Mining Geology: Exploration and Management

Mining Geology: Exploration and Management

Edited by **Beth Thorpe**

⬚SYRAWOOD
PUBLISHING HOUSE

New York

Published by Syrawood Publishing House,
750 Third Avenue, 9th Floor,
New York, NY 10017, USA
www.syrawoodpublishinghouse.com

Mining Geology: Exploration and Management
Edited by Beth Thorpe

International Standard Book Number: 978-1-68286-180-6 (Hardback)

Printed in the United States of America.

Contents

Preface

Mining is a significant economic activity. The recent advances in technology have aided the tools and techniques used in mining processes. This book is an extensive source of knowledge which discusses the technological advances in mining and mineral engineering. It also throws light on new minerals, gem deposits, biomineralogy, etc. This text will prove beneficial to students, geologists, researchers and professionals engaged in this field.

Various studies have approached the subject by analyzing it with a single perspective, but the present book provides diverse methodologies and techniques to address this field. This book contains theories and applications needed for understanding the subject from different perspectives. The aim is to keep the readers informed about the progress in the field; therefore, the contributions were carefully examined to compile novel researches by specialists from across the globe.

Indeed, the job of the editor is the most crucial and challenging in compiling all chapters into a single book. In the end, I would extend my sincere thanks to the chapter authors for their profound work. I am also thankful for the support provided by my family and colleagues during the compilation of this book.

Editor

Arsenic Adsorption onto Minerals: Connecting Experimental Observations with Density Functional Theory Calculations

Heath D. Watts [1,*], Lorena Tribe [2] and James D. Kubicki [1,3,*]

[1] Department of Geosciences, The Pennsylvania State University, University Park, PA 16802, USA

[2] Division of Science, The Pennsylvania State University, Berks, Reading, PA 19610, USA;
E-Mail: lut1@psu.edu

[3] Earth and Environmental Systems Institute, The Pennsylvania State University, University Park, PA 16802, USA

* Authors to whom correspondence should be addressed; E-Mails: hdw115@psu.edu (H.D.W.); jdk7@psu.edu (J.D.K.)

Abstract: A review of the literature about calculating the adsorption properties of arsenic onto mineral models using density functional theory (DFT) is presented. Furthermore, this work presents DFT results that show the effect of model charge, hydration, oxidation state, and DFT method on the structures and adsorption energies for As^{III} and As^{V} onto Fe^{3+}-(oxyhydr)oxide cluster models. Calculated interatomic distances from periodic planewave and cluster-model DFT are compared with experimental data for As^{III} and As^{V} adsorbed to Fe^{3+}-(oxyhydr)oxide models. In addition, reaction rates for the adsorption of As^{V} on α-FeOOH (goethite) (010) and Fe^{3+} (oxyhydr)oxide cluster models were calculated using planewave and cluster-model DFT methods.

Keywords: arsenic; density functional theory (DFT); kinetics; thermodynamics; adsorption; computational chemistry; planewave DFT; reaction rates; As—Fe bond distances

1. Introduction

1.1. Arsenic Chemistry, Geochemistry, Prevalence, and Toxicity

The study of arsenic (As) adsorption on mineral surfaces is necessary to understand both the distribution and mobility of As species in nature as well as to develop remediation strategies for As waste sites. Arsenic is found in a variety of geochemical environments at aqueous concentrations varying from <0.5 to >5000 µg/L, and is found in a variety of geochemical environments [1,2]. Natural and anthropogenically-mediated biogeochemical interactions among arsenic species, biota, and minerals can affect the distribution, mobility, and toxicity of As in the environment [2–8]. Although recent work has posited that arsenic could be a potential biochemical and astrobiological proxy for phosphorus during biological evolution [9], this hypothesis is controversial [10].

Arsenic can occur in both inorganic (iAs) and organic (oAs) forms; the chemical form of As, or species, as well as the concentration of As, affects the solubility, mobility, reactivity, bioavailability, and toxicity of As [11,12]. iAs occurs predominately as either arsenious acid ($H_nAs^{III}O_3{}^{n-3}$; sometimes called arsenous acid) in reducing environments, or arsenic acid ($H_nAs^{V}O_4{}^{n-3}$) in oxidizing environments, where $n = 0$, 1, 2, or 3 [12–14]. Both the oxidation and protonation states of iAs depend on the physiochemical conditions of the sample environment [2]. For example, the three pK_a values for arsenic acid are 2.2., 6.9, and 11.5 [15]; therefore, the pH of the environment will affect the protonation state of H_3AsO_4, which will, in turn, affect the mobility, reactivity, and bioavailability of iAs^V. In addition to iAs, methylated As^{III} and As^V compounds such as monoarsinic acid (MMA^V) and dimethylarsinic acid (DMA^V) occur both naturally and due to anthropogenic sources [12,13]. Arsenic toxicity depends on the species present; a general trend of decreasing toxicity is: $R_3As > H_3AsO_3 > H_3AsO_4 > R_4As^+ > As^0$, where R is an alkyl group or a proton [12,16].

Arsenic originating from natural water-rock interactions of surface and groundwater [2,6] and from anthropogenic sources such as acid mine drainages [17,18] can lead to drinking water contamination. Biogeochemical processes result in organic As compounds accumulating in oil, shale, and coal [19]. DMA and MMA have been used as herbicides, pesticides, and defoliants and pose a contamination problem to surface and groundwater [20]. Roxarsone ($C_6AsNH_6O_6$) is used as a supplement in chicken feed and ends up in waste from poultry operations [21]. In addition to natural and anthropogenic groundwater contamination by As, anthropogenic As is contaminating the oceans, and subsequently seafood [16], which could have human health implications.

Human diseases caused by As contamination include various cancers, liver disease, cardiovascular disease, and an increase in mortality from pulmonary tuberculosis [22–28]. Groundwater that contains As concentrations >10 µg/L limit set by the World Health Organization (WHO) places more than ten million people at risk from arsenicosis. Locales ranging from Southern Bangladesh, India, Argentina, Chile, and Vietnam have groundwater sources with As concentrations >10 µg/L [2,3,6,25,26,28–33].

The remediation of arsenic in aquifers can be challenging due to the size of contaminated groundwater systems and varying biogeochemical conditions. For instance, the shallow groundwater of the Chaco Pampean Plain of Argentina spans 10^6 km^2 and contains 10 to 5300 µg As/L [34]. A variety of As remediation methods were reviewed recently [35,36]; these methods have shown varying success rates. For example, an *in situ* study of an alkaline aquifer showed relatively low adsorption of iAs [37]. Conversely, a permeable reactive barrier study showed As removal to <5 μg/L due to induced sulfate reduction and the presence of zero-valent Fe [38], and experiments with household sand filters in Vietnam were able to reduce As concentrations in drinking water to <10 μg/L with a 40% success rate [39]. Laboratory and field experiments with household zero-valent Fe filters showed similarly effective results in Bangladesh [40]. Field tests with Fe-coated zeolites showed 99% removal of As [41]. Mn-Fe oxide-coated diatomites were able to reduce iAs from >40 μg/L to <10 μg/L in a pilot field-scale study [42]. Although many of these studies show promise for field-scale remediation of As, the biogeochemistry, aqueous geochemistry, and mineralogy of the surface and groundwater can affect the efficiency of the methods; therefore, it is necessary to understand the As adsorption process more thoroughly, so that the remediation methods can be applied more effectively.

1.2. Arsenic Treatment Methods

Due to the prevalence of As contamination worldwide and the threat of arsenicosis [43], research and the development of methods to understand As chemistry and to attenuate As in water are imperative [44]. Methods that have been developed to attenuate As such as electrocoagulation and electrodialysis [45,46], treatment with microorganisms to affect the biogeochemical cycling of As [7], and the adsorption of As onto a variety sorbents [47–49]. Among the sorbents used for As attenuation are organic polymers [50], and minerals such as dolomite [51], zeolites [52], and Al minerals such as alumina or gibbsite [53,54]. Alumina has been found to effectively remove iAs^V [55], but it is necessary to use activated alumina to efficiently remove iAs^{III} from solution [56]. A multitude of studies on As adsorbents have been conducted using Fe-based mineral sorbents [57,58], such as magnetite [54,59–63], ferrihydrite [53,54,64–66], goethite [48,54,66,67], and zero-valent iron [62].

The focus of the current work is on the chemistry of iAs species adsorbed to Fe-oxide and Fe-hydroxide mineral surfaces. Prior experiments have found that inorganic [64] and organic [64,68–70] ligands may adsorb to Fe minerals and compete with As for adsorption sites, but Zn cations [59] may augment As adsorption to Fe sorbents. Therefore, if ligands that inhibit As adsorption can be removed or precipitated and ligands that enhance As adsorption can be added to As-containing site, it could be possible to develop improved As-remediation techniques. Arsenic adsorption may also be enhanced by the addition of magnetite to agricultural waste such as wheat straw [61], but adsorption of As can be reduced by microorganisms [7,66]. If it is possible to control the growth and metabolism of microorganisms present in As-containing water, then it could be possible to enhance As adsorption.

1.3. Studying As Adsorption with Experimental and Modeling Methods

Because of the complex and often uncharacterized biogeochemistry of arsenic-rich environments, it is useful and necessary to use a variety of experimental data and modeling results to characterize these complex matrices. For example, Figure 1 shows models of the monodentate, bidentate, and outer-sphere adsorption of $HAsO_4^{2-}$ to Fe-(oxyhyr)oxide clusters (Figure 2A–C) and periodic models (Figures 2D–F) models. Experimental and modeling techniques have been employed to determine the geometries, energetics, and spectroscopic properties exhibited by As species adsorbed to mineral surfaces; improved knowledge about the adsorption chemistry of As species can aid in the development of methods for attenuating As in the environment. Although the focus of our work is on the use of quantum mechanics (QM) techniques for studying the properties of As adsorption, it is necessary to frame these results in relationship to experimental and other modeling methods, because QM results can be useful for interpreting experimental data and for parameterizing other modeling methods such as classical force fields [71] and surface complexation models [72,73].

Figure 1. (**A**) Ferric iron (oxyhydr)oxide cluster model of $Fe_2(OH)_6(H_2O)_4 \cdot 4H_2O$; (**B**) monodentate mononuclear (MM) cluster model of $Fe_2(OH)_5(H_2O)_4H_2AsO_4 \cdot 4H_2O$; and (**C**) binuclear bidentate (BB) cluster model of $Fe_2(OH)_4(H_2O)_4HAsO_4 \cdot 4H_2O$.

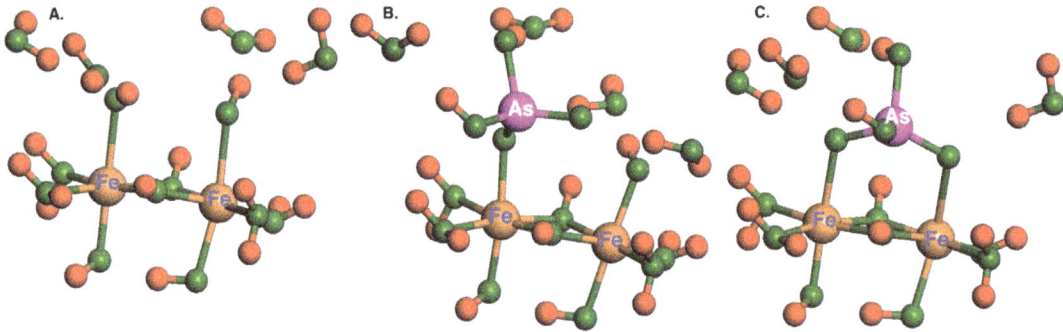

Figure 2. (**A**) and (**D**) monodentate mononuclear adsorption of $HAsO_4^{2-}$; (**B**) and (**E**) bidentate binuclear adsorption of $HAsO_4^{2-}$; and (**C**) and (**F**) outer-sphere complex of $HAsO_4^{-}$, on clusters of Fe^{3+}-(oxyhydr)oxide (**A–C**) and periodic α-goethite (010) (**D–F**) models.

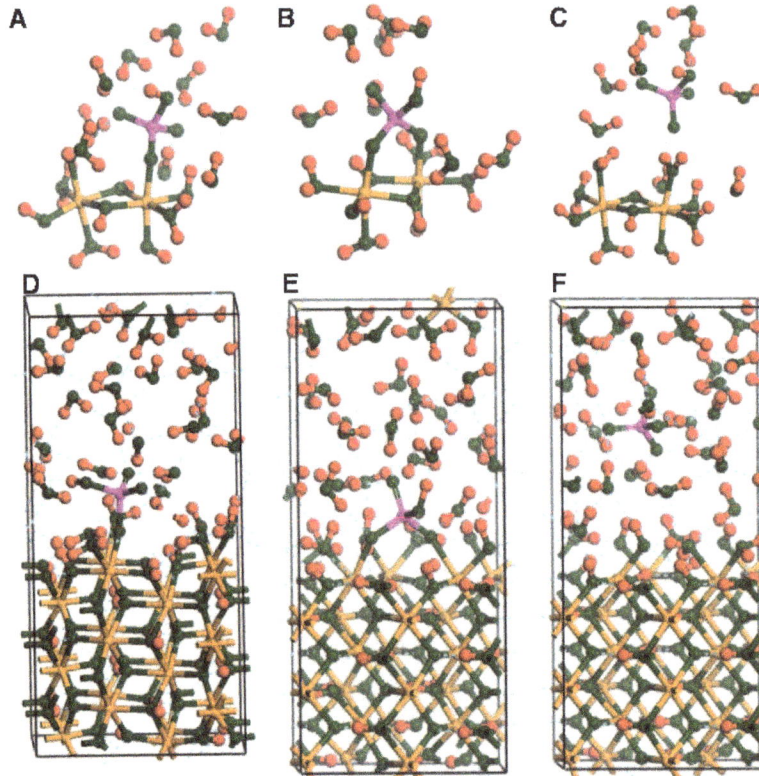

1.4. Studying As Adsorption with Experiments

Vibrational spectroscopy (e.g., Fourier transform infrared spectroscopy (FTIR) or Raman), X-ray absorption near edge structure (XANES), extended X-ray absorption fine structure (EXAFS) spectroscopies, as well as adsorption isotherm and kinetics experiments are useful for determining the chemistry of As adsorption. FTIR and Raman studies are useful for determining the bonding configurations between As species and mineral surfaces. When As adsorbs as an inner-sphere complex, characteristic vibrational frequency shifts are observable [74–79]. XANES spectroscopy can provide information about the oxidation state of As that is adsorbed to mineral surfaces [80–84], and can determine if the oxidation state of As or the surface changes during the adsorption process [78,85]. EXAFS spectroscopy provides information about the coordination chemistry of adsorbed As [46,80–82,86–92]. Coordination state information is useful for determining whether the As adsorbs as a monodentate, bidentate, or outer-sphere complex (Figure 2). Moreover, studies that use two or more instrumental methods such as XANES/FTIR [78] or XANES/EXAFS [82] can increase the reliability of data interpretation. Furthermore, kinetics and isotherm studies provide information about the rates and energetics of As adsorption onto mineral surfaces that further help constrain adsorption mechanisms [55,93–97].

Although experimental techniques have contributed greatly to the understanding of As adsorption chemistry, computational chemistry can be used to help interpret experimental data on the mechanisms of As adsorption to mineral surfaces. In addition, computational chemistry can fill in missing information on the details of adsorption mechanisms and kinetics.

1.5. Studying As Adsorption with Mathematical Models

Surface complexation modeling (SCM) techniques such as the charge distribution multi-site complexation (CD-MUSIC) [76,95,98–106], the extended triple-layer model [107], isotherm modeling [108,109], and the ligand and charge distribution model (LCD) [110,111] have been used to model As species in solution and adsorbed to mineral surfaces. In general, SCM models use experimental data or quantum mechanics results such as equilibrium constants, bond lengths, and surface charges to calculate the adsorption isotherms of As on Fe mineral surfaces. When integrated, the CD-MUSIC and LCD SCMs are able to model humic substances interacting with Fe surfaces and the effect they have on As adsorption [111,112]. SCM can be used to interpret interactions among As, mineral surfaces, and organic and inorganic ligands, as well as charge and protonation effects. However, the precision of these models depends on the quality of their parameterization that is obtained from experimental data, which can be difficult to interpret, or with QM results [72,73].

1.6. Studying As Adsorption with Quantum Mechanics Modeling Methods

Numerous studies applying QM, or specifically density functional theory (DFT) methods [113,114], use models similar to those shown in Figures 2A–C. For example, DFT methods have been used to study the thermodynamics [91,115–117], vibrational frequencies [75,118–121], kinetics [120–122], ligand effects [87], oxidation-reduction reactions [123], and the coordination of adsorbed As [90,91,117,119,124–132].

Many of the previous computational chemistry studies relied on cluster models such as those in Figures 2A–C, but some groups have used periodic models to capture the chemistry of the mineral surface sorbates more precisely [124,125,127]. In this paper, we report structures and kinetics comparisons for the results from both cluster models (Figures 2A–C) and periodic models (Figures 2D–F).

These comparisons are necessary to determine if and how the results from the molecular cluster and periodic model calculations differ.

In addition to the use of cluster *versus* periodic models, other factors can affect the results obtained from DFT calculations. These factors include surface charge, hydration, model convergence criteria, and potentially the software used for the calculation. For example, prior studies used highly charged models [91] to study the adsorption of As to Fe clusters; however, the work that we are presenting used models for the surface clusters with neutral charges, because localized high charges are unlikely. Prior work using the implicit solvation using the conductor reaction field (COSMO) [133,134] suggests that a single water molecule can produce activation energy results that are more precise than the addition of multiple water molecules can [122]. Conversely, other studies have used anhydrous surfaces to model As adsorption to Fe clusters [91]. The work we present herein used both hydrated cluster models and hydrated periodic models in an attempt to model the natural environment of As adsorption more accurately and realistically.

As the processing power and speed of computers increases, the size and complexity of As adsorption models increases, as illustrated by two Fe cluster models from papers dating from 2001 and 2006 [90,129]. The Fe cluster model in the latter paper is larger and likely more realistic than the model in the former paper. Results from prior DFT calculations provide contradictory results about the coordination state of As adsorption. For example, there are DFT results that predict that monodentate [128], bidentate [90,91,119,129], or a mixture of As coordination states are occurring on Fe-mineral surface models [130]. Because both the experimental data and the DFT results are providing ambiguous information about As coordination state, further calculations and experiments are necessary to clarify this topic. Alternatively, As may adsorb in a variety of configurations depending upon which surfaces are present on a given mineral sample [135].

The convergence criteria of the energy minimization calculations for the models could also have an effect on the precision of the calculated results and the ability of the results to reproduce experimental data and provide insight about As adsorption chemistry. For example we use a minimization convergence criteria of 0.03 kJ/mol, whereas, for example, Sherman and Randall [91] used higher tolerance energy minimization criteria 5 kJ/mol; however, tighter convergence criteria again reflects the availability and evolution of computational resources.

In the studies presented here, the thermodynamics, geometries, and kinetics of inner-sphere iAs^{III} and iAs^{V} adsorbed as monodentate mononuclear (MM) or bidentate binuclear (BB) complexes to solvated Fe clusters were evaluated. The molecular cluster results show how hydration and the initial Fe cluster charge affect iAs^{III} and iAs^{V} adsorption for BB models, and the results compare the of adsorption energies of BB iAs^{III} and iAs^{V} onto hydrated neutral Fe clusters. The DFT calculations on the cluster models also compare the calculated As—Fe distances for the BB models with pertinent experimental observations. In addition, inner- and outer-sphere As-Fe complexes were used to determine the activation energies (ΔE_a) of the adsorption/desorption process; these calculations were performed using both molecular cluster models and periodic models of the α-FeOOH (goethite) (010) surface.

2. Methods

2.1. Applied Quantum Mechanics Background

The application of QM with quantum chemistry software allows one to calculate chemical properties such as thermodynamics, kinetics, molecular geometries, spectroscopic parameters, transition states structures that might not be experimentally observable, and potentially hazardous chemistries [136–138].

Quantum chemistry calculations begin with an initial input of Cartesian coordinates for the molecule of interest, and then these coordinates are subsequently allowed to change during energy minimization calculations. The bond lengths, bond angles, and dihedral angles of the model are systematically perturbed, followed by the calculation of the relative energy of the model. After each energy calculation, subsequent systematic perturbations of the model geometry take place until the force (F), where $F = dE/dr$ and dE and dr are the change in the energy and change in the model coordinates, respectively, converges to at or near "zero" (*i.e.*, "zero" is defined as the convergence criteria by the modeler). When F is zero, the model then resides at a stationary point on a potential energy surface (PES). The user of computational chemistry software can specify the criterion for energy minimization convergence [136,139].

Subsequent calculation of the vibrational frequencies for the model determines the second derivative of energy with respect to atomic coordinates (d^2E/dr^2). If the calculated vibrational frequencies are all real (positive), then the model is at a PES minimum; if one vibrational frequency is imaginary (negative), then the model is at a transition state or PES maximum; if the model exhibits > 1 imaginary frequency, then the model is unstable and a new input geometry should be used. Obtaining a PES minimum does not guarantee that the model is at a global minimum, only that the model is at a local minimum. Using multiple input models obtained from a conformational analysis can aid in the attainment of a globally minimized final geometry [136,139].

The calculation of the vibrational frequencies (d^2E/dr^2) also allows the calculation of thermodynamic properties such as enthalpy, Gibbs free energy, and entropy. If one imaginary frequency is present, the model is at a transition state and the results from this model, the initial structure of the model, and the PES minimum model can be used to calculate rate constants for reactions. Note that throughout the manuscript the output from QM calculations is referred to as results and not data, we refer to the output from experiments as data. The vibrational frequency calculation provides infrared and Raman frequencies for the model, and further calculations can provide NMR chemical shifts, UV-Visible wavelengths, isotope effects, and temperature effects [136].

There are numerous methods available for calculating energies and other chemical properties with QM, among the most widely used are Hartree-Fock (HF) method [140], Møller-Plesset perturbation (MP) theory [141], and density functional theory (DFT) [113,114]. All of these methods arose from the development of quantum mechanics and the Schrödinger equation ($\hat{H}\Psi(r) = E\Psi(r)$), where $\hat{H}$ is a Hamiltonian operator, $\Psi(r)$ is the wave function, and E is the energy of the model. If $\Psi(r)$ is known for a model, it is possible to solve for E and thus obtain the molecular properties for any model of interest. The Schrödinger equation has only to solve for the electronic energy of the model, because the nuclear energy and positions are relatively low and stationary compared to those of the electrons [142]. Thus far, it has been impossible to solve the Schrödinger equation for models that are more complex than H_2 because $\Psi(r)$ makes the solution of the equation untenable for larger molecules with a greater number of electrons; therefore, it has been necessary to develop approximations for the Schrödinger equation such as HF and MP theory.

HF theory minimizes the energy of each electron iteratively with respect to the average energy of the other electrons in a model [140]. The shortcoming HF is that it does not account for electron correlation (repulsion) between each electron in the model. Using HF leads to an overestimation of model stability. MP theory accounts for electron correlation by systematically perturbing the molecular Hamiltonian [141]; however, the cost of using MP theory is prohibitive for models of geochemical interest. Furthermore, unlike HF theory, MP theory is not variational [143], so, the calculated energy could be lower than the ground state energy.

Unlike HF and MP theories that calculate electron interaction to obtain molecular energies, DFT calculates the electron density of the molecule to determine the energy [113,114,144]. The shortcoming of DFT is that the theory lacks a method for calculating the exact energy of the electron correlation and exchange term (E_{xc}). The inability to calculate the E_{xc} results in an underestimation of the total energy. Neglecting E_{xc}, as HF theory does, or underestimating E_{xc}, as DFT does, results in an underestimation of the total energy of a given model. For DFT, the lack of a precise correlation energy results because DFT does not account for the columbic interaction (repulsion) between electrons with anti-parallel (opposite) spin. Imprecise exchange energy results because DFT does not account precisely for the fact that electrons with parallel (same) spins cannot reside in the same orbital. Therefore, neglecting or underestimating the E_{xc} violates the Pauli Exclusion Principle. However, a variety of DFT methods have been developed to approximate E_{xc} and these methods can produce precise results [138,145]. Significantly, the computational cost of DFT calculations is substantially less than the cost of MP theory calculations, and the precision of DFT calculations is greater than that of HF calculations. Therefore, although DFT methods account imprecisely for E_{xc}, they are preferable to HF and MP methods.

Many DFT methods are available, and a particular method could be useful for calculating a particular chemical property (e.g., energy) but not for calculating other properties (e.g., NMR chemical shifts); these differences in precision are due to the parameters used for E_{xc} in the DFT methods [138,145]. Additional work is necessary to evaluate the efficacy of DFT methods for calculating properties such as adsorption energies, rate constants, and structures. Although DFT methods such as B3LYP [146,147] can provide accurate results for a variety of chemical properties, we suggest that computational geochemists begin to explore the use of other DFT methods that could provide improved results.

In addition to using an electron correlation method such as MP or DFT, it is necessary to specify a basis set when using molecular orbital (MO) calculations [136,148–151]; however, when using planewave calculations [152–158], basis sets are not used.

For DFT calculations on the clusters, the basis sets are equations that define atomic orbitals and are used in linear combinations to create molecular orbitals. When using basis sets, each atomic orbital for a given atom contains one electron [149], so for the C atom, which has six electrons and an electronic configuration of $1s^2 2s^2 2p^2$, it is necessary to have a minimum of five basis functions (*i.e.*, 1s, 2s, $2p_x$, $2p_y$, and $2p_z$). Note that because the p-orbitals are energetically degenerate, the $2p_x$, $2p_y$, and $2p_z$ are present as basis functions. The variational principle of quantum mechanics states that $E_g \leq \langle \Psi(r) | \hat{H} | \Psi(r) \rangle$, where E_g is the ground state (lowest) energy of the molecule. The variational principle shows that increasing the accuracy of $\Psi(r)$ will increase the calculated accuracy E relative to E_g; increased basis set size can increase the accuracy of $\Psi(r)$.

Basis set size can be increased by using double zeta (DZ) or triple zeta (TZ) basis sets which double or triple the number of basis sets used, relative to the minimal basis set [148]. For C, the DZ and TZ basis sets have ten and fifteen basis functions, respectively. Another method for increasing the accuracy of a basis set is to use split basis sets, where more basis sets are used for the bond-forming valence electrons, while the non-bond-forming core electrons are treated with minimal basis sets. For C, this would involve one 1s orbital and two each of the 2s, $2p_x$, $2p_y$, and $2p_z$ basis sets for a total of nine basis sets.

Further increases in basis set size and precision can be obtained by the addition of polarization and diffuse basis functions [148]. Polarization functions increase the angular momentum basis sets on a particular atomic orbital; for example, in methane each H atom has a 1s orbital, but in the C-H bonds, the H atoms receives some p-orbital character from the C atom; therefore, including p-orbital polarized functions on the H atoms increases the accuracy of the orbital description. In addition, diffuse basis sets are used increase the electronic radius where electrons can reside. Diffuse basis sets are useful for calculations with anions and for weak interactions such as van der Waals interactions.

2.2. Molecular Orbital Theory Calculations with Fe Clusters

For the cluster model DFT calculations, all models were constructed in Materials Studio (Accelrys Inc., San Diego, CA, USA) and the energy minimization, Gibbs free energy, and transition-state calculations were performed in the gas phase using Gaussian 09 software [136]. All energy minimization calculations on the cluster models were performed without symmetry or atomic constraints. Energy minimizations, frequency, and kinetics (transition state) calculations were performed using the hybrid density functional B3LYP with the 6-31G(d) basis set [148–151]. B3LYP accounts for E_{xc} and the 6-31G(d) basis set is a DZ basis set with p-polarization functions on the non-H atoms. Energy convergence was set to 0.03 kJ/mol during the energy minimization calculations. The frequency calculations using B3LYP/6-31G(d) ensured that each model was at either a potential energy minimum (no imaginary frequencies) or at a transition state (one imaginary frequency) [136]; however, the frequency calculation does not ensure the model is at a global energy minimum.

For Gaussian calculations, it is necessary to specify an electron correlation method (e.g., B3LYP), a basis set, the type of calculation (e.g., optimization), the Cartesian coordinates of the atoms in the model, the charge of the model, and the spin multiplicity of the model [136]. The electron configuration of Fe^{3+} is $[Ar]3d^5$. For the energy minimization calculations, we used high-spin Fe^{3+}, where each 3d electron occupies one of the five d-orbitals, and where the electrons are all either spin up or spin down; this means that each of the two Fe atoms in the cluster has five unpaired electrons. The multiplicity is $= 2S + 1$, where S is the spin of an unpaired electron and can be $+\frac{1}{2}$ or $-\frac{1}{2}$. Therefore, for our high-spin clusters there are 10 unpaired electrons, each having a spin of $+\frac{1}{2}$, so the multiplicity $= 2(10/2) + 1$ or 11. For the rate constant calculations, we experimented by using high-spin multiplicity for both Fe atoms (i.e., 11), and a combination of up spin for one Fe atoms (i.e., +5) and down spin for the other (i.e., −5).

Fe model surface complex clusters were designed to minimize surface charge because charge buildup on the actual mineral surfaces is believed to be relatively small. However, surface models with charges were also calculated to demonstrate the effect of model charge on calculated energetics, and to compare with prior studies that used charged models [91]. In addition, explicit hydrating H_2O molecules were included for the aqueous species, the surface Fe-OH groups, and the adsorbed arsenic acid (iAs^V) molecules. Figure 1 shows examples of a tetrahydrated Fe (oxyhydr)oxide cluster, (Fe$_2$(OH)$_6$(H$_2$O)$_4$, A, without adsorbed HAsO$_4^{2-}$, B, monodentate mononuclear (MM) adsorbed HAsO$_4^{2-}$, and C, bidentate binuclear (BB) adsorbed HAsO$_4^{2-}$. Hydration can be a key in accurately predicting structures and frequencies of anionic surface complexes because they are both a function of the hydration state of the sample [75,157]. The interatomic distances calculated in each model surface complex were compared to As-O bond lengths and As—Fe distances obtained from EXAFS spectra [46,80–82,86–92].

Gibbs free energies (G) of each species were estimated by calculating G in a polarized continuum the permittivity for water of 78.4 at 298.15 K. The integral equation formalism of the polarized continuum model (IEFPCM) [159] was used to calculate the Gibbs free energy in solution. The polarized continuum places the model in the cavity of a field with a constant permittivity, in this case for water; this field is used as a proxy for the solvent of interest. Single-point energy calculations were performed with the B3LYP functional and the 6-31+G(d,p) basis set [148–151]; the TZ basis set with augmented functions on the non-H atoms, and d-polarized and p-polarized functions for the non-H atoms and H-atoms, respectively, was used in order to improve the accuracy of the energy that was calculated using B3LYP/6-31(d) during the energy minimizations; this is a standard practice. The ΔG_{ads} was then determined from stoichiometrically balanced reactions. Configurational entropy terms are neglected in this approach; hence, we emphasize that these are Gibbs free energy estimates. We do not expect the precision of ΔG_{ads} to be better than ±10 kJ/mol.

To illustrate the potential effect that a chosen DFT method can have on geometry and thermodynamics, we report the As—Fe, As-OH, As=O, and As-OFe bond distances for the Fe$_2$(OH)$_4$(OH$_2$)$_4$HAsO$_4$·4H$_2$O model and ΔG_{ads} for the reaction:

$$H_2AsO_4^-·8H_2O + Fe_2(OH)_6(OH_2)_4·8H_2O \rightarrow Fe_2(OH)_4(OH_2)_4HAsO_4·4H_2O + OH^-·4H_2O + 5H_2O \quad (1)$$

For this reaction, we used either B3LYP, PBE0 [160–162], and M06-L [163] DFT methods to minimize each structure in the reaction with the 6-31G(d) basis set, and the 6-311+G(d,p) basis with the self-consistent reaction field (SCRF) IEFPCM and the solvent water to calculate the single-point energy of each structure. The PBE0 and M06-L methods were chosen to compare with the results from the often-used B3LYP method, because the PBE0 function was the method used for the periodic planewave calculations for this work, and the M06-L method was specifically parameterized for use with transition metals such as Fe [162].

For the transition-state calculations on the clusters, the outer-sphere complexes of AsO$_4^{3-}$ on an Fe^{3+}-(oxyhydr)oxide clusters were obtained by constraining one Fe—As distance and allowing all other atoms to relax. The constrained distance was increased incrementally, allowing for energy minimization of the system at each new distance. The reaction path was graphically visualized. The change in energy for the adsorption reactions (ΔE_{ads}) were inferred by using the total electronic energy plus the zero-point correction obtained from the inner-sphere frequency calculations [163].

2.3. Planewave Calculations Using α-FeOOH (010)

The starting configuration for the periodic bidentate, binuclear $HAsO_4^{2-}$ on the goethite (α-FeOOH) (010) surface was taken from previous simulations of HPO_4^{2-} on the same surface [135]. Phosphate and arsenate structures and chemistries are similar, so this starting configuration is a reasonable approximation. An energy minimization was performed on this starting configuration to allow the atoms to relax as necessary for the As for P substitution. Energy minimizations were carried out with the lattice parameters constrained the experimental values (9.24 × 9.95 Å^2) [164] and with a vacuum gap between surface slabs of 10 Å. The model stoichiometry was 24FeOOH, $HAsO_4^{2-}$, $29H_2O$, and $2H_3O^+$ ($Fe_{24}O_{83}H_{89}As$). The small model system size and the high percentage of H^+ per H_2O molecules severely limits the realism of the model compared to experimental systems, so the results from these calculations should be considered exploratory of model system behavior rather than an accurate portrayal of arsenate adsorption thermodynamics and kinetics.

Projector-augmented planewave calculations [152,165] with the Perdew-Erzenhof-Burke exchange correlation functional [160] were performed with the Vienna Ab-initio Simulation Package (VASP 5.2) [153–157]. The PBE0 exchange correlation functionals were Fe pv (14 valence e^-), O (6 valence e^-), H (1 valence e^-) and As (5 valence e^-) as labeled in the VASP exchange correlation functional library. Energy cut-offs (ENCUT in VASP input files) of 500 eV and 400 eV were used for energy minimizations and molecular dynamics simulations, respectively. The precision of the self-consistent field calculation of electron density was (PREC = Accurate = 700 eV/ROPT = −2.5 × 10^{-4}) for energy minimizations and (PREC = Medium − 700 eV/ROPT − −2.5 × 10^{-3}) for molecular dynamics simulations. The PREC-flag determines the energy cutoff (ENCUT) when no value is given for ENCUT in the central input file of VASP, INCAR, and the ROPT-tag controls the real-space optimization. The lower accuracies of the molecular dynamics (MD) simulations were chosen for practical reasons. Thousands of configurations and their energies need to be calculated for MD simulations, so the less stringent electron density grid speeds up the energy calculation at each step. The assumption here is that although the MD simulations are less accurate, they are not dramatically in error for predicting atomic structure. Thus, the MD simulations can be used to relax the atomic positions to achieve an approximate configuration and energy, and then energy minimizations can be performed to obtain structures and energies that are more precise. Without the MD simulations, the likelihood of the energy minimizations becoming trapped in a local potential energy minimum is much greater.

Periodic DFT calculations were run with 1 k-point created with the Monkhorst-Pack mesh. The DFT+U correction [166,167] was employed with a $U = 4$ eV for Fe and 0 eV for all other elements. Spin states were ordered according to the experimental observed magnetic ordering of goethite [168]. These selections have worked reasonably well in a previous study on goethite and goethite-water [73]. No dispersion-corrections were employed in these calculations although the results of DFT calculations of adsorption onto mineral surfaces may be affected by Van der Waals forces and how they are approximated [86]. The MD simulations were run at a temperature of 300 K maintained by the Nosé-Hoover thermostat [169]. Time steps were 0.5 fs (POTIM = 0.5). POTIM is a time-step variable and its value depends on the type of calculation being performed. Note that because some DFT methods may over structure and freeze water at 300 K, some authors have used higher temperatures to overcome this problem [170]. Another method is to use D instead of H [171], so that a 1 fs time step can be used instead of a 0.5 fs time step. Both are accepted practices, but we prefer to use the actual temperature and H atoms. This may cause error, but these errors are intrinsic to the method and as such not different in character from other computational uncertainties. Introducing errors by giving the atoms extra kinetic energy or mass may mask the discrepancies with experiment. Instead, highlighting these discrepancies is important and points to the need to improve computational methods.

Calculation of the periodic surface complex structures also allows for creation of more realistic molecular clusters that are surface specific. Figure 3 shows how an extended cluster (Figure 3B) can be extracted from the periodic model (Figure 3A). By selecting all the O atoms bonded to the Fe atoms connected to the As surface complex, a molecular cluster is created that retains surface-specific structure. The O atoms at the edge of the cluster are then terminated with H^+ in order to satisfy valence and adjust the overall charge of the cluster as desired. Positions of hydration H_2O molecules can be included in the extended cluster (Figure 3B) to better mimic the aqueous phase. The combination of periodic and molecular cluster DFT results can take advantage of the strengths of each approach. For example, the periodic calculations should provide more accurate surface structures and adsorption energies, but molecular cluster models can be used to predict IR, Raman, and NMR spectra [135].

Figure 3. (**A**) Periodic model of goethite and (**B**) an extended cluster extracted from the periodic model.

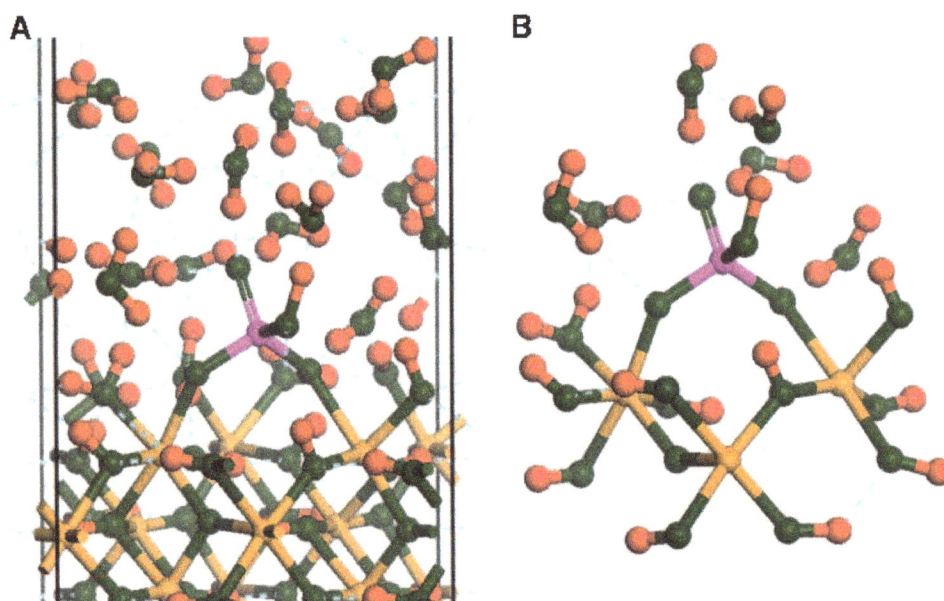

3. Results and Discussion

The molecular cluster results show how the initial cluster charges affect Gibbs free energy of adsorption (ΔG_{ads}) of iAs^V and iAs^{III}, the effect of hydration on ΔG_{ads} for the iAs^V models, and compare the of adsorption energies of triprotic iAs^{III}, and monoprotic or diprotic iAs^V onto hydrated neutral Fe clusters. The molecular cluster calculations also compare the calculated As—Fe distances with EXAFS data. In addition, calculations to determine the rate constant for iAs^V adsorption on Fe clusters and periodic models are discussed.

3.1. Effect of Cluster Charge on ΔG_{ads}

Reactions (1)–(7) in Table 1 show the ΔG_{ads} of H_3AsO_3 (Reaction (1) and (2)), $HAsO_4^{2-}$ (Reaction (3) and (4)), and $H_2AsO_4^{-}$ (Reaction (5)–(7)) onto either a neutral ($Fe_2(OH)_6(OH_2)_4^{0}$) or +4 charged ($Fe_2(OH)_2(OH_2)_8^{4+}$). Reaction (7) is stoichiometrically equivalent to one previously reported; however, for our calculation we used an energy minimization convergence criteria of 0.03 kJ/mol, whereas Sherman and Randall [91] used 5 kJ/mol. Significantly, Sherman and Randall [91] reported that the reaction:

$$Fe_2(OH)_2(OH_2)_6H_2AsO_4^{3+} (BB) + H_2O \rightarrow Fe_2(OH)_2(OH_2)_7H_2AsO_4^{3+} (MM) \qquad (2)$$

was endothermic and required +95 kJ/mol of energy; however, our results predict that this conversion would require +17 kJ/mol of Gibbs free energy. Both calculations predict that the BB structure is energetically favorable, but our results show energy difference between the BB and MM structures is not as large, and that these structures could co-exist in nature. The possibility of the presence of both BB and MM agree with prior research [130], but the lower ΔG_{ads} of the BB is inconsistent with the claim that MM adsorption is dominant [128]. The methodologies used to calculate the conversion of BB to MM could account for the calculated energy differences; the methods differ by:

- Model convergence criteria;
- Implicit solvation (our work) and gas-phase results from Sherman and Randall [91];
- Electronic energies (ΔE) from Sherman and Randall [91] and ΔG for our work;
- Use of a single, gas-phase minimized H_2O model to balance Equation (2) [91], whereas we used 1/8 the energy of an implicitly solvated model with eight H_2O molecules.

Throughout this work, unless otherwise noted, we used explicitly hydrated models for all of the reactants and products for the energy minimization calculations, and those same models for the implicitly solvated (*i.e.*, IEFPCM) single-point energy calculations.

Table 1. Effect of Fe cluster charge on the ΔG_{ads} of iAs^{III} and iAs^{V}.

Reaction #	Reaction	ΔG_{ads} (kJ/mol)
(1)	$H_3AsO_3 + Fe_2(OH)_6(OH_2)_4^{0} \rightarrow Fe_2(OH)_4(OH_2)_4HAsO_3 + 2H_2O$	−60
(2)	$H_3AsO_3 + Fe_2(OH)_2(OH_2)_8^{4+} + 16H_2O \rightarrow Fe_2(OH)_2(OH_2)_6HAsO_3^{2+} + 2H_3O^{+}\cdot 8H_2O$	−159
(3)	$HAsO_4^{2-} + Fe_2(OH)_6(OH_2)_4^{0} + 8H_2O \rightarrow Fe_2(OH)_4(OH_2)_4HAsO_4 + 2OH^{-}\cdot 4H_2O$	+14
(4)	$HAsO_4^{2-} + Fe_2(OH)_2(OH_2)_8^{4+} \rightarrow Fe_2(OH)_2(OH_2)_6HAsO_4^{2+} + 2H_2O$	−263
(5)	$H_2AsO_4^{-} + Fe_2(OH)_6(OH_2)_4^{0} + 3H_2O \rightarrow Fe_2(OH)_4(OH_2)_4HAsO_4 + OH^{-}\cdot 4H_2O$	−309
(6)	$H_2AsO_4^{-} + Fe_2(OH)_2(OH_2)_6^{4+} + 7H_2O \rightarrow Fe_2(OH)_2(OH_2)_6HAsO_4^{2+} + H_3O^{+}\cdot 8H_2O$	−336
(7)	$H_2AsO_4^{-} + Fe_2(OH)_2(OH_2)_8^{4+} \rightarrow Fe_2(OH)_2(OH_2)_6H_2AsO_4^{3+} + 2H_2O$	−338

Comparing Reactions (1) with (2), (3) with (4), and (5) with (6) shows that the adsorption of iAs onto the more highly charged surfaces is more energetically favorable. However, it is unlikely that a +4 localized charged would occur in nature, and the results for ΔG_{ads} for the 0 charged Fe clusters are more realistic for exergonic adsorption of H_3AsO_3 and the endergonic adsorption of $HAsO_4^{2-}$. The adsorption of $H_2AsO_4^{-}$ is energetically favorable, regardless of the initial Fe cluster charge (Reactions (5)–(7)).

3.2. Effect of Fe Cluster Hydration on ΔG_{ads} for Anhydrous and Octahydrated $H_2AsO_4^-$

Reactions (8)–(13) in Table 2 give examples of the effect of neutral Fe cluster hydration on the ΔG_{ads} of $H_2AsO_4^-$. The reactions differ only by the number of H_2O molecules of hydration that are present on the initial Fe cluster (*i.e.*, 0, 4, 8, or 8 for Reactions (8)–(11), respectively) and the $HAsO_4^{2-}$-Fe cluster (*i.e.*, 0, 4, 4, and 8, for Reactions (8)–(11), respectively). Reactions (12) and (13) used octahydrated $H_2AsO_4^-$ as the reactant. Note that we used a tetrahydrated hydroxide model to mass and charge balance Reactions (8)–(11).

Table 2. Effect of Fe cluster hydration (Reactions (8)–(11)) and iAs^V hydration (Reactions (12) and (13)) on ΔG_{ads} of iAs^V.

Reaction #	Reaction	ΔG_{ads} (kJ/mol)
(8)	$H_2AsO_4^- + Fe_2(OH)_6(OH_2)_4 + 3H_2O \rightarrow Fe_2(OH)_4(OH_2)_4HAsO_4 + OH^-\cdot4H_2O$	−186
(9)	$H_2AsO_4^- + Fe_2(OH)_6(OH_2)_4\cdot4H_2O + 3H_2O \rightarrow Fe_2(OH)_4(OH_2)_4HAsO_4\cdot4H2O + OH^-\cdot4H_2O$	−195
(10)	$H_2AsO_4^- + Fe_2(OH)_6(OH_2)_4\cdot8H_2O \rightarrow Fe_2(OH)_4(OH_2)_4HAsO_4\cdot4H_2O + OH^-\cdot4H_2O + H_2O$	−217
(11)	$H_2AsO_4^- + Fe_2(OH)_6(OH_2)_4\cdot8H_2O + 3H_2O \rightarrow Fe_2(OH)_4(OH_2)_4HAsO_4\cdot8H_2O + OH^-\cdot4H_2O$	−223
(12)	$H_2AsO_4^-\cdot8H_2O + Fe_2(OH)_6(OH_2)_4\cdot4H_2O \rightarrow Fe_2(OH)_4(OH_2)_4HAsO_4\cdot4H_2O + OH^-\cdot4H_2O + 5H_2O$	−64
(13)	$H_2AsO_4^-\cdot8H_2O + Fe_2(OH)_6(OH_2)_4\cdot8H_2O \rightarrow Fe_2(OH)_4(OH_2)_4HAsO_4\cdot4H_2O + OH^-\cdot4H_2O + 5H_2O$	−86

As the H_2O molecules of hydration increases from $Fe_2(OH)_6(OH_2)_4$ to $Fe_2(OH)_6(OH_2)_4\cdot8H_2O$ (Reactions (8)–(11)), the ΔG_{ads} of $H_2AsO_4^-$ becomes more negative. Reactions (10) and (11) differ only by the numbers of H_2O molecules present on the cluster product, the $HAsO_4^{2-}$-Fe cluster, four for Reaction (10) and eight for Reaction (11). The ΔG_{ads} for Reactions (10) and (11) differ by 6 kJ/mol, which is less than the ±10 kJ/mol error associated with thermodynamics calculations, so the results are indistinguishable. Using eight hydrating H_2O molecules on the $HAsO_4^{2-}$-Fe cluster rather than four significantly increased the time need to minimize the model. Significantly, Reactions (12) and (13), where iAs^V is present as octahydrated $H_2AsO_4^-$ exhibit ΔG_{ads} that are likely more realistic than those seen for ΔG_{ads} of anhydrous $H_2AsO_4^-$ (Reactions (8)–(11)). Reactions (8)–(11) are shown here to emphasize the cluster hydration results. Therefore, we used the anhydrous $H_2AsO_4^-$ reactant in Reactions (8)–(11) because these calculations are focusing on the hydration state of the clusters and are used here as a teaching tool. To attain results that are meaningful with respect to nature and experimental conditions, all of the products and reactants should be hydrated.

Recently [122], a claim was made that using a single explicit H_2O molecule and implicit solvation with the self-consistent reaction field (SCRF) COSMO [133,134] could produce superior results to using multiple H_2O molecules. This argument is based on calculations that used small, monohydrated organic and inorganic molecules in the COSMO SCRF that showed better agreement with experiment when a single, rather than multiple H_2O molecules were used to hydrate the models [172]. However, the work that modeled iAs^V interacting with ferric hydroxide clusters assumed that the result from simple organic and inorganic molecules would be applicable for the cluster calculations [122]; they did not test models with more than one H_2O molecule.

One argument for the addition of multiple H_2O molecules is that the model would better approximate the aqueous environment in which As adsorption occurs. However, if the multiple H_2O molecules are arranged in a way that does not lead to an observable PES minimum (*i.e.*, becoming trapped in a local minimum), then using more than one H_2O molecule could lead to errors. On the other hand, the results for Reactions (12) and (13) suggest that multiple H_2O of hydration for all products and reactants could lead to results that are chemically more realistic than reactions with anhydrous reactants and products. Furthermore, chemical properties such as vibrational frequencies [75] are dependent on hydrogen bonding, so the inclusion of additional explicit H_2O molecules could be necessary to calculate precise spectroscopic and structural results.

Our work used the IEFPCM, and not the COSMO reaction field used by Farrell and Chaudhary [122]. A particular quantum chemistry method such as the SCRF, DFT method, or basis set can be useful for precisely calculating particular chemical properties, such as energies, but may produce other chemical properties that are imprecise. In this instance, the COSMO SCRF was parameterized to work most successfully with limited explicit hydration, but other SCRF such as IEFPCM may require the addition of more H_2O molecules to obtain precise results. Furthermore, particular DFT methods have been developed that are useful for calculating energies, geometries, and kinetics [138,173,174], whereas other DFT methods are useful for calculating spectroscopic properties such as NMR chemical shifts for H and C [175,176]. Because the exact E_{xc} for DFT is unknown, it is not yet possible to use one DFT method to calculate every chemical property. Therefore, it is necessary when doing DFT calculations to read the literature to find methodologies that are efficient for calculating the chemical properties of interest and to be willing to experiment with variations of those methods, if the calculated results are imprecise when compared with experimental data. This procedure is similar to those an experimentalist takes when deciding how to study a chemistry of interest.

3.3. Effect of As Oxidation State and DFT Method on ΔG_{ads}

Reactions (14)–(19) in Table 3 show the ΔG_{ads} of octahydrated H_3AsO_3, $HAsO_4^{2-}$, and $H_2AsO_4^{-}$ onto tetra- and octahydrated $Fe_2(OH)_6(OH_2)_4^{0}$. The ΔG_{ads} for iAs^{III} is more favorable than it is for iAs^V; this is fortunate, if correct, because iAs^{III} is more toxic than iAs^V is. The results for the preferential adsorption of iAs^{III} over iAs^V at the point-of-zero charge for the Fe cluster are supported by experimental data that shows the same trend [101]. Under acidic or basic conditions, the Fe clusters would have charges, and neutral H_3AsO_3 might not adsorb to Fe surfaces as favorably. Conversely, although these reactions show less favorable adsorption of octahydrated $HAsO_4^{2-}$ and $H_2AsO_4^{2-}$ to the neutral Fe cluster than for iAs^{III}, under acidic or basic conditions when the cluster is charged, the charged iAs^V ions could adsorb more favorably than uncharged iAs^{III}. Although the products and reactants are anhydrous, Reactions (1)–(7) support this assertion, where neutral H_3AsO_3 adsorbs more strongly to the neutral cluster (Reaction (1)) than to the +4-charged cluster (Reaction (2)). Conversely, the charged iAs^V reactants in Reactions (3)–(7) adsorb more strongly to the +4-charged clusters than they do to the neutral clusters. These results that show weaker adsorption of H_3AsO_3 and stronger adsorption of charged iAs^V to charged Fe clusters are also supported by experimental data [101]. Furthermore, the results for Reactions (14) and (15) show that $HAsO_4^{2-}$ would adsorb less favorable to the neutral Fe cluster than does $H_2AsO_4^{-}$. Note that a tetrahydrated hydroxide model was used to mass and charge balance Reactions (14)–(17); the hydroxide was not necessary to mass charge balance Reactions (12) and (13).

Table 3. Effect of As oxidation state on BB adsorption to neutral Fe clusters. For Reaction (16), the density functional theory (DFT) methods used to calculate ΔG_{ads} were: [a], B3LYP; [b], PBE0; and [c], M06-L.

Reaction #	Reaction	ΔG_{ads} (kJ/mol)
(12)	$H_3AsO_3 \cdot 8H_2O + Fe_2(OH)_6(OH_2)_4 \cdot 4H_2O \rightarrow Fe_2(OH)_4(OH_2)_4HAsO_3 \cdot 4H_2O + 10H_2O$	-124
(13)	$H_3AsO_3 \cdot 8H_2O + Fe_2(OH)_6(OH_2)_4 \cdot 8H_2O \rightarrow Fe_2(OH)_4(OH_2)_4HAsO_3 \cdot 4H_2O + 10H_2O$	-146
(14)	$HAsO_4^{2-} \cdot 8H_2O + Fe_2(OH)_6(OH_2)_4 \cdot 4H_2O \rightarrow Fe_2(OH)_4(OH_2)_4HAsO_4 \cdot 4H_2O + 2OH^- \cdot 4H_2O$	$+15$
(15)	$HAsO_4^{2-} \cdot 8H_2O + Fe_2(OH)_6(OH_2)_4 \cdot 8H_2O \rightarrow Fe_2(OH)_4(OH_2)_4HAsO_4 \cdot 4H_2O + 2OH^- \cdot 4H_2O + 4H_2O$	-6
(16)	$H_2AsO_4^- \cdot 8H_2O + Fe_2(OH)_6(OH_2)_4 \cdot 4H_2O \rightarrow Fe_2(OH)_4(OH_2)_4HAsO_4 \cdot 4H_2O + OH^- \cdot 4H_2O + 5H_2O$	-64 [a], -35 [b], -3 [c]
(17)	$H_2AsO_4^- \cdot 8H_2O + Fe_2(OH)_6(OH_2)_4 \cdot 8H_2O \rightarrow Fe_2(OH)_4(OH_2)_4HAsO_4 \cdot 4H_2O + OH^- \cdot 4H_2O + 5H_2O$	-86

For Reaction (16), we compared the ΔG_{ads} calculated with the B3LYP, PBE0, or M06-L DFT methods. The ΔG_{ads} calculated with B3LYP (-64 kJ/mol), M06-L (-3 kJ/mol), and PBE0 (-35 kJ/mol) results all predict favorable, exergonic reactions. These results show that calculated thermodynamic results are dependent on the chosen DFT method. Because the adsorption of iAs^V is experimentally observed over a wide pH range [101], the results from B3LYP, PBE0, or M06-L could be correct. Thermodynamic results from DFT calculations are typically precise within ±10 kJ/mol; therefore, the B3LYP results would range from -74 to -54 kJ/mol and the PBE0 results would range from -45 to -25 kJ/mol; these results do not overlap, so the precision of the results from B3LYP and PBE0 differ. Within the ±10 kJ/mol error range of DFT methods, the M06-L results would range from -13 to $+7$ kJ/mol; therefore, because iAs^V adsorption is favorable, the M06-L results could be erroneous.

3.4. As—Fe Distance and As-O Bond Length Data from Experiments Compared with Cluster and Periodic Model Results

Table 4 shows the BB As—Fe distances and As-O bond distances calculated from this work for As^{III} and As^V, the calculated BB results of Sherman and Randall [91] for As^V, and EXAFS data for As^{III} and As^V on four mineral surfaces.

Notably, that the $Fe_2(OH)_2(OH_2)_6H_2As^VO_4^{3+}$ (BB) model [91] exhibits four As-O bonds, two of which are As-OH bonds that are 1.73 Å. Sherman and Randall [91] observed a 1.62–1.64 Å As-O bond with EXAFS that is not present in their models where $H_2AsO_4^-$ has two As-OH single bonds to the As atom. We argue that the 1.62–1.64 Å As-O bond is an As-O double bond (As=O) that our calculations predict to be 1.63 Å due to $HAsO_4^{2-}$ being adsorbed to the Fe surface (model $Fe_2(OH)_4(OH_2)_4HAs^VO_4$ (BB)), rather than a As-OH single bond, and that iAs^V is not present on Fe surfaces as $H_2AsO_4^-$ with two As-OH single bonds. However, we also note that the $Fe_2(OH)_4(OH_2)_4HAs^VO_4$ (BB) model is not hydrated with explicit H_2O molecules and that when the model is hydrated with either four or eight explicit H_2O molecules, the As=O bond length increases from 1.63 to 1.67 Å, which is slightly longer than the observed 1.62–1.64 Å As=O length, but is still shorter than the 1.73 Å bond lengths from the model of Sherman and Randall [91]. In Table 4, we report the results for the +3-charged model that was energy minimized using the methods described in this work. The As-OFe and As-OH bond distances calculated here differ by 0.01 Å from those reported by Sherman and Randall [91], whereas the As—Fe distances calculated here are both 0.05 Å shorter than those reported by Sherman and Randall [91]. The difference in As—Fe distances likely occur due to differences in methodology, but the results from both models lie within the range of experimental uncertainty.

Table 4. As—Fe interatomic distance, As-OFe bond distance, and As-O bond distance results from this work compared to previous calculations and extended X-ray adsorption fine structure (EXAFS) data for As^V and As^{III} adsorption onto ferrihydrite (Fh), hematite (Hm), goethite (Gt), and lepidocrocite (Lp). For $Fe_2(OH)_4(OH_2)_4HAs^VO_4 \cdot 4H_2O$ (BB), the DFT methods used to calculate ΔG_{ads} were: [x], B3LYP; [y], PBE0; and [z], M06-L.

As^V Complex	As—Fe (Å)	As—Fe (Å)	As-OFe (Å)	As-OFe (Å)	As-OH (Å)	As-OH (Å)	As=O (Å)
$Fe_2(OH)_4(OH_2)_4HAs^VO_4$ (BB)	3.13	3.25	1.71	1.73	1.83		1.63
$Fe_2(OH)_4(OH_2)_4HAs^VO_4 \cdot 4H_2O$ (BB)	3.20 [x], 3.19 [y], 3.08 [z]	3.28 [x], 3.24 [y], 3.29 [z]	1.69 [x], 1.68 [y], 1.69 [z]	1.72 [x], 1.71 [y], 1.72 [z]	1.76 [x], 1.75 [y], 1.79 [z]		1.67 [x], 1.66 [y], 1.65 [z]
$Fe_2(OH)_4(OH_2)_4HAs^VO_4 \cdot 8H_2O$ (BB)	3.30	3.30	1.70	1.70	1.76		1.67
Goethite (010) periodic model (BB)	<u>3.56</u>	<u>3.68</u>	<u>1.72</u>	<u>1.72</u>	<u>1.78</u>		<u>1.73</u>
$Fe_2(OH)_2(OH_2)_6H_2As^VO_4^{3+}$ (BB)	3.24	3.24	1.70	1.70	1.72	1.72	
$Fe_2(OH)_2(OH_2)_6H_2As^VO_4^{3+}$ (BB) [a]	3.29	3.29	1.71	1.71	1.73	1.73	
As^V on Fh (BB) [a]	3.27	3.38	1.70	1.70	1.67		1.64
As^V on Gt (BB) [a]	3.30	3.30	1.70	1.70	1.70		1.63
As^V on Lp (BB) [a]	3.30	3.32	1.71	1.71	1.66		1.63
As^V on Hm (BB) [a]	3.24	3.35	1.70	1.70	1.70		1.62
As^V on Fh (BB) [b]	3.25 (+0.02?)						
As^V on Gt (BB) [b]	3.28 (±0.01)						
As^V on Lp (BB) [c]	3.31 (±0.014)		1.69 (±0.004)				
As^V on Gt (BB) [c]	3.30 (±0.008)		1.69 (±0.004)				
As^V on Fh (BB) [d]	3.27						
Goethite (010) periodic model (MM)	3.54	5.00 †	1.78	1.75	1.71 ‡		1.68
AsV on Gt [e]	<u>3.25 §</u>		<u>1.689</u>				<u>1.679</u>
As^{III} Complex	As—Fe (Å)	As—Fe (Å)	As-O (Å)	As-O (Å)		As-O (Å)	As=O (Å)
$Fe_2(OH)_4(OH_2)_4HAs^{III}O_3 \cdot 4H_2O$ (BB)	3.26	3.41	1.77	1.74		1.90	na
$Fe_2(OH)_4(OH_2)_4HAs^{III}O_3 \cdot 8H_2O$ (BB)	3.29	3.39	1.78	1.72		1.90	
As^{III} on Lp (BB) [c]	3.41 (±0.013)		1.78 (±0.014)				
As^{III} on Gt (BB) [c]	3.31 (±0.013)		1.78 (±0.012)				
As^{III} on Fh (BB) [d]	3.41–3.44						
As^{III} on Fh and Hm (BB) [f]	3.35 (±0.05)						
As^{III} on Gt and Lp (BB) [f]	3.3–3.4						
As^{III} on Gt (BB) [g]	3.378 (±0.014)						

Notes: [a] Sherman and Randall [91]; [b] Waychunas et al. [89]; [c] Farquhar et al. [82]; [d] Gao et al. [87]; [e] Loring et al. [128]; [f] Ona-Nguema et al. [92]; [g] Manning et al. [81]. † This As—Fe distance does not agree with that reported by Loring et al. [128] for a MM. ‡ For the MM periodic model, there is one As-OH bond and two As partial double bonds (1.71 and 1.68 Å), because $HAsO_4^{2-}$ is adsorbed to the surface. § For the Loring et al. [128] model, both As-O bonds are aprotic and should have partial double bonds.

For the iAs^V models, the two calculated As—Fe distances within each configuration differ by 0.12, 0.08, and 0.00 Å from each other for the $Fe_2(OH)_4(OH_2)4HAs^VO_4$, $Fe_2(OH)_4(OH_2)_4HAs^VO_4 \cdot 4H_2O$, and $Fe_2(OH)_4(OH_2)4HAs^VO_4 \cdot 8H_2O$ models, respectively. Sherman and Randall [91] reported two As—Fe distances for ferrihydrite (Fh), lepidocrocite (Lp), hematite (Hm), and goethite (Gt), whereas the other data report a single As—Fe distance for the adsorption onto Fh, Gt, or Lp. The As^V—Fe distance results from the $Fe_2(OH)_4(OH_2)_4HAs^VO_4 \cdot 8H_2O$ model agree within experimental uncertainty with the Gt and Lp data of Sherman and Randall [91], the Gt and Fh data of Waychunas *et al.* [89], and the Gt and Lp data of Farquhar *et al.* [82]. The data for the As^V—Fe distances overlap for the minerals used for these studies; therefore, determining the type of mineral to which the iAs^V is bonding could be difficult; however, we can state that the Fe cluster models are predicting As—Fe distances that are indicative of BB adsorption of As^V. Similarly, the calculated experimental and As^V—OFe bond distances agree precisely.

For the BB results for the periodic structures of α-FeOOH (010), the As-OFe bonds (1.72 Å) show precise agreement with experiment and correlate well with the cluster results. However, the As—Fe distances (3.56 and 3.68 Å) both overestimate the experimental data, and the As-OH and As=O bonds both overestimate the experimentally observed bond distances. For the MM periodic model, the calculated As—Fe distance is overestimating the 3.25 Å distance measured distance of Loring *et al.* [128] by 0.29 Å. The errors associated with the periodic calculations could be due to systematic errors in the planewave calculations or could be because the α-FeOOH (010) surface is not the surface where As adsorption predominately occurs.

The two As^{III}—Fe bond distances in the $Fe_2(OH)_4(OH_2)_4HAs^{III}O_3 \cdot 4H_2O$ model differ by 0.15 Å and by 0.10 for the octahydrated version of that model., The longer calculated As—Fe distances (*ca.* 3.4 Å) agree with most of the As^{III} data within uncertainty, whereas the shorter calculated As—Fe distance agrees well with the data of Gao *et al.* [87]. Again, because of the data overlap and due to the uncertainty in the As—Fe distances observed for As^{III} adsorption onto Hm, Gt, Lp, and Fh, it is difficult to resolve different sorption mechanisms of various Fe-oxide and Fe-hydroxide minerals; however, it is possible to state the BB adsorption is occurring. Furthermore, it is possible to differentiate As^{III}—Fe and As^V—Fe BB adsorption, because the former distances are approximately 3.4 Å, whereas the latter are approximately 3.3 Å. Significantly, these distances are seen both experimentally and computationally. Moreover, there is good agreement between the calculated As^{III}-OFe distance and the EXAFS data of Sherman and Randall [91].

For the $Fe_2(OH)_4(OH_2)_4HAs^VO_4 \cdot 4H_2O$ (BB) model, we compared the As—Fe distance results obtained from the B3LYP, PBE0, and M06-L methods. For the As—Fe distances, the B3LYP (3.20 Å) and PBE0 (3.19 Å) results correlate for one of the As—Fe distance, and the B3LYP (3.28 Å) and M06-L (3.29 Å) results agree for the other As—Fe distance. The As-O and As=O bond lengths agree well among the DFT methods and agree precisely with the experimental data in Table 4. This type of DFT method testing helps eliminate the possible effects of exchange-correlation functional errors on the results.

The $Fe_2(OH)_4(OH_2)_4HAs^VO_4 \cdot 4H_2O$ (BB) model used for these DFT method comparisons is the adsorption product of Reaction (16), and the M06-L ΔG_{ads} results for Reaction (16) ranged from −13 to +7 kJ/mol, suggesting thermodynamic adsorption results from M06-L that are potentially unfavorable, relative to the results from B3LYP and PBE0 (Table 3). In addition, the 3.08 Å As—Fe distance from the M06-L minimized model (Table 4) is predicts is significantly shorter than the experimental data and the results from B3LYP and PBE0. Therefore, because the As—Fe distance calculated with M06-L is imprecise, it is likely that the ΔG_{ads} results from M06-L is also imprecise.

Notably, the PBE0 As—Fe distance results from the cluster and periodic calculations differed distinctly. The cluster results underestimated the As—Fe distance data by approximately 0.1 Å, whereas the periodic calculation results overestimated those data by approximately 0.25 Å. The PBE0 method, like many DFT methods, may contain different parameters, depending on which software package implements it (e.g., VASP, Gaussian 09, *etc.*), so the results obtain with a particular DFT method using different software packages might not be directly comparable. In addition to the potential differences in the DFT methods, model sizes could also contribute to the discrepancies in the calculated distances obtained from the periodic and cluster models. One would presume that the larger periodic models would provide results that are more precise relative to the data than the smaller cluster models do; however, neither model size produced precise As—Fe distances. Differences between periodic and cluster model results have been discussed previously [73,135].

3.5. Sorption Kinetics for iAsV on Cluster and Periodic Models

Calculations were completed to show a possible reaction pathways for desorption of the monodentate inner-sphere complex of $HAsO_4^{2-}$ from Fe^{3+}-(oxyhydr)oxide cluster and from the periodic goethite (010) surface. Although the bidentate binuclear complex is likely to be more stable [88,91], the monodentate species is an intermediate between the bidentate and outer-sphere species. Figure 4 shows desorption of iAsV from a $Fe_2(OH)_4(H_2O)_5$-$HAsO_4$ cluster model, where the model begins as a MM structure with a As—Fe distance of 3.27 Å (Figure 4A), moves through a transition state structure (Figure 4B), and reaches the outer-sphere structure (Figure 4C) where the As—Fe distance is 4.36 Å. The Fe—As distances were increased manually and then held constant in each calculation until there ceased to be a bonding interaction. The energies of the monodentate reaction pathway are portrayed as a function of the Fe—As distance in the Figure 4 for both periodic and cluster models.

Figure 4. Desorption of iAs^V from Fe clusters showing (**A**) the initial MM model, (**B**) the transition state model, and (**C**) the outer-sphere, final structure model.

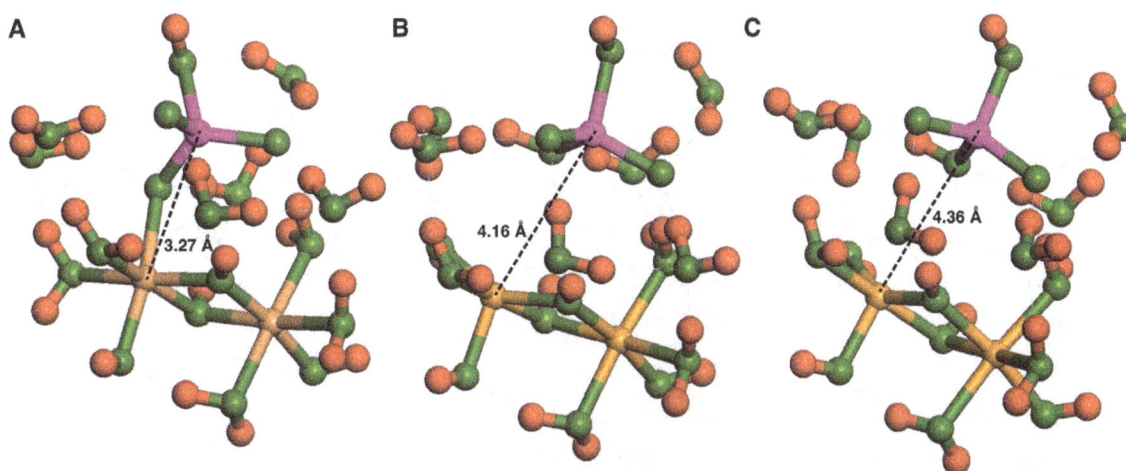

Based on the results shown in Figure 5, ΔE_a for the breaking of the first bond in monodentate complex requires approximately +133 and +70 kJ/mol in the periodic and cluster models, respectively. (Note that there is a small increase in energy of the model system near a Fe—As distance of 4.2 Å, but this energy increase is insignificant compared to the first barrier.) The energy barrier for the reverse reaction is higher, +148 kJ/mol, in the periodic model because the outer-sphere complex is lower in energy in this model. The higher energy of the inner-sphere complex and the high-energy barrier of adsorption suggest that adsorption would not occur to the goethite (010) surface under these conditions; this result corroborates with the discussion the long As—Fe distances reported in the previous section for the periodic models. In the molecular cluster models, the adsorption reaction barrier is insignificant, *i.e.*, +1 kJ/mol. We strongly remind the reader, however, that the conditions of this model are not realistic compared to experimental conditions where lower H^+ activity, lower arsenate concentrations and greater volume of water would exist and affect the results. On the other hand, the cluster models exhibit almost no energy barrier to adsorption from outer-sphere to monodentate with the inner-sphere complex having a lower energy (Figure 5).

Figure 5. Constrained scan of Fe—As distances starting from monodentate configuration to outer-sphere using periodic and molecular cluster DFT calculations results in ΔE_a of adsorption of +148 and +1 kJ/mol, respectively and desorption of +133 and +70 kJ/mol, respectively.

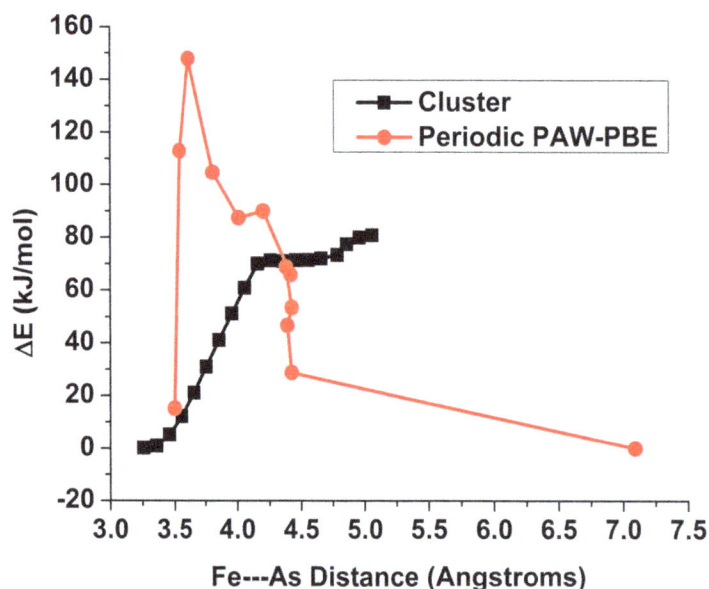

The discrepancies between the two types of models are illustrative of some of the problems that can be encountered by each type of approach. First, although the periodic models were run for short (*i.e.*, 6000 steps × 0.5 fs = 3 ps) molecular dynamics simulations at 300 K to relax the atoms, the complex nature of the periodic model all but ensures that a global minimum configuration will not be obtained. This is an example of a general problem, *i.e.*, adding more atoms to the simulation may make it more realistic but increases the number of potential energy minima dramatically. Thus, the transition state may overestimate the ΔE_a because the system is not in the lowest possible potential energy configuration, especially with respect to the configuration of H_2O molecules.

There are numerous possibilities for overcoming the metastable minimum problem. Longer simulation runtimes are one option. These longer simulations could be performed with tight-binding DFT (*i.e.*, DFT-B; see REF for a review of DFT-B) or classical force fields. However, one problem with classical simulations is that it is difficult to created accurate parameterizations that allow for bond-making and bond-breaking, especially for configurations far from equilibrium such as transition states (see [71] for a review). Replicate MD simulations are another option for exploring configuration space. These require multiple simulations at different temperatures to be run simultaneously such that higher and lower temperature configurations can be switched with potential energies overlap. Again, however, this method requires a significant expansion of computational effort to run the multiple MD simulations.

Alternatively, the cluster model allows the "surface" atoms to relax without constraint from the remainder of the crystal, which may help explain the lower ΔE_a. The "outer-sphere" configuration is higher in energy in this case because the $HAsO_4^{2-}$ is not completely solvated by the six extra H_2O molecules in the model. In addition to the loss of solvation energy, protonation of the $HAsO_4^{2-}$ does not occur in the cluster whereas in the periodic model $H_2AsO_4^-$ is the product.

Even with these discrepancies in the energies and products between the periodic and cluster models, the structures of the reactants and transition states are similar. A combined approach using insights from both periodic and cluster models is useful at this time because each has strengths and weaknesses. These first simulations of the reaction path can be reiterated by performing longer MD simulations at various steps and by searching for lower energy points along the reaction path. In addition, lower energy transition states determined via the cluster approach can be used to guide construction of transition states in the more realistic (but more complex) periodic simulations.

The main problem in this case, however, is likely to be that the (010) surface is not the dominant face responsible for adsorbing arsenate onto goethite. Thus, extensive calculations of any type could prove futile in terms of reproducing the observed ΔE_a of adsorption/desorption. Other surfaces could be examined, because the (010) surface may not be the most preferred surface for adsorption of arsenate onto goethite. We had selected the (010) surface for arsenate adsorption onto goethite based on the analogy with chromate adsorption [177,178]. Recent DFT calculations have been used to suggest that the (210) surface adsorbs phosphate more strongly than the (010) surface [135]. None of the three just-mentioned studies included arsenate, however, so we are using the other oxyanions as analogs for arsenate. Our future computational work will focus on arsenic sorption reaction mechanisms onto the (210) surface, and this work would benefit from experiments similar to those that used chromate [177]. This point emphasizes that realistic model construction is one of the most important considerations in performing computational geochemistry. Too often, missing components are some inaccuracy in the original model creation leads to discrepancies with observation that cannot be resolved with even the most accurate quantum mechanical calculations.

4. Conclusions

This work explored the effect of cluster charge, hydration, As oxidation state, and DFT methods on the Gibbs free energy of adsorption (ΔG_{ads}) of inorganic arsenic (iAs) species onto Fe^{3+}-(oxyhydr)oxide models. In general neutral clusters and hydrated models produced ΔG_{ads} results that are likely more realistic than models with charged clusters and anhydrous models. As shown with experiments [101], iAs^{III} adsorption onto neutral Fe^{3+}-(oxyhydr)oxide cluster models was more exergonic than iAs^V adsorption onto the same cluster models. For the DFT calculations on the clusters, the results showed that both ΔG_{ads} and As—Fe distances depend on the DFT method used to calculate those properties; however the As—Fe distance results from these calculations generally agreed precisely with the experimental data cited.

The cluster model As—Fe distance and As-O bond distance results showed relatively precise agreement with the experimental data. Conversely, the periodic planewave calculation results for iAsV adsorption onto α-FeOOH (010) generally overestimated the As—Fe distance and As-O bond length data for iAs^V adsorption onto goethite. Other α-FeOOH surfaces could produce results that are more precise.

Sorption kinetics calculations using DFT with cluster and a periodic model of α-goethite (010) showed discrepancies in the calculated activation energies of iAs^V adsorption. One major difference for the discrepant results could be that the relatively large periodic model did not reach an energy minimum during the DFT MD simulation, whereas the cluster model was smaller than the periodic model and did attain a PES minimum. Although the calculated activation energies for the two methods differed, the initial and transition state structures for both calculations were structurally similar. Longer DFT MD simulations and periodic structures other than the (010) surface of α-FeOOH could produce results that are more precise.

The calculated reaction rates, thermodynamics, and structural results presented in this work provide results that could lead to a better understanding of the adsorption of arsenic to Fe (oxyhydr)oxide minerals. However, further studies are necessary to better determine which DFT methods produce the most precise results, the effect of model size on model precision, and the effects of model hydration and surface charge on As adsorption to Fe (oxyhydr)oxide models. Furthermore, basis set size, which was not addressed herein, could potentially affect the precision of the results for the cluster models; therefore, future studies should include the evaluation of basis set effects. Furthermore, increased collaborative efforts among experimental and computational (geo)chemists could lead to improved knowledge about arsenic adsorption on Fe minerals.

Acknowledgments

The authors would like to thank Hind A. Al-Abadleh, Associate Director at the Laurier Centre for Women in Science (WinS) for her help with editing and improving this paper. We also thank Benjamin Tutolo for running preliminary calculations on arsenate desorption who was supported by a research experience for undergraduates grant from the National Science Foundation under Grant No. CHE-0431328 (the Center for Environmental Kinetics Analysis, an NSF-DOE environmental molecular sciences institute). Computational support was provided by the Research Computing and Cyberinfrastructure group at The Pennsylvania State University. Lorena Tribe acknowledges support of the Research Collaboration Fellowship funded by The Pennsylvania State University.

Author Contributions

Heath D. Watts did the work reported in Sections 3.1–3.4. James D. Kubicki and Lorena Tribe did the work reported in Section 3.5. All authors contributed to the preparation and writing of the manuscript.

Conflicts of Interest

The authors declare no conflict of interest.

References and Notes

1. Kim, K.-W.; Chanpiwat, P.; Hanh, H.T.; Phan, K.; Sthiannopkao, S. Arsenic geochemistry of groundwater in Southeast Asia. *Front. Med.* **2011**, *5*, 420–433.

2. Smedley, P.L.; Kinniburgh, D.G. A review of the source, behaviour and distribution of arsenic in natural waters. *Appl. Geochem.* **2002**, *17*, 517–568.

3. Anawar, H.M.; Akai, J.; Mihaljevič, M.; Sikder, A.M.; Ahmed, G.; Tareq, S.M.; Rahman, M.M. Arsenic contamination in groundwater of bangladesh: Perspectives on geochemical, microbial and anthropogenic issues. *Water* **2011**, *3*, 1050–1076.

4. Drewniak, L.; Maryan, N.; Lewandowski, W.; Kaczanowski, S.; Sklodowska, A. The contribution of microbial mats to the arsenic geochemistry of an ancient gold mine. *Environ. Pollut.* **2012**, *162*, 190–201.

5. Signes-Pastor, A.; Burló, F.; Mitra, K.; Carbonell-Barrachina, A.A. Arsenic biogeochemistry as affected by phosphorus fertilizer addition, redox potential and pH in a west Bengal (India) soil. *Geoderma* **2007**, *137*, 504–510.

6. Oremland, R.S.; Stolz, J.F. The ecology of arsenic. *Science* **2003**, *300*, 939–944.

7. Dhuldhaj, U.P.; Yadav, I.C.; Singh, S.; Sharma, N.K. Microbial interactions in the arsenic cycle: Adoptive strategies and applications in environmental management. *Rev. Environ. Contam. Toxicol.* **2013**, *224*, 1–38.

8. Moreno-Jiménez, E.; Esteban, E.; PeñaloBa, J.M The fate of arsenic in soil-plant systems. *Rev. Environ. Contam. Toxicol.* **2012**, *215*, 1–37.

9. Wolfe-Simon, F.; Blum, J.S.; Kulp, T.R.; Gordon, G.W.; Hoeft, S.E.; Pett-Ridge, J.; Stolz, J.F.; Webb, S.M.; Weber, P.K.; Davies, P.C.W.; *et al.* A bacterium that can grow by using arsenic instead of phosphorus. *Science* **2011**, *332*, 1163–1166.

10. Wolfe-Simon, F.; Blum, J.S.; Kulp, T.R.; Gordon, G.W.; Hoeft, S.E.; Pett-Ridge, J.; Stolz, J.F.; Webb, S.M.; Weber, P.K.; Davies, P.C.W.; *et al.* Response to comments on "A bacterium that can grow using arsenic instead of phosphorus." *Science* **2011**, *332*, 1149–1149.

11. Masscheleyn, P.H.; Delaune, R.D.; Patrick, W.H. Arsenic and selenium chemistry as affected by sediment redox potential and pH. *J. Environ. Qual.* **1991**, *20*, 522–527.

12. Cullen, W.R.; Reimer, K.J. Arsenic speciation in the environment. *Chem. Rev.* **1989**, *89*, 713–764.

13. Greenwood, N.N.; Earnshaw, A. Arsenic, Antimony and Bismuth. In *Chemistry of the Elements*; Pergamon Press: Oxford, UK, 1997; pp. 547–599.

14. Masscheleyn, P.H.; Delaune, R.D.; Patrick, W.H. Effect of redox potential and pH on arsenic speciation and solubility in a contaminated soil. *Environ. Sci. Technol.* **1991**, *25*, 1414–1419.

15. Flis, I.E.; Mishchenko, K.P.; Tumanova, T.A.; Russ, J. Dissociation of arsenic acid. *J. Inorg. Chem.* **1959**, *4*, 120–124.

16. Francesconi, K.A. Arsenic species in seafood: Origin and human health implications. *Pure Appl. Chem.* **2010**, *82*, 373–381.

17. Beauchemin, S.; Fiset, J.-F.; Poirier, G.; Ablett, J. Arsenic in an alkaline AMD treatment sludge: Characterization and stability under prolonged anoxic conditions. *Appl. Geochem.* **2010**, *25*, 1487–1499.

18. Cheng, H.; Hu, Y.; Luo, J.; Xu, B.; Zhao, J. Geochemical processes controlling fate and transport of arsenic in acid mine drainage (AMD) and natural systems. *J. Hazard. Mater.* **2009**, *165*, 13–26.

19. Cramer, S.P.; Siskin, M.; Brown, L.D.; George, G.N. Characterization of arsenic in oil shale and oil shale derivatives by X-ray absorption spectroscopy. *Energy Fuels* **1988**, *2*, 175–180.

20. Pelley, J. Commonarsenical pesticide under scrutiny. *Environ. Sci. Technol.* **2005**, *39*, 122–123.

21. Arai, Y.; Lanzirotti, A.; Sutton, S.; Davis, J.A.; Sparks, D.L. Arsenic speciation and reactivity in poultry litter. *Environ. Sci. Technol.* **2003**, *37*, 4083–4090.

22. Argos, M.; Kalra, T.; Rathouz, P.J.; Chen, Y.; Pierce, B.; Parvez, F.; Islam, T.; Ahmed, A.; Rakibuz-Zaman, M.; Hasan, R.; *et al.* Arsenic exposure from drinking water, and all-cause and chronic-disease mortalities in Bangladesh (HEALS): A prospective cohort study. *Lancet* **2010**, *376*, 252–258.

23. Chen, Y.; Graziano, J.H.; Parvez, F.; Liu, M.; Slavkovich, V.; Kalra, T.; Argos, M.; Islam, T.; Ahmed, A.; Rakibuz-Zaman, M.; *et al.* Arsenic exposure from drinking water and mortality from cardiovascular disease in Bangladesh: Prospective cohort study. *Br. Med. J.* **2011**, *342*, d2431, doi:10.1136/bmj.d2431.

24. Das, N.; Paul, S.; Chatterjee, D.; Banerjee, N.; Majumder, N.S.; Sarma, N.; Sau, T.J.; Basu, S.; Banerjee, S.; Majumder, P.; *et al.* Arsenic exposure through drinking water increases the risk of liver and cardiovascular diseases in the population of West Bengal, India. *BMC Public Health* **2012**, *12*, 639, doi:10.1186/1471-2458-12-639.

25. Ferreccio, C.; Smith, A.H.; Durán, V.; Barlaro, T.; Benítez, H.; Valdés, R.; Aguirre, J.J.; Moore, L.E.; Acevedo, J.; Vásquez, M.I.; *et al.* Case-control study of arsenic in drinking water and kidney cancer in uniquely exposed Northern Chile. *Am. J. Epidemiol.* **2013**, *178*, 813–818.

26. Meliker, J.R.; Slotnick, M.J.; AvRuskin, G.A.; Schottenfeld, D.; Jacquez, G.M.; Wilson, M.L.; Goovaerts, P.; Franzblau, A.; Nriagu, J.O. Lifetime exposure to arsenic in drinking water and bladder cancer: A population-based case-control study in Michigan, USA. *Cancer Causes Control* **2010**, *21*, 745–757.

27. Paul, S.; Bhattacharjee, P.; Mishra, P.K.; Chatterjee, D.; Biswas, A.; Deb, D.; Ghosh, A.; Mazumder, D.N.G.; Giri, A.K. Human urothelial micronucleus assay to assess genotoxic recovery by reduction of arsenic in drinking water: A cohort study in West Bengal, India. *Biometals* **2013**, *26*, 855–862.

28. Smith, A.H.; Marshall, G.; Yuan, Y.; Liaw, J.; Ferreccio, C.; Steinmaus, C. Evidence from Chile that arsenic in drinking water may increase mortality from pulmonary tuberculosis. *Am. J. Epidemiol.* **2011**, *173*, 414–420.

29. Concha, G.; Broberg, K.; Grandér, M.; Cardozo, A.; Palm, B.; Vahter, M. High-level exposure to lithium, boron, cesium, and arsenic via drinking water in the Andes of northern Argentina. *Environ. Sci. Technol.* **2010**, *44*, 6875–6880.

30. Kinniburgh, D.G.; Smedley, P.L.; Davies, J.; Milne, C.J.; Gaus, I.; Trafford, J.M.; Ahmed, K.M. The Scale and Causes of the Groundwater Arsenic Problem in Bangladesh. In *Arsenic in Ground Water*; Springer: Berlin, Germany, 2003; pp. 211–257.

31. Mondal, D.; Banerjee, M.; Kundu, M.; Banerjee, N.; Bhattacharya, U.; Giri, A.K.; Ganguli, B.; Sen Roy, S.; Polya, D.A. Comparison of drinking water, raw rice and cooking of rice as arsenic exposure routes in three contrasting areas of West Bengal, India. *Environ. Geochem. Health* **2010**, *32*, 463–477.

32. Sun, G.; Li, X.; Pi, J.; Sun, Y.; Li, B.; Jin, Y.; Xu, Y. Current research problems of chronic arsenicosis in China. *J. Health Popul. Nutr.* **2011**, *24*, 176–181.

33. Nickson, R.; McArthur, J.; Burgess, W.; Ahmed, K.M.; Ravenscroft, P.; Rahman, M. Arsenic poisoning of Bangladesh groundwater. *Nature* **1998**, *395*, 338.

34. Nicolli, H.B.; Bundschuh, J.; Blanco, M.C.; Tujchneider, O.C.; Panarello, H.O.; Dapeña, C.; Rusansky, J.E. Arsenic and associated trace-elements in groundwater from the Chaco-Pampean plain, Argentina: Results from 100 years of research. *Sci. Total Environ.* **2012**, *429*, 36–56.

35. Sharma, V.K.; Sohn, M. Aquatic arsenic: Toxicity, speciation, transformations, and remediation. *Environ. Int.* **2009**, *35*, 743–759.

36. Malik, A.H.; Khan, Z.M.; Mahmood, Q.; Nasreen, S.; Bhatti, Z.A. Perspectives of low cost arsenic remediation of drinking water in Pakistan and other countries. *J. Hazard. Mater.* **2009**, *168*, 1–12.

37. Welch, A.H.; Stollenwerk, K.G.; Maurer, D.K.; Feinson, L.S. *In Situ* Arsenic Remediation in a Fractured, Alkaline Aquifer. In *Arsenic in Ground Water*; Welch, A.H., Stollenwerk, K.G., Eds.; Kluwer Academic Publishers: Dordrecht, The Netherlands, 2003; pp. 403–419.

38. Beaulieu, B.; Ramirez, R.E. Arsenic remediation field study using a sulfate reduction and zero-valent iron PRB. *Groundw. Monit. Remediat.* **2013**, *33*, 85–94.

39. Berg, M.; Luzi, S.; Trang, P.T.K.; Viet, P.H.; Giger, W.; Stüben, D. Arsenic removal from groundwater by household sand filters: Comparative field study, model calculations, and health benefits. *Environ. Sci. Technol.* **2006**, *40*, 5567–5573.

40. Neumann, A.; Kaegi, R.; Voegelin, A.; Hussam, A.; Munir, A.K.M.; Hug, S.J. Arsenic removal with composite iron matrix filters in Bangladesh: A field and laboratory study. *Environ. Sci. Technol.* **2013**, *47*, 4544–4554.

41. Jeon, C.-S.; Park, S.-W.; Baek, K.; Yang, J.-S.; Park, J.-G. Application of iron-coated zeolites (ICZ) for mine drainage treatment. *Korean J. Chem. Eng.* **2012**, *29*, 1171–1177.

42. Wu, K.; Liu, R.; Liu, H.; Chang, F.; Lan, H.; Qu, J. Arsenic species transformation and transportation in arsenic removal by Fe-Mn binary oxide–coated diatomite: Pilot-scale field Study. *J. Environ. Eng.* **2011**, *137*, 1122–1127.

43. Mudhoo, A.; Sharma, S.K.; Garg, V.K.; Tseng, C.-H. Arsenic: An overview of applications, health, and environmental concerns and removal processes. *Crit. Rev. Environ. Sci. Technol.* **2011**, *41*, 435–519.

44. Ng, K.-S.; Ujang, Z.; Le-Clech, P. Arsenic removal technologies for drinking water treatment. *Rev. Environ. Sci. Biotechnol.* **2004**, *3*, 43–53.

45. Ali, I.; Khan, T.A.; Asim, M. Removal of arsenic from water by electrocoagulation and electrodialysis techniques. *Sep. Purif. Rev.* **2011**, *40*, 25–42.

46. Van Genuchten, C.M.; Addy, S.E.A.; Peña, J.; Gadgil, A.J. Removing arsenic from synthetic groundwater with iron electrocoagulation: An Fe and As K-edge EXAFS study. *Environ. Sci. Technol.* **2012**, *46*, 986–994.

47. Ali, I. Water treatment by adsorption columns: Evaluation at ground level. *Sep. Purif. Rev.* **2014**, *43*, 175–205.

48. Mohan, D.; Pittman, C.U. Arsenic removal from water/wastewater using adsorbents—A critical review. *J. Hazard. Mater.* **2007**, *142*, 1–53.

49. Ali, I.; Gupta, V.K. Advances in water treatment by adsorption technology. *Nat. Protoc.* **2006**, *1*, 2661–2667.

50. Wei, Y.-T.; Zheng, Y.-M.; Chen, J.P. Uptake of methylated arsenic by a polymeric adsorbent: Process performance and adsorption chemistry. *Water Res.* **2011**, *45*, 2290–2296.

51. Salameh, Y.; Al-Lagtah, N.; Ahmad, M.N.M.; Allen, S.J.; Walker, G.M. Kinetic and thermodynamic investigations on arsenic adsorption onto dolomitic sorbents. *Chem. Eng. J.* **2010**, *160*, 440–446.

52. Chutia, P.; Kato, S.; Kojima, T.; Satokawa, S. Arsenic adsorption from aqueous solution on synthetic zeolites. *J. Hazard. Mater.* **2009**, *162*, 440–447.

53. Adra, A.; Morin, G.; Ona-Nguema, G.; Menguy, N.; Maillot, F.; Casiot, C.; Bruneel, O.; Lebrun, S.; Juillot, F.; Brest, J. Arsenic scavenging by aluminum-substituted ferrihydrites in a circumneutral pH river impacted by acid mine drainage. *Environ. Sci. Technol.* **2013**, *47*, 12784–12792.

54. Giles, D.E.; Mohapatra, M.; Issa, T.B.; Anand, S.; Singh, P. Iron and aluminium based adsorption strategies for removing arsenic from water. *J. Environ. Manag.* **2011**, *92*, 3011–3022.

55. Manning, B.A.; Goldberg, S. Adsorption and stability of arsenic(III) at the clay mineral–water interface. *Environ. Sci. Technol.* **1997**, *31*, 2005–2011.

56. Singh, T.S.; Pant, K.K. Equilibrium, kinetics and thermodynamic studies for adsorption of As(III) on activated alumina. *Sep. Purif. Technol.* **2004**, *36*, 139–147.

57. Gallegos-Garcia, M.; Ramírez-Muñiz, K.; Song, S. Arsenic removal from water by adsorption using iron oxide minerals as adsorbents: A review. *Miner. Process. Extr. Metall. Rev.* **2012**, *33*, 301–315.

58. Miretzky, P.; Cirelli, A.F. Remediation of arsenic-contaminated soils by iron amendments: A review. *Crit. Rev. Environ. Sci. Technol.* **2010**, *40*, 93–115.

59. Yang, W.; Kan, A.T.; Chen, W.; Tomson, M.B. pH-dependent effect of zinc on arsenic adsorption to magnetite nanoparticles. *Water Res.* **2010**, *44*, 5693–5701.

60. Zhang, S.; Niu, H.; Cai, Y.; Zhao, X.; Shi, Y. Arsenite and arsenate adsorption on coprecipitated bimetal oxide magnetic nanomaterials: $MnFe_2O_4$ and $CoFe_2O_4$. *Chem. Eng. J.* **2010**, *158*, 599–607.

61. Tian, Y.; Wu, M.; Lin, X.; Huang, P.; Huang, Y. Synthesis of magnetic wheat straw for arsenic adsorption. *J. Hazard. Mater.* **2011**, *193*, 10–16.

62. Kanel, S.R.; Manning, B.; Charlet, L.; Choi, H. Removal of arsenic(III) from groundwater by nanoscale zero-valent iron. *Environ. Sci. Technol.* **2005**, *39*, 1291–1298.

63. Yavuz, C.T.; Mayo, J.T.; Suchecki, C.; Wang, J.; Ellsworth, A.Z.; D'Couto, H.; Quevedo, E.; Prakash, A.; Gonzalez, L.; Nguyen, C.; *et al.* Pollution magnet: Nano-magnetite for arsenic removal from drinking water. *Environ. Geochem. Health* **2010**, *32*, 327–334.

64. Zhu, J.; Pigna, M.; Cozzolino, V.; Caporale, A.G.; Violante, A. Sorption of arsenite and arsenate on ferrihydrite: Effect of organic and inorganic ligands. *J. Hazard. Mater.* **2011**, *189*, 564–571.

65. Villalobos, M.; Antelo, J. A unified surface structural model for ferrihydrite: Proton charge, electrolyte binding, and arsenate adsorption. *Rev. Int. Contam. Ambie* **2011**, *27*, 139–151.

66. Huang, J.-H.; Voegelin, A.; Pombo, S.A.; Lazzaro, A.; Zeyer, J.; Kretzschmar, R. Influence of arsenate adsorption to ferrihydrite, goethite, and boehmite on the kinetics of arsenate reduction by Shewanella putrefaciens strain CN-32. *Environ. Sci. Technol.* **2011**, *45*, 7701–7709.

67. Mamindy-Pajany, Y.; Hurel, C.; Marmier, N.; Roméo, M. Arsenic adsorption onto hematite and goethite. *Comptes Rendus Chim.* **2009**, *12*, 876–881.

68. Bowell, R.J. Sorption of arsenic by iron oxides and oxyhydroxides in soils. *Appl. Geochem.* **1994**, *9*, 279–286.

69. Ko, I.; Davis, A.P.; Kim, J.-Y.; Kim, K.-W. Effect of contact order on the adsorption of inorganic arsenic species onto hematite in the presence of humic acid. *J. Hazard. Mater.* **2007**, *141*, 53–60.

70. Simeoni, M.A.; Batts, B.D.; McRae, C. Effect of groundwater fulvic acid on the adsorption of arsenate by ferrihydrite and gibbsite. *Appl. Geochem.* **2003**, *18*, 1507–1515.

71. Aryanpour, M.; van Duin, A.C.T.; Kubicki, J.D. Development of a reactive force field for iron-oxyhydroxide systems. *J. Phys. Chem. A* **2010**, *114*, 6298–6307.

72. Fitts, J.P.; Machesky, M.L.; Wesolowski, D.J.; Shang, X.; Kubicki, J.D.; Flynn, G.W.; Heinz, T.F.; Eisenthal, K.B. Second-harmonic generation and theoretical studies of protonation at the water/α-TiO2 (110) interface. *Chem. Phys. Lett.* **2005**, *411*, 399–403.

73. Kubicki, J.D.; Paul, K.W.; Sparks, D.L. Periodic density functional theory calculations of bulk and the (010) surface of goethite. *Geochem. Trans.* **2008**, *9*, doi:10.1186/1467-4866-9-4.

74. Arts, D.; Sabur, M.A.; Al-Abadleh, H.A. Surface interactions of aromatic organoarsenical compounds with hematite nanoparticles using ATR-FTIR: Kinetic studies. *J. Phys. Chem. A* **2013**, *117*, 2195–2204.

75. Bargar, J.R.; Kubicki, J.D.; Reitmeyer, R.; Davis, J.A. ATR-FTIR spectroscopic characterization of coexisting carbonate surface complexes on hematite. *Geochim. Cosmochim. Acta* **2005**, *69*, 1527–1542.

76. Goldberg, S.; Johnston, C.T. Mechanisms of arsenic adsorption on amorphous oxides evaluated using macroscopic measurements, vibrational spectroscopy, and surface complexation modeling. *J. Colloid Interface Sci.* **2001**, *234*, 204–216.

77. Sun, X.; Doner, H. An investigation of arsenate and arsenite bonding structures on goethite by FTIR. *Soil Sci.* **1996**, *161*, 865–872.

78. Zhao, K.; Guo, H. Behavior and mechanism of arsenate adsorption on activated natural siderite: Evidences from FTIR and XANES analysis. *Environ. Sci. Pollut. Res.* **2014**, *21*, 1944–1953.

79. Müller, K.; Ciminelli, V.S.T.; Dantas, M.S.S.; Willscher, S. A comparative study of As(III) and As(V) in aqueous solutions and adsorbed on iron oxy-hydroxides by Raman spectroscopy. *Water Res.* **2010**, *44*, 5660–5672.

80. Illera, V.; Rivera, N.A.; O'Day, P.A. Spectroscopic Characterization of Co-Precipitated Arsenic- and Iron-Bearing Sulfide Phases at Circum-Neutral pH. In Proceedings of the 2009 American Geophysical Union Fall Meeting, San Francisco, CA, USA, 14–18 December 2009.

81. Manning, B.A.; Fendorf, S.E.; Goldberg, S. Surface structures and stability of arsenic(III) on goethite: Spectroscopic evidence for inner-sphere complexes. *Environ. Sci. Technol.* **1998**, *32*, 2383–2388.

82. Farquhar, M.L.; Charnock, J.M.; Livens, F.R.; Vaughan, D.J. Mechanisms of arsenic uptake from aqueous solution by interaction with goethite, lepidocrocite, mackinawite, and pyrite: An X-ray absorption spectroscopy study. *Environ. Sci. Technol.* **2002**, *36*, 1757–1762.

83. Ona-Nguema, G.; Morin, G.; Wang, Y.; Foster, A.L.; Juillot, F.; Calas, G.; Brown, G.E. XANES evidence for rapid arsenic(III) oxidation at magnetite and ferrihydrite surfaces by dissolved O_2 via Fe^{2+}-mediated reactions. *Environ. Sci. Technol.* **2010**, *44*, 5416–5422.

84. Tu, Y.-J.; You, C.-F.; Chang, C.-K.; Wang, S.-L. XANES evidence of arsenate removal from water with magnetic ferrite. *J. Environ. Manag.* **2013**, *120*, 114–119.

85. Xu, L.; Zhao, Z.; Wang, S.; Pan, R.; Jia, Y. Transformation of arsenic in offshore sediment under the impact of anaerobic microbial activities. *Water Res.* **2011**, *45*, 6781–6788.

86. Couture, R.-M.; Rose, J.; Kumar, N.; Mitchell, K.; Wallschläger, D.; van Cappellen, P. Sorption of arsenite, arsenate, and thioarsenates to iron oxides and iron sulfides: A kinetic and spectroscopic investigation. *Environ. Sci. Technol.* **2013**, *47*, 5652–5659.

87. Gao, X.; Root, R.A.; Farrell, J.; Ela, W.; Chorover, J. Effect of silicic acid on arsenate and arsenite retention mechanisms on 6-L ferrihydrite: A spectroscopic and batch adsorption approach. *Appl. Geochem.* **2013**, *38*, 110–120.

88. Waychunas, G.A.; Davis, J.A.; Fuller, C.C. Geometry of sorbed arsenate on ferrihydrite and crystalline FeOOH: Re-evaluation of EXAFS results and topological factors in predicting sorbate geometry, and evidence for monodentate complexes. *Geochim. Cosmochim. Acta* **1995**, *59*, 3655–3661.

89. Waychunas, G.A.; Rea, B.A.; Fuller, C.C.; Davis, J.A. Surface chemistry of ferrihydrite: Part 1. EXAFS studies of the geometry of coprecipitated and adsorbed arsenate. *Geochim. Cosmochim. Acta* **1993**, *57*, 2251–2269.

90. Ladeira, A.C.Q.; Ciminelli, V.S.T.; Duarte, H.A.; Alves, M.C.M.; Ramos, A.Y. Mechanism of anion retention from EXAFS and density functional calculations: Arsenic(V) adsorbed on gibbsite. *Geochim. Cosmochim. Acta* **2001**, *65*, 1211–1217.

91. Sherman, D.M.; Randall, S.R. Surface complexation of arsenic(V) to iron(III) (hydr)oxides: Structural mechanism from *ab initio* molecular geometries and EXAFS spectroscopy. *Geochim. Cosmochim. Acta* **2003**, *67*, 4223–4230.

92. Ona-Nguema, G.; Morin, G.; Juillot, F.; Calas, G.; Brown, G.E. EXAFS analysis of arsenite adsorption onto two-line ferrihydrite, hematite, goethite, and lepidocrocite. *Environ. Sci. Technol.* **2005**, *39*, 9147–9155.

93. Fuller, C.C.; Davis, J.A.; Waychunas, G.A. Surface chemistry of ferrihydrite: Part 2. Kinetics of arsenate adsorption and coprecipitation. *Geochim. Cosmochim. Acta* **1993**, *57*, 2271–2282.

94. Goldberg, S. Competitive adsorption of arsenate and arsenite on oxides and clay minerals. *Soil Sci. Soc. Am. J.* **2002**, *66*, 413–421.

95. Jain, A.; Raven, K.P.; Loeppert, R.H. Arsenite and arsenate adsorption on ferrihydrite: Surface charge reduction and net OH-release stoichiometry. *Environ. Sci. Technol.* **1999**, *33*, 1179–1184.

96. Maji, S.K.; Kao, Y.-H.; Liao, P.-Y.; Lin, Y.-J.; Liu, C.-W. Implementation of the adsorbent iron-oxide-coated natural rock (IOCNR) on synthetic As(III) and on real arsenic-bearing sample with filter. *Appl. Surf. Sci.* **2013**, *284*, 40–48.

97. Raven, K.P.; Jain, A.; Loeppert, R.H. Arsenite and arsenate adsorption on ferrihydrite: Kinetics, equilibrium, and adsorption envelopes. *Environ. Sci. Technol.* **1998**, *32*, 344–349.

98. Antelo, J.; Avena, M.; Fiol, S.; López, R.; Arce, F. Effects of pH and ionic strength on the adsorption of phosphate and arsenate at the goethite–water interface. *J. Colloid Interface Sci.* **2005**, *285*, 476–486.

99. Hiemstra, T.; van Riemsdijk, W.H. A surface structural approach to ion adsorption: The charge distribution (CD) model. *J. Colloid Interface Sci.* **1996**, *179*, 488–508.

100. Weng, L.; van Riemsdijk, W.H.; Hiemstra, T. Effects of fulvic and humic acids on arsenate adsorption to goethite: Experiments and modeling. *Environ. Sci. Technol.* **2009**, *43*, 7198–7204.

101. Dixit, S.; Hering, J.G. Comparison of arsenic(V) and arsenic(III) sorption onto iron oxide minerals: Implications for arsenic mobility. *Environ. Sci. Technol.* **2003**, *37*, 4182–4189.

102. Ngantcha, T.A.; Vaughan, R.; Reed, B.E. Modeling As(III) and As(V) removal by an iron oxide impregnated activated carbon in a binary adsorbate system. *Sep. Sci. Technol.* **2011**, *46*, 1419–1429.

103. Que, S.; Papelis, C.; Hanson, A.T. Predicting arsenate adsorption on iron-coated sand based on a surface complexation model. *J. Environ. Eng.* **2013**, *139*, 368–374.

104. Jeppu, G.P.; Clement, T.P.; Barnett, M.O.; Lee, K.-K. A scalable surface complexation modeling framework for predicting arsenate adsorption on goethite-coated sands. *Environ. Eng. Sci.* **2010**, *27*, 147–158.

105. Jessen, S.; Postma, D.; Larsen, F.; Nhan, P.Q.; Hoa, L.Q.; Trang, P.T.K.; Long, T.V.; Viet, P.H.; Jakobsen, R. Surface complexation modeling of groundwater arsenic mobility: Results of a forced gradient experiment in a Red River flood plain aquifer, Vietnam. *Geochim. Cosmochim. Acta* **2012**, *98*, 186–201.

106. Sharifa, S.U.; Davisa, R.K.; Steelea, K.F.; Kima, B.; Haysa, P.D.; Kresseb, T.M.; Fazioc, J.A. Surface complexation modeling for predicting solid phase arsenic concentrations in the sediments of the Mississippi River Valley alluvial aquifer, Arkansas, USA. *Appl. Geochem.* **2011**, *26*, 496–504.

107. Pakzadeh, B.; Batista, J.R. Surface complexation modeling of the removal of arsenic from ion-exchange waste brines with ferric chloride. *J. Hazard. Mater.* **2011**, *188*, 399–407.

108. Kanematsu, M.; Young, T.M.; Fukushi, K.; Green, P.G.; Darby, J.L. Arsenic(III,V) adsorption on a goethite-based adsorbent in the presence of major co-existing ions: Modeling competitive adsorption consistent with spectroscopic and molecular evidence. *Geochim. Cosmochim. Acta* **2013**, *106*, 404–428.

109. Selim, H.; Zhang, H. Modeling approaches of competitive sorption and transport of trace metals and metalloids in soils: A review. *J. Environ. Qual.* **2013**, *42*, 640–653.

110. Wan, J.; Simon, S.; Deluchat, V.; Dictor, M.-C.; Dagot, C. Adsorption of As(III), As(V) and dimethylarsinic acid onto synthesized lepidocrocite. *J. Environ. Sci. Health Part A. Tox Hazard. Subst. Environ. Eng.* **2013**, *48*, 1272–1279.

111. Cui, Y.; Weng, L. Arsenate and phosphate adsorption in relation to oxides composition in soils: LCD modeling. *Environ. Sci. Technol.* **2013**, *47*, 7269–7276.

112. Weng, L.; van Riemsdijk, W.H.; Koopal, L.K.; Hiemstra, T. Ligand and Charge Distribution (LCD) model for the description of fulvic acid adsorption to goethite. *J. Colloid Interface Sci.* **2006**, *302*, 442–457.

113. Hohenberg, P.; Kohn, W. Inhomogeneous electron gas. *Phys. Rev.* **1964**, *136*, 864–871.

114. Kohn, W.; Sham, L.J. Self-consistent equations including exchange and correlation effects. *Phys. Rev.* **1965**, *140*, 1133–1138.

115. Adamescu, A.; Hamilton, I.P.; Al-Abadleh, H.A. Thermodynamics of dimethylarsinic acid and arsenate interactions with hydrated iron-(oxyhydr)oxide clusters: DFT calculations. *Environ. Sci. Technol.* **2011**, *45*, 10438–10444.

116. He, G.; Zhang, M.; Pan, G. Influence of pH on initial concentration effect of arsenate adsorption on TiO_2 surfaces: Thermodynamic, DFT, and EXAFS interpretations. *J. Phys. Chem. C* **2009**, *113*, 21679–21686.

117. Zhang, N.; Blowers, P.; Farrell, J. Evaluation of density functional theory methods for studying chemisorption of arsenite on ferric hydroxides. *Environ. Sci. Technol.* **2005**, *39*, 4816–4822.

118. Adamescu, A.; Mitchell, W.; Hamilton, I.P.; Al-Abadleh, H.A. Insights into the surface complexation of dimethylarsinic acid on iron (oxyhydr)oxides from ATR-FTIR studies and quantum chemical calculations. *Environ. Sci. Technol.* **2010**, *44*, 7802–7807.

119. Kubicki, J.D.; Kwon, K.D.; Paul, K.W.; Sparks, D.L. Surface complex structures modelled with quantum chemical calculations: Carbonate, phosphate, sulphate, arsenate and arsenite. *Eur. J. Soil Sci.* **2007**, *58*, 932–944.

120. Tofan-Lazar, J.; Al-Abadleh, H. ATR-FTIR studies on the adsorption/desorption kinetics of dimethylarsinic acid on iron-(oxyhydr)oxides. *J. Phys. Chem. A* **2012**, *116*, 1596–1604.

121. Tofan-Lazar, J.; Al-Abadleh, H. Kinetic ATR-FTIR studies on phosphate adsorption on iron (oxyhydr)oxides in the absence and presence of surface arsenic: Molecular-level insights into the ligand exchange mechanism. *J. Phys. Chem. A* **2012**, *116*, 10143–10149.

122. Farrell, J.; Chaudhary, B.K. Understanding arsenate reaction kinetics with ferric hydroxides. *Environ. Sci. Technol.* **2013**, *47*, 8342–8347.

123. Zhu, M.; Paul, K.W.; Kubicki, J.D.; Sparks, D.L. Quantum chemical study of arsenic(III,V) adsorption on Mn-oxides: Implications for arsenic(III) oxidation. *Environ. Sci. Technol.* **2009**, *43*, 6655–6661.

124. Blanchard, M.; Morin, G.; Lazzeri, M.; Balan, E.; Dabo, I. First-principles simulation of arsenate adsorption on the (1Ī2) surface of hematite. *Geochim. Cosmochim. Acta* **2012**, *86*, 182–195.

125. Blanchard, M.; Wright, K.; Gale, J.D.; Catlow, C.R.A. Adsorption of $As(OH)_3$ on the (001) surface of FeS_2 pyrite: A quantum-mechanical DFT Study. *J. Phys. Chem. C* **2007**, *111*, 11390–11396.

126. Duarte, G.; Ciminelli, V.S.T.; Dantas, M.S.S.; Duarte, H.A.; Vasconcelos, I.F.; Oliveira, A.F.; Osseo-Asare, K. As(III) immobilization on gibbsite: Investigation of the complexation mechanism by combining EXAFS analyses and DFT calculations. *Geochim. Cosmochim. Acta* **2012**, *83*, 205–216.

127. Goffinet, C.J.; Mason, S.E. Comparative DFT study of inner-sphere As(III) complexes on hydrated α-Fe_2O_3 (0001) surface models. *J. Environ. Monit.* **2012**, *14*, 1860–1871.

128. Loring, J.; Sandström, M.; Norén, K.; Persson, P. Rethinking arsenate coordination at the surface of goethite. *Chem. Eur. J.* **2009**, *15*, 5063–5072.

129. Oliveira, A.F.; Ladeira, A.C.Q.; Ciminelli, V.S.T.; Heine, T.; Duarte, H.A. Structural model of arsenic(III) adsorbed on gibbsite based on DFT calculations. *J. Mol. Struct. Theochem.* **2006**, *762*, 17–23.

130. Otte, K.; Schmahl, W.W.; Pentcheva, R. DFT+ U study of arsenate adsorption on FeOOH surfaces: Evidence for competing binding mechanisms. *J. Phys. Chem. C* **2013**, *117*, 15571–15582.

131. Stachowicz, M.; Hiemstra, T.; van Riemsdijk, W.H. Surface speciation of As(III) and As(V) in relation to charge distribution. *J. Colloid Interface Sci.* **2006**, *302*, 62–75.

132. Tanaka, M.; Takahashi, Y.; Yamaguchi, N. A study on adsorption mechanism of organoarsenic compounds on ferrihydrite by XAFS. *J. Phys. Conf. Ser.* **2013**, *430*, 012100, doi:10.1088/1742-6596/430/1/012100.

133. Klamt, A.; Jonas, V.; Bürger, T.; Lohrenz, J.C.W. Refinement and parametrization of COSMO-RS. *J. Phys. Chem. A* **1998**, *102*, 5074–5085.

134. Delley, B. The conductor-like screening model for polymers and surfaces. *Mol. Simul.* **2006**, *32*, 117–123.

135. Kubicki, J.D.; Paul, K.W.; Kabalan, L.; Zhu, Q.; Mrozik, M.K.; Aryanpour, M.; Pierre-Louis, A.-M.; Strongin, D.R. ATR-FTIR and density functional theory study of the structures, energetics, and vibrational spectra of phosphate adsorbed onto goethite. *Langmuir* **2012**, *28*, 14573–14587.

136. Frisch, M.J.; Trucks, G.W.; Schlegel, H.B.; Scuseria, G.E.; Robb, M.A.; Cheeseman, J.R.; Scalmani, G.; Barone, V.; Mennucci, B.; Petersson, G.A.; Nakatsuji, H.; Caricato, M.; Li, X.; Hratchian, H.P.; Izmaylov, A.F.; Bloino, J.; Zheng, G.; Sonnenberg, J.L.; Hada, M.; Ehara, M.; Toyota, K.; Fukuda, R.; Hasegawa, J.; Ishida, M.; Nakajima, T.; Honda, Y.; Kitao, O.; Nakai, H.; Vreven, T.; Montgomery, J.A., Jr.; Peralta, J.E.; Ogliaro, F.; Bearpark, M.; Heyd, J.J.; Brothers, E.; Kudin, K.N.; Staroverov, V.N.; Kobayashi, R.; Normand, J.; Raghavachari, K.; Rendell, A.; Burant, J.C.; Iyengar, S.S.; Tomasi, J.; Cossi, M.; Rega, N.; Millam, N.J.; Klene, M.; Knox, J.E.; Cross, J.B.; Bakken, V.; Adamo, C.; Jaramillo, J.; Gomperts, R.; Stratmann, R.E.; Yazyev, O.; Austin, A.J.; Cammi, R.; Pomelli, C.; Ochterski, J.W.; Martin, R.L.; Morokuma, K.; Zakrzewski, V.G.; Voth, G.A.; Salvador, P.; Dannenberg, J.J.; Dapprich, S.; Daniels, A.D.; Farkas, Ö.; Foresman, J.B.; Ortiz, J.V.; Cioslowski, J.; Fox, D.J. Gaussian 09, Revision B.01; Gaussian, Inc.: Wallingford CT, USA, 2009. Available online: http://www.gaussian.com/g_tech/g_ur/m_citation.htm (accessed on 10 March 2014).

137. Curtiss, L.; Redfern, P.; Raghavachari, K. Gaussian-3X (G3X) theory: Use of improved geometries, zero-point energies, and Hartree–Fock basis sets. *J. Chem. Phys.* **2001**, *114*, 108–117.

138. Zhao, Y.; Truhlar, D.G. Density functionals with broad applicability in chemistry. *Acc. Chem. Res.* **2008**, *41*, 157–167.

139. Leach, A.R. Energy Minimisation and Related Methods for Exploring the Energy Surface. In *Molecular Modelling: Principles and Applications*; Prentice Hall: Upper Saddle River, NJ, USA, 2001; pp. 253–302.

140. Szabo, A.; Ostlund, N.S. The Hartree-Fock Approximation. In *Modern Quantum Chemistry*; Dover Publications: Mineola, NY, USA, 1989; pp. 108–230.

141. Møller, C.; Plesset, M.S. Note on an approximation treatment for many-electron systems. *Phys. Rev.* **1934**, *46*, 618–622.

142. Levine, I.N. Electronic Structure of Diatomic Molecules. In *Quantum Chemistry*; Pearson: Upper Saddle River, NJ, USA, 2009; pp. 369–373.

143. Levine, I.N. The Variation Method. In *Quantum Chemistry*; Pearson: Upper Saddle River, NJ, USA, 2009; pp. 211–247.

144. Matta, C.F.; Boyd, R.J. An Introduction to the Quantum Theory of Atoms in Molecules. In *A Chemist's Guide to Density Functional Theory*; Koch, W., Holthausen, M.C., Eds.; John Wiley & Sons: Hoboken, NJ, USA, 2001; pp. 1–34.

145. Korth, M.; Grimme, S. "Mindless" DFT benchmarking. *J. Chem. Theory Comput.* **2009**, *5*, 993–1003.

146. Becke, A.D. A new mixing of Hartree–Fock and local density-functional theories. *J. Chem. Phys.* **1993**, *98*, 1372, doi:10.1063/1.464304.

147. Lee, C.; Yang, W.; Parr, R.G. Development of the Colle-Salvetti correlation-energy formula into a functional of the electron density. *Phys. Rev. B* **1988**, *37*, 785–789.

148. Bachrach, S.M. Quantum Mechanics for Organic Chemistry. In *Computational Organic Chemistry*; John Wiley & Sons: Hoboken, NJ, USA, 2007; pp. 8–11.

149. Clark, T.; Chandrasekhar, J.; Spitznagel, G.W.; Schleyer, P.V.R. Efficient diffuse function-augmented basis sets for anion calculations. III. The 3-21+G basis set for first-row elements, Li-F. *J. Comput. Chem.* **1983**, *4*, 294–301.

150. Krishnan, R.; Brinkley, J.S.; Seeger, R.; Pople, J.A. Self-consistent molecular orbital methods. XX. A basis set for correlated wave functions. *J. Chem. Phys.* **1980**, *72*, 650–654.

151. Papajak, E.; Zheng, J.; Xu, X.; Leverentz, H.R.; Truhlar, D.G. Perspectives on basis sets beautiful: Seasonal plantings of diffuse basis functions. *J. Chem. Theory Comput.* **2011**, *7*, 3027–3034.

152. Blöchl, P.E. Projector augmented-wave method. *Phys. Rev. B* **1994**, *50*, 17953–17979.

153. Kresse, G.; Hafner, J. *Ab initio* molecular-dynamics simulation of the liquid-metal-amorphous-semiconductor transition in germanium. *Phys. Rev. B* **1994**, *49*, 14251–14269.

154. Kresse, G.; Hafner, J. *Ab initio* molecular dynamics for open-shell transition metals. *Phys. Rev. B.* **1993**, *48*, 13115–13118.

155. Kresse, G.; Furthmüller, J. Efficiency of ab-initio total energy calculations for metals and semiconductors using a plane-wave basis set. *Comput. Mater. Sci.* **1996**, *6*, 15–50.

156. Kresse, G.; Furthmüller, J. Efficient iterative schemes for *ab initio* total-energy calculations using a plane-wave basis set. *Phys. Rev. B* **1996**, *54*, 11169–11186.

157. Kresse, G.; Furthmüller, J.; Hafner, J. Theory of the crystal structures of selenium and tellurium: The effect of generalized-gradient corrections to the local-density approximation. *Phys. Rev. B* **1994**, *50*, 13181–13185.

158. Myneni, S.C.B.; Traina, S.J.; Waychunas, G.A.; Logan, T.J. Experimental and theoretical vibrational spectroscopic evaluation of arsenate coordination in aqueous solutions, solids, and at mineral-water interfaces. *Geochim. Cosmochim. Acta* **1998**, *62*, 3285–3300.

159. Cancès, E.; Mennucci, B.; Tomasi, J. A new integral equation formalism for the polarizable continuum model: Theoretical background and applications to isotropic and anisotropic dielectrics. *J. Chem. Phys.* **1997**, *107*, 3032–3041.

160. Perdew, J.; Burke, K.; Ernzerhof, M. Errata: Generalized gradient approximation made simple. *Phys. Rev. Lett.* **1996**, *78*, 1396, doi:10.1103/PhysRevLett.78.1396.

161. Perdew, J.; Burke, K.; Ernzerhof, M. Generalized gradient approximation made simple. *Phys. Rev. Lett.* **1996**, *77*, 3865–3868.

162. Adamo, C.; Barone, V. Toward reliable density functional methods without adjustable parameters: The PBE0 model. *J. Chem. Phys.* **1999**, *110*, 6158–6170.

163. Zhao, Y.; Truhlar, D.G. A new local density functional for main-group thermochemistry, transition metal bonding, thermochemical kinetics, and noncovalent interactions. *J. Chem. Phys.* **2006**, *125*, 194101, doi:10.1063/1.2370993.

164. Szytuła, A.; Burewicz, A.; Dimitrijević, Ž.; Kraśnicki, S.; Rżany, H.; Todorović, J.; Wanic, A.; Wolski, W. Neutron diffraction studies of α-FeOOH. *Phys. Status Solidi* **1968**, *26*, 429–434.

165. Kresse, G.; Joubert, D. From ultrasoft pseudopotentials to the projector augmented-wave method. *Phys. Rev. B* **1999**, *59*, 1758–1775.

166. Dudarev, S.L.; Botton, G.A.; Savrasov, S.Y.; Humphreys, C.J.; Sutton, A.P. Electron-energy-loss spectra and the structural stability of nickel oxide: An LSDA+ U study. *Phys. Rev. B* **1998**, *57*, 1505–1509.

167. Rollmann, G.; Rohrbach, A.; Entel, P.; Hafner, J. First-principles calculation of the structure and magnetic phases of hematite. *Phys. Rev. B* **2004**, *69*, 165107, doi:10.1103/PhysRevB.69.165107.

168. Coey, J.M.D.; Barry, A.; Brotto, J.; Rakoto, H.; Brennan, S.; Mussel, W.N.; Collomb, A.; Fruchart, D. Spin flop in goethite. *J. Phys. Condens. Matter* **1995**, *7*, 759–768.

169. Nosé, S. A unified formulation of the constant temperature molecular dynamics methods. *J. Chem. Phys.* **1984**, *81*, 511, doi:10.1063/1.447334.

170. Leung, K.; Nielsen, I.M.B.; Criscenti, L.J. Elucidating the bimodal acid-base behavior of the water-silica interface from first principles. *J. Am. Chem. Soc.* **2009**, *131*, 18358–18365.

171. Liu, L.; Zhang, C.; Thornton, G.; Michaelides, A. Structure and dynamics of liquid water on rutile TiO$_2$(110). *Phys. Rev. B* **2010**, *82*, 161415, doi:10.1103/PhysRevB.82.161415.

172. Kelly, C.P.; Cramer, C.J.; Truhlar, D.G. Adding explicit solvent molecules to continuum solvent calculations for the calculation of aqueous acid dissociation constants. *J. Phys. Chem. A* **2006**, *110*, 2493–2499.

173. Felipe, M.A.; Xiao, Y.; Kubicki, J.D. Molecular orbital modeling and transition state theory in geochemistry. *Rev. Mineral. Geochem.* **2001**, *42*, 485–531.

174. Zhao, Y.; Truhlar, D.G. Design of density functionals that are broadly accurate for thermochemistry, thermochemical kinetics, and nonbonded interactions. *J. Phys. Chem. A* **2005**, *109*, 5656–5667.

175. Sarotti, A.M.; Pellegrinet, S.C. Application of the multi-standard methodology for calculating ^{1}H NMR chemical shifts. *J. Org. Chem.* **2012**, *77*, 6059–6065.

176. Sarotti, A.M.; Pellegrinet, S.C. A multi-standard approach for GIAO ^{13}C NMR calculations. *J. Org. Chem.* **2009**, *74*, 7254–7260.

177. Villalobos, M.; Pérez-Gallegos, A. Goethite surface reactivity: A macroscopic investigation unifying proton, chromate, carbonate, and lead(II) adsorption. *J. Colloid Interface Sci.* **2008**, *326*, 307–323.

178. Villalobos, M.; Cheney, M.A.; Alcaraz-Cienfuegos, J. Goethite surface reactivity: II. A microscopic site-density model that describes its surface area-normalized variability. *J. Colloid Interface Sci.* **2009**, *336*, 412–422.

Particle Size-Specific Magnetic Measurements as a Tool for Enhancing Our Understanding of the Bulk Magnetic Properties of Sediments

Robert G. Hatfield

College of Earth, Ocean, and Atmospheric Science, Oregon State University, 104 CEOAS Admin Building, Corvallis, OR 97331, USA; E-Mail: rhatfiel@coas.oregonstate.edu

External Editor: Dimitrina Dimitrova

Abstract: Bulk magnetic properties of soils and sediments are often sensitive proxies for environmental change but commonly require interpretation in terms of the different sources of magnetic minerals (or components) that combine to generate them. Discrimination of different components in the bulk magnetic record is often attempted through endmember unmixing and/or high resolution measurements that can require intensive measurement plans, assume linear additivity, and sometimes have difficulty in discriminating a large number of sources. As an alternative, magnetic measurements can be made on isolated sediment fractions that constitute the bulk sample. When these types of measurements are taken, heterogeneity is frequently observed between the magnetic properties of different fractions, suggesting different magnetic components often associate with different physical grain sizes. Using a particle size-specific methodology, individual components can be isolated and studied and bulk magnetic properties can be linked to, and isolated from, sedimentological variations. Deconvolving sedimentary and magnetic variability in this way has strong potential for increased understanding of how magnetic fragments are carried in natural systems, how they vary with different source(s), and allows for a better assessment of the effect environmental variability has in driving bulk magnetic properties. However, despite these benefits, very few studies exploit the information they can provide. Here, I present an overview of the different sources of magnetic minerals, why they might associate with different sediment fractions, how bulk magnetic measurements have been used to understand the contribution of different components to the bulk magnetic record, and outline how particle size-specific magnetic measurements can assist in their better understanding. Advantages and disadvantages of this methodology, their role alongside bulk magnetic measurements, and potential future directions of research are also discussed.

Keywords: magnetic properties; environmental magnetism; magnetic susceptibility; magnetic grain size; sediment; particle size

1. Introduction

Magnetic properties of soils and sediments are among the most commonly made physical property measurements to understand records of past environmental change [1–12]. A simple suite of magnetic measurements can rapidly and non-destructively characterize the concentration, mineralogy, and magnetic grain size of magnetic minerals present in samples. These properties are strongly sensitive to the abundance and type of Fe-bearing minerals, which are both ubiquitous in environmental systems and sensitive to environmental change, such that magnetic measurements offer the opportunity to study temporal and spatial environmental variability in a range of settings. As a result, magnetic properties have been exploited in a variety of terrestrial, lacustrine, and marine environments to map anthropogenic pollution [13–15], investigate erosion, sediment sourcing and historical land use change [1,11,16–18], reconstruct paleorainfall estimates [19–21], map the incidence of ice rafted debris (IRD) [4,22,23] and dust [12,23,24] to the open ocean, understand the response of icesheets to climate [25], investigate ocean circulation [9,26–28], and document the cyclicity of Pleistocene climate in loess [29] and ocean sediments [24].

While sensitivity to multiple processes makes bulk (whole sample) magnetic measurements an attractive tool, it can also complicate interpretation. Identification of the process(es) driving the magnetic response in any particular record can be a non-trivial undertaking as magnetic concentration, grain size and mineralogy can be sensitive to geological variations [2,23], sediment particle size [11,30–32], sediment transport, delivery, and flux [4,7,9,25,26], anthropogenic pollution [13–15], and diagenetic [33–37], and authigenic [38–40] changes. When bulk magnetic properties are measured they often reflect this competition of different processes and magnetic mineral sources. If one process or source dominates, then interpretation of the drivers of magnetic properties is often easier than one experiencing multiple influences which can present a rather complex record. Interpretation is often achieved through correlation to independent variables, e.g., the abundance of lithic grains (e.g., >150 μm) if used as a proxy for IRD [4,41], heavy metals if tracking pollution [15,42], or sediment geochemistry for sediment fingerprinting and diagenesis [43–46]. However, the use of correlation in this manner often negates the advantages of exploiting magnetic properties in the first place; for example, why use magnetic susceptibility (MS) as a proxy for IRD if you have to physically count the IRD grains to qualify the magnetic proxy for them? Furthermore, inter-property correlations often sidestep addressing the more fundamental questions about what specifically drives magnetic property variations in any particular record. Lack of this understanding may be a major reason why magnetic measurements are often poorly understood and overlooked as potential investigative tools by the wider geoscience community.

When interpreting bulk magnetic properties, it is important to remember that they are rarely controlled by a single process or source. Instead, they are often the average of different magnetic mineral sources (termed components), each of which can potentially possess its own individual magnetic fingerprint. The aim for any environmental magnetic study should therefore be to try and identify the presence of different magnetic components and understand how in combination they influence bulk magnetic properties. In the literature, two main approaches have been utilized to try and unravel this complexity. The first approach attempts to unmix the signatures of different components using end members and/or through detailed examination of coercivity variances between components [47–55]. The second approach aims to characterize heterogeneity in the magnetic record and isolate different components through measurement of the magnetic properties of different sediment particle size constituents of the bulk sediment [10–12,17,30–32]. Both approaches have provided greater insights into what drives bulk magnetic properties in a variety of different environments, however, they are rarely employed in routine magnetic studies. Here, I discuss the origins of some different magnetic components in the bulk sediment record and how they can potentially be discriminated using both techniques. Then, as relatively underused, yet simple techniques, I shift to focus on the particle size dependence of magnetic properties and outline some important considerations when designing a particle size specific magnetic study, discuss their role alongside bulk magnetic property measurements, before finally providing an outlook and direction for future research.

2. Sources of Magnetic Minerals in the Environment

2.1. Sedimentary and Magnetic Properties

While all materials exhibit a response to an externally applied magnetic field ferrimagnetic minerals (e.g., magnetite, maghemite, titano-magnetite, pyrrhotite, and greigite) produce the strongest reactions and often dominate the magnetic properties of natural samples [2,56]. In their absence, weaker (canted-)antiferromagnetic minerals (e.g., hematite and goethite) and paramagnetic minerals (e.g., ilmenite, olivine, pyrite) can be important components in driving bulk magnetic properties [2,56]. These minerals largely originate from three sources; primary geological contributions, secondary neoformation and alteration through processes of diagenesis and authigenesis, and tertiary contributions relating to anthropogenic pollution. Not to be confused with physical grain size, magnetic grain size is a fundamental magnetic property related to domain state. For magnetite (Fe_3O_4), one of the most common ferrimagnetic minerals in environmental materials, magnetic grain sizes are given as super-paramagnetic (SP < 0.03 μm), single domain (SD, 0.03–0.1 μm), pseudo single domain (PSD, 0.1–20 μm) and multi-domain (MD > 20 μm) for cubic and/or slightly elongated grains [57,58]. These size bounds can be expected to change with grain shape and mineralogy; for example, the upper SD limit for hematite is around 100 μm [59]. Differences in magnetic grain size and mineralogy are often discriminable in terms of the coercive force needed to (de)magnetize samples [2,56,60] and as such a variety of magnetic parameters have been developed to characterize and discriminate between them (e.g., [2,47,56,58,60]).

In terms of sediment grain size, magnetic grain size boundaries for magnetite naturally fall in the clay (<2 μm; SP/SD, SD/PSD) and medium silt (16–32 μm; PSD/MD) fractions while larger MD grains fall in the coarse silt (32–63 μm) and sand (>63 μm) fractions. For discrete magnetite particles, it might be expected that differences in magnetic grain size exist between different sediment size fractions. However, magnetic minerals do not always exist as discrete entities and can often occur as inclusions within polycrystalline clasts. For example, basaltic glasses can be dominated by SP-SD grains [61] and medium to coarse silts can be dominated by relatively fine PSD size grains [32] complicating the seemingly simple and often assumed [9,27,28,62] scaling of magnetic and sedimentary granulometry. Understanding how sediment and magnetic grain size relate to each other is therefore important for interpretation, however, it is rarely considered in routine magnetic studies. Before discussing how to investigate this relationship, I will first discuss the origins of some common magnetic components that often contribute to bulk magnetic properties in environmental materials. This overview is intended to give an indication of the potential magnetic heterogeneity in environmental samples and how they might affect different sedimentary grain size ranges, not to evaluate the entire breadth of magnetic minerals in the environment. For further information and examples, see the excellent and detailed texts of Thompson and Oldfield [2], Maher and Thompson [63], and Evans and Heller [56].

2.2. Origins and Properties of Different Magnetic Components

Parent geology can provide a large proportion of the magnetic minerals to a soil or sediment often making it the dominant control on magnetic concentration, mineralogy, and magnetic grain size. Compounds of iron comprise roughly 5% of the Earth's crust and depending on fundamental controls such as geochemical composition and cooling rate [64,65], different iron-bearing minerals can grow to various sizes driving variances in magnetic mineralogy, concentration, and magnetic grain size. For example, red beds can get their distinctive coloring from hematite staining, iron sulphides can be the dominant magnetic minerals in slates, and ferrimagnetic iron oxides (e.g., magnetite, titano-magnetite) can constitute 2%–6% of basalts [2]. These magnetic mixtures are broken down via chemical, biological and, mechanical weathering to release both discrete and included magnetic minerals in a range clay, silt, and sand fractions into the environment. As a result, geological sources of magnetic minerals can often strongly dictate the magnetic properties of soils [3,66–68] and suspended [11], lake [1,69], and marine sediments [4,7,22–25]. If magnetic properties are assumed to be conservative [70] and maintain their geological fingerprints during sediment erosion, transport, and deposition, then magnetic discrimination of geological units can be exploited to facilitate sediment provenance studies [1,11,23,31,69].

While magnetic minerals are often assumed to be conservative in a range of settings, variation in environmental conditions can alter the reactive iron species within the mineral magnetic assemblage, potentially modifying their magnetic properties. Organic matter, pH, Eh, and sediment and porewater geochemistry are important factors that can promote reductive and/or oxidative diagenesis and authigenic (re)precipitation of magnetic mineral species that may differ significantly from those originally deposited. For example, given strongly reducing conditions, and excesses in porewater sulphate and organic matter [33,34,43] iron monosulphides (FeS), pyrite (FeS_2), greigite (Fe_3S_4), and ultimately pyrrhotite (Fe_7S_8), can form authigenically at the expense of magnetite and other reactive iron-(oxyhydr)oxides (e.g., goethite [FeO(OH)]). Commonly of relatively fine, magnetic grain size (SP/SD) [37,71–73], these species can be ferrimagnetic (e.g., greigite and pyrrhotite) or paramagnetic (e.g., pyrite) and constitute a significant portion of the magnetic mineral assemblage under favorable conditions. Similarly, oxygen rich environments can promote oxidation of magnetite to maghemite [74].

Dynamic sensitivity to variations in environmental conditions means that magnetic properties can reflect the non-stationarity of different processes, e.g., the formation of ferrimagnetic iron oxides on the surface of sapropels that were previously deposited under anoxic conditions [35,75]. Indeed, non-steady state conditions promoted during repeated wetting and drying cycles in soils are particularly important for the pedogenic formation and ferrimagnetic enhancement of soils in SP grains [38,67,76,77]. It is important to note that although redox state is an important factor dictating the nature of any diagenetic (or authigenic) change, the prevailing redox conditions do not necessarily result in alteration of the magnetic mineral population. The nature of any alteration (and its products) is constrained by the magnetic mineral species present in the sediment, their redox stability, biogenic activity and/or sediment geochemistry. However, significant alteration can strongly affect down core sediment bulk magnetic properties. For instance, during reductive diagenesis consumption of fine grained ferrimagnets and formation of paramagnetic pyrite can result in decreased magnetic concentration, an increase in paramagnetic concentration, and a coarsening of the magnetic remanence carrying components [33,44].

In addition to diagenetic consumption, oxidation and/or precipitation of magnetic minerals biological processes are capable of generating ferrimagnetic minerals. Magnetotactic bacteria can synthesize SD size magnetite and greigite into elongate chains and have been observed living throughout the sediment column or soil profile [39,40,78–80] sometimes forming a large remanence carrying component [79–83]. Possessing a relatively narrow coercivity range [49,50,52] magnetosomes are relatively well characterized magnetic components with First Order Reversal Curve (FORC) analyses particularly well suited to their identification [84–87]. However, as SD grain sizes are not unique to magnetosomes, it is still often necessary to confirm their presence using complementary techniques, e.g., transmission electron microscopy (TEM). Other natural processes including fire, lightning strikes, and in soils waterlogging and gleying can also have important influences on the stability and/or formation of different magnetic minerals resulting in changes in magnetic concentration, mineralogy, and magnetic grain size [3,66,67,88].

Tertiary contributions originate from a relatively narrow and specific range of sources related to ferrimagnetic particulates produced by industrial and vehicular exhausts. Often confirmed through scanning electron microscopy (SEM) and energy dispersive X-ray spectroscopy (EDS) discrete MD size spherules of magnetite can accumulate proximal to point sources [14,15,89]. These particles are often generated alongside, and strongly correlate with, other heavy metals and pollutants [43,88] making them a useful proxy for understanding the atmospheric dispersal of contaminants. Through measurement of the magnetic properties of tree leaves [15,89] and *in situ* surface sediments and soils [13,14] the extent and concentration of pollutants has been mapped from the individual road scale [14], through urban areas [90] and entire counties [91].

Different magnetic mineral species of SP, SD, PSD, and MD grains sizes can be generated by multiple processes and originate from multiple sources. Unfortunately, sources of different minerals are often not unique and/or confined to a single process or source, e.g., geology and bacterial magnetosomes can both produce ferrimagnetic SD size magnetite/maghemite. Therefore, for most environmental magnetic studies, the aim is often to discriminate between different minerals (e.g., magnetite and hematite) and grain sizes (e.g., SD/PSD/MD) that might relate to different processes. Aggregation of different components in the bulk sediment is responsible for the generation of the measured bulk magnetic properties of the sample. Discriminating the relative contribution of different components is therefore important for understanding the drivers of and processes involved in the generation of the bulk magnetic record.

3. Unmixing Different Components Using Bulk Magnetic Measurements

If magnetic properties are assumed to be linearly additive and the different magnetic components that are present in the bulk sediment are distinguishable from one another, then potential exists to unmix their contributions within the bulk magnetic signal. Bivariate plots of two magnetic parameters provide a popular and relatively rapid method to visualize component mixtures. For instance, the Day plot [92] combines the ratios of the coercivity of remanence and coercivity (Hcr/Hc) to saturation remanence and saturation magnetization (Mrs/Ms) and is frequently used to help provide a semi-quantitative estimate of the average magnetic grain size of magnetite and the potential mixing of SP, SD, and MD contributions [93,94]. Similarly, the ratio of the mass normalized susceptibility of the Anhysteretic Remanent Magnetization (ARM) to mass normalized magnetic susceptibility (χ_{ARM}/χ) and the χ_{ARM} to mass normalized frequency dependence of susceptibility (χ_{ARM}/χ_{fd}) has been used to discriminate SD chains of bacterial magnetosomes from detrital catchment sources [31,95–97], while combinations of concentration independent ratios can be used to discriminate between different magnetic mineralogies [98]. If end-members of different sources or components can be defined then bivariate or multivariate unmixing can provide a quantitative estimate of the combination of different components [11,17,69,99]. For example, using catchment samples as source endmembers, Yu and Oldfield [69] used linear programming in an attempt to unmix source contributions in lake sediments [69]. In this example, lake sediments fall outside of the source endmembers which was interpreted to reflect inadequate source characterization and highlights the importance of endmember characterization for the linear unmixing of sediments [69,99,100]. In order to characterize end members for MD, interacting SD (ISD), uniaxial non-interacting SD (UNISD), and super-paramagnetic (SP) components, Lascu *et al.* [53] measured Ms, Mrs/Ms, the ratio of χ_{ARM} to the Saturation Isothermal Remanent Magnetization ($\chi_{ARM}/SIRM$), and the mass normalized susceptibility of the ferrimagnetic fraction (χ_f/Ms) [53]. Using the discrimination afforded by these magnetic grain size end members to represent different detrital, biogenic, and pedogenic components, they were then able to unmix the flux of different sources in lake sediments and discuss the potential processes responsible for driving temporal change [53,101].

Despite the breadth of different magnetic parameters, they can only discriminate the magnetic fraction in terms of their concentration, mineralogy, and magnetic grain size. This presents issues for linear unmixing as increasing the number of potential sources in mixtures increases the uncertainty in the result [98,100,102], with those results being increasingly non-unique [103]. As an alternative, high resolution measurements of ARM and/or IRM can be decomposed to reveal discriminable coercivity distributions that potentially relate to different components. Generation and optimization of the fit of a number of cumulative log Gaussian (CLG) functions to the changes in the slope of magnetic remanence acquisition or demagnetization curves can then be used to represent different components (e.g., [47,48,104,105]). Combinations of these distributions, each of which possessing a modal coercivity, dispersal range, and a relative proportion, can then be used to model the measured bulk magnetic properties (e.g., Figure 1). For example, the gradient of the IRM acquisition curve in Figure 1a is best modeled using three components (Figure 1d) which exhibits improved fit over the one- (Figure 1b) or two-component (Figure 1c) solution. This technique benefits from requiring no prior knowledge about a sample, however, the resulting distributions (*i.e.*, those in Figure 1d) then require interpretation in terms of the magnetic components they physically represent. While this approach is efficient at identifying high coercivity components (e.g., hematite or goethite), from multiple lower and intermediate coercivity components (e.g., detrital and bacterial magnetite), discrimination of components with similar coercivity distributions can be more difficult [48,105,106].

Figure 1. (a) Isothermal Remanent Magnetization (IRM) acquisition curve (orange dots and line) and the gradient of the IRM (blue dots). One **(b)**, two **(c)**, and three **(d)** cumulative log Gaussian (CLG) distribution fits are shown to the data with their associated R^2 values. Note the greater fit of three CLG distributions to the IRM gradient data suggesting that the bulk magnetic dataset is best described as having contributions from three coercivity distinct components. Redrawn from Heslop *et al.* [48].

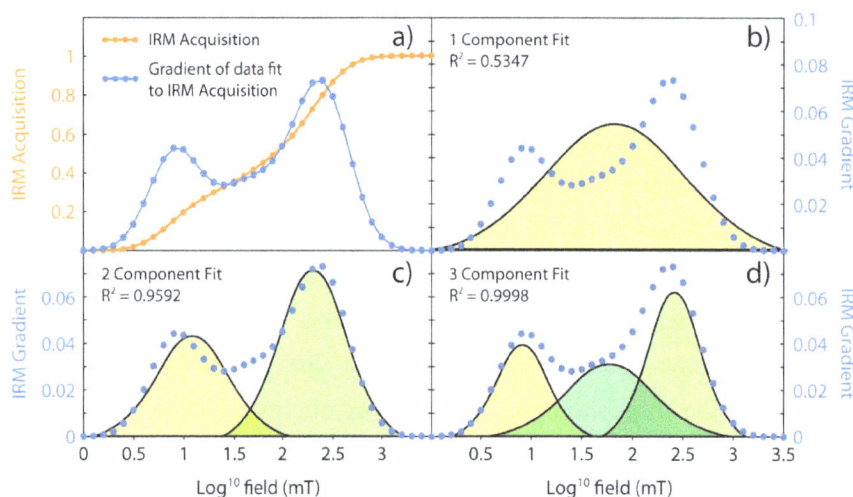

Furthermore, subsequent studies found CLG functions might not be the most appropriate representations of some magnetic components [49,50,106] which can often display characteristics that deviate from idealized Gaussian distributions [49,50]. Development of model functions that could incorporate non-Gaussian variation have shown to better fit and describe commonly occurring low-coercivity magnetic components [49–51] and performed better in identification of components with overlapping coercivity spectra [50,51]. These components can be summarized and discriminated using simple rock magnetic parameters (e.g., the median destructive field of the ARM (MDF$_{ARM}$) and volume normalized susceptibility of ARM to IRM (κ_{ARM}/IRM), Figure 2) providing the basis for discrimination of bacterial magnetosomes, extracellular and pedogenic ferrimagnets, detrital, eolian, and loessic components, and those signatures related to urban pollution from a range of environmental samples (Figure 2; [50–52,107–110]). Similar high resolution modelling approaches have been employed to unmix different shapes within hysteresis loops that relate to different magnetic components [55,111] and quantitative analysis of FORCs [84–86,112–114] are a useful tool to assess the nature, degree of magnetostatic interaction, and the abundance of different magnetic grain sizes within bulk samples [40,114].

Modeling and linear unmixing of the bulk magnetic signal is dependent on magnetic properties being linearly additive [48,50,53,99]. This may not always be a valid assumption in natural samples [99], particularly given the strong interacting nature of some components [40,53,101], that can promote strong non-linearity. However, the generally low concentration of magnetic minerals—and therefore greater dispersal of potentially interacting particles—in sediments compared to synthetic mixtures [99,100] or concentrated magnetic extracts often makes this less of a concern in natural samples [50,115,116]. Interpretation of identified components often assumes a link between sediment and magnetic granulometry. For example, SD components are commonly assumed to be bacterial in origin, while MD grains are often interpreted as detrital contributions. This may be the correct assumption if magnetic fragments exist as discrete particles, but does less well in accounting for the potential of fine magnetic grains to be carried within larger polycrystalline clasts. These techniques also poorly characterize PSD grains often considering them as linear mixtures of SD and MD contributions in the bulk sediment [93,94] despite many natural samples being dominated by these magnetic grains [28,32,117–120]. Furthermore, detailed coercivity measurements can often require an intensive measurement program, sometimes utilizing specialized equipment that can make processing a large number of samples difficult [53]. However, despite these assumptions and limitations these techniques are extremely valuable for characterizing the magnetic heterogeneity inherent in bulk sediment samples.

Figure 2. Bi-plot summary of the median destructive field of the Anhysteretic Remanent Magnetization (MDF$_{ARM}$) and κ_{ARM}/IRM clusters for different magnetic components as identified by Egli [50]. Letters classify and discriminate the different components into low-coercivity magnetosomes (biogenic soft, BS: blue dots and rectangle), high-coercivity magnetosomes (biogenic hard, BH: green squares and rectangle), ultra-fine extra cellular magnetite (EX: brown circles), pedogenic magnetite (PD: brown diamonds) and fluvially transported detrital particles (D: brown diamonds and brown ellipse), wind-blown particles (eolian dust, ED: orange squares), urban pollution particulate matter (UP: black crosses and ellipse), and maghemite in loess (L: red triangles and ellipse). The open rectangle intersecting BS relates to cultured magnetotactic bacteria (blue triangles); magnetosomes with intermediate coercivity fall between BS and BH clusters (open light blue half-filled squares). Arrows indicate the decrease of κ_{ARM}/IRM observed during anoxic lake sediment conditions. All data redrawn from Egli [50].

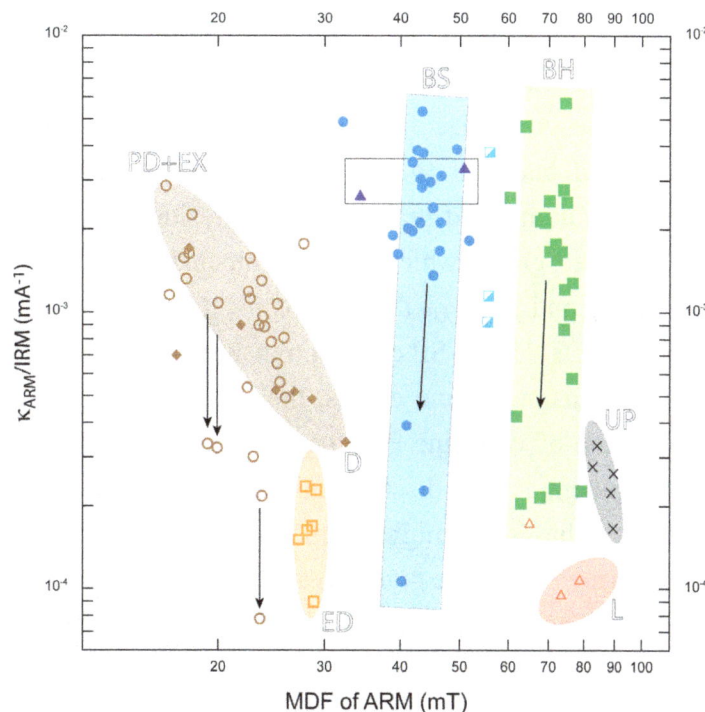

4. Particle Size-Specific Measurements

4.1. Origins of Particle Size Dependence

In contrast to unmixing of the bulk magnetic properties which exploit variations in concentration and/or coercivity, particle size-specific measurements inform on magnetic heterogeneity through the natural preferential association of different components with different physical grain sizes. When magnetic measurements are made on different size fractions they often reveal differences in the magnetic properties of different sediment size fractions [3,11,17,30–32,121–123]. Discrete magnetic grains might be expected to fractionate according to their physical size such that SP, SD, and fine PSD grains will likely reside in the clay fraction [31,121] and coarser PSD and MD grains in the silt and sand fractions [11,17,32]. If we assume different magnetic grain size populations reflect different components, then fractionation of the bulk sediment into different grain size bins can potentially isolate different processes and/or sources. This is neatly illustrated in the two soil profiles in Figure 3. Bulk MS is higher in the A horizon of both the brown earth and the brown sand suggesting ferrimagnetic enrichment in surficial layers relative to the sub soil, and high bulk frequency dependent susceptibility suggests the brown earth has a higher proportion of SP grains than the brown sand soil (Figure 3). Fractionation and measurement of different sediment particle sizes shows that ferrimagnetic grains are carried differently between fractions and between soil profiles. In the brown earth, highest MS and $\chi_{ARM}/SIRM$ occur in the clay fraction consistent with discrete SP/SD size ($\chi_{ARM}/SIRM > 100 \times 10^{-5}$ Am^{-1}; [3]) pedogenic magnetite [3,67]. High bulk MS in the surficial layers of the brown sand soil is associated with relatively coarse MD size ($\chi_{ARM}/SIRM \sim 20 \times 10^{-5}$ Am^{-1}; [3]) magnetic grains that are concentrated in the coarse silt fraction and relate to the atmospheric deposition of pollutants from combustion point sources [3,67]. In these soils, isolation and measurement of different size fractions quickly characterizes and discriminates how magnetic properties are carried in the bulk sediment, links their magnetic properties to sedimentology, and provides a method to discriminate them in the bulk sediment.

While PSD, MD and SP/SD domain boundaries occur within distinct sedimentological grain size classes the SP/SD domain boundary resides in the clay size fraction, over an order of magnitude below the fine silt—clay boundary (~2 µm). It might therefore be anticipated that mixtures of SP/SD grains cannot be isolated by routine grain size fractionation. However, in an analysis of different pedogenic (dominantly SP-size) and bacterial (dominantly SD-size) magnetite dominated samples Oldfield *et al.* [31] showed that SD dominated grains regularly resided in a particle size class (e.g., 2–4 µm) larger than the SP dominated grains which often dominated the finest grades (e.g., <2 µm). Discriminable on a bi-plot of χ_{ARM}/χ and χ_{ARM}/χ_{fd} the reasons for this apparent differential association are not immediately clear, especially as the SD/SP domain transition is below 0.05 µm, but may relate to the aggregation of SD grains in magnetosome chains, their attachment to clay particles affecting settling velocity, and/or the inclusion of very fine PSD particles in the 0.2–2.5 µm range [31]. These results highlight not only the potential for discrimination of different sources using magnetic measurements—the authors suggest that 3 µm might be the optimum cut-off to isolate the pedogenic contribution—but that although discrete magnetic grain sizes largely scale with physical grain size, other factors (that are still not fully determined) can cause deviation from domain size bounds [31].

Figure 3. IRM, ARM, and χ_{ARM}/SIRM as a function of physical grain size for a Brown Earth (Cambisol) soil (**a,b**) and a Brown Sand Soil (**c,d**). Also shown are bulk MS and bulk frequency dependence of susceptibility with depth for the two soil profiles (**e,f**). Figure redrawn from Maher [3]. Note the χ_{ARM}/SIRM scale changes between profiles. The Exmoor Brown Earth soil has a relatively high percentage of ultra-fine pedogenic SP grains throughout the profile concentrated in the clay size fraction. In contrast, the brown sand soil carries its relatively coarse magnetic grains in the silt fraction and is restricted to the upper few centimeters of the profile resulting from atmospheric fall out of MD size ferrimagnetic pollution spherules.

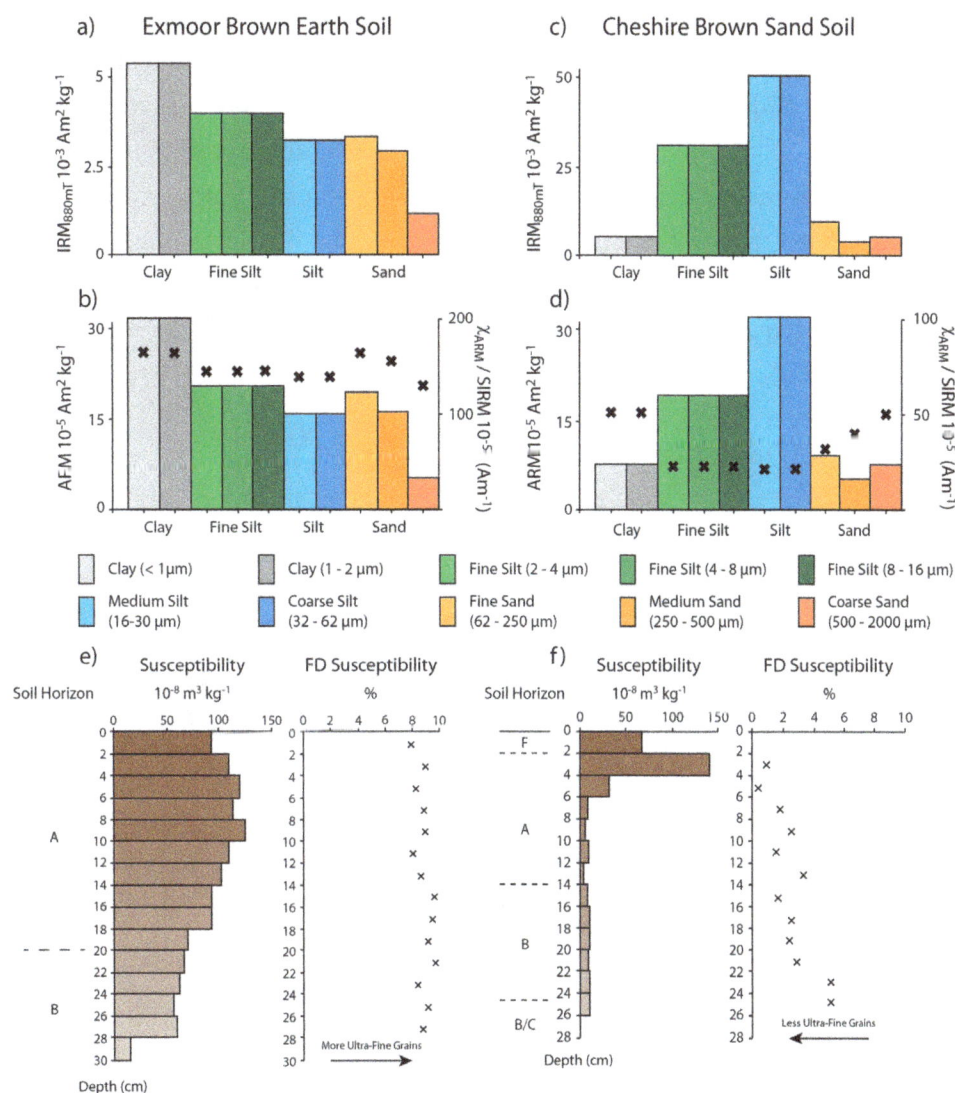

Discrete magnetic components, e.g., magnetic spherules and bacterial and pedogenic magnetite, have relatively well defined magnetic properties and potentially preferential fractions they reside in. However, not all processes that affect the magnetic mineral population act in defined sedimentological ranges nor within certain magnetic domain bounds. The inclusion of magnetic minerals within polycrystalline clasts has significant implications for the interpretation of the magnetic record as it further decouples magnetic grain size from physical grain size. For example, in a comparison of glaciogenic sediments draining the Precambrian continental crust terranes of southern Greenland and oceanic basalts from Iceland magnetic properties vary strongly both with source and sediment particle size (Figure 4, [32]). In both regions, ferrimagnetic minerals are concentrated in the silt fraction which possesses 2–5 times the MS of either the sand or clay fraction (Figure 4a, [32]). A Day plot shows that Icelandic clays, silts, and sands consistently contain relatively fine PSD sized magnetic grains (Figure 4b) while the magnetic grain-size of Greenlandic sediments scale with physical grain-size to a greater extent, so that the magnetic grain-size of the sand fraction is coarser than silt, which is coarser than clay. Only clay from Greenland contains magnetic grain sizes as fine as those sediments from Iceland [32]. Concentration of relatively fine PSD size grains in the Icelandic silt and sand fractions suggests that these grains occur as inclusions within larger polycrystalline clasts. As Iceland is dominated by extrusive mid-ocean ridge basalts [124,125], relatively rapid cooling rates may restrict the growth of larger magnetic grains. In contrast the intruded and metamorphosed continental crust of southern Greenland cooled over a longer period [126,127] potentially permitting growth of different magnetic grain sizes up to and including relatively coarse MD grains [32]. Upon weathering and erosion, this mixture of magnetic grains scales to a greater extent with physical grain size resulting in greater particle size dependence of magnetic properties. Incorporation of these different sources into sedimentary records has implications for their interpretation. For example, while magnetic grain size coarsening would accompany sedimentary grain size coarsening in Greenland sourced records and may be a useful paleo-current strength indicator [8,9,27,28], magnetic grain size proxies of Icelandic sourced sediment records would be less sensitive to changes in sediment grain size.

Figure 4. Variation of terrestrial magnetic properties with grain size: (**a**) Magnetic susceptibility (MS) (one standard deviation range about the mean) of six particle size fractions from Greenland and Iceland. Note how the MS of silts (10–63 μm) are 2–5 times higher than fine silt/clay (<10 μm) or sand (>63 μm) fractions. (**b**) Day plot (Day *et al.* [92]) of clay (<3 μm), silt (3–63 μm), and sand (>63 μm) fractions from Greenland (green data) and Iceland (black data). Note that all Icelandic fractions possess relatively similar fine pseudo single domain (PSD) magnetic grain sizes. In contrast, only Greenland clays possess the same magnetic grain size as Iceland samples, Greenland silts and sands fall further along the SD/MD (super-paramagnetic/multi-domain) mixing line possessing higher concentrations of coarser PSD and MD grains. (**c**) Silt data from (**b**) highlight the discrimination of source afforded by this fraction. Boxes on the *y*-axis show the range of Mrs/Ms values from different sources. All data from Hatfield *et al.* [32].

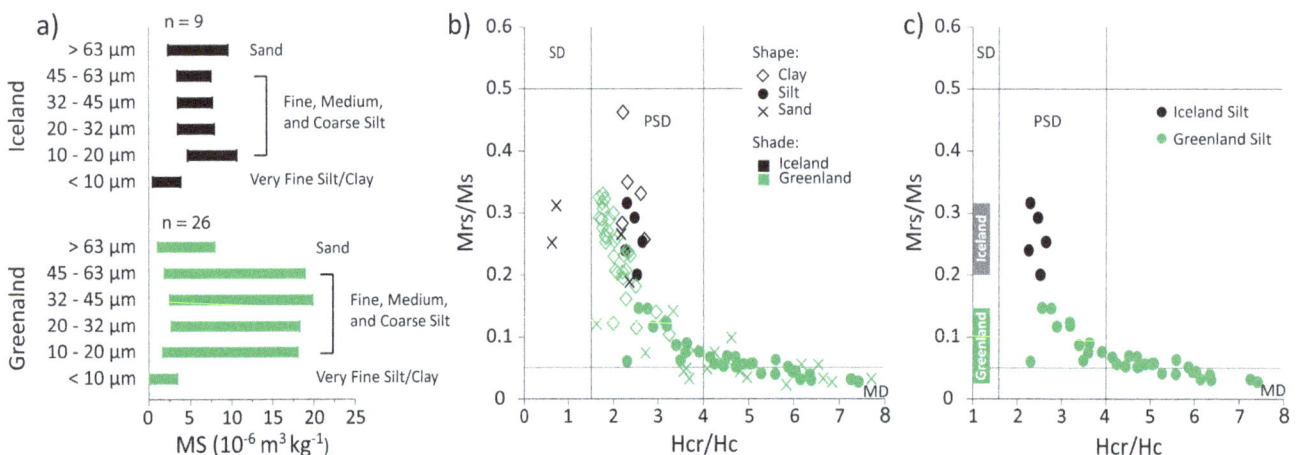

In addition to magnetic grain size variations, magnetic mineralogy can also possess strong particle size dependence especially if different fractions have different sources and/or transport mechanisms (e.g., [12]) The only way to investigate how the magnetic fraction is carried in the bulk sediment is through characterization of the different size fractions that constitute the bulk sediment sample. This methodology provides a way to examine what processes might be important in driving the bulk magnetic signal, which fraction it is held in, and how best to interpret changes in bulk properties that might occur both spatially and temporally. Furthermore, as sediment size in many environments is influenced by sediment transport processes, e.g., dust, bottom currents, or fluvial discharge, bulk magnetic properties can be sensitive not only to sediment source, but also to changes in sediment transport that control the availability and abundance of different fractions.

4.2. The Influence of Sediment Particle Size on Bulk Magnetic Properties

In many environments, different processes dictate the source, transport, and deposition of different physical grain sizes. For example, ice rafting during glacial periods can deliver otherwise absent terrigenous sand to the deep ocean, winnowing can dictate the spatial (and temporal) distribution of sediment particle sizes along current flow paths, turbidites have a characteristic upwards fining clastic signature, storms and catchment flooding can result in mobilization of large quantities of coarse sediment, and eolian sorting can often prescribe the physical grain size of loess. This information alone can be useful for understanding process, but if strong differences exist between the magnetic properties of different particle size fractions, then bulk magnetic properties can be strongly sensitive to changes in abundance of different sediment fractions within the bulk sediment. Assessment of the particle size dependence of magnetic properties and the sediment particle size distribution should be important considerations for the interpretation of bulk magnetic records where potential exists for strong variances in source and/or transport processes. For instance, let us consider a North Atlantic marine sediment core that receives sediment from both Greenland and Iceland. Terrestrial endmembers suggest that ferrimagnetic minerals are concentrated in the silt fraction (Figure 4). If these terrestrial properties are conservative through transport and deposition in the ocean, then small changes in the proportion of silt in the bulk sediment would be amplified in measurement of bulk MS potentially making MS a sensitive proxy for the amount of silt. This is the case in Upper Klamath Lake in Oregon where glacial flour with a median grain size of ~5 μm is estimated to contain ~9 times the MS and a finer magnetic grain size than the non-glacially derived silt fraction (median grain size 24 μm) (Figure 5a, [10,128]). Concentration of ferrimagnetic minerals in the finest fractions explains the strong coherence between changes in the proportion of fine grained silt and clay and bulk IRM (Figure 5b(i)) and helps establish the use of IRM as a proxy for the abundance of glaciogenic silt [128]. Magnetic mineralogy and magnetic grain size of the bulk sediment is also strongly biased by the ferrimagnetically strong, fine-grained, glaciogenic fraction. For example, changes in the proportion of the <10 μm fraction in Klamath Lake sediments drives changes in bulk magnetic mineralogy and magnetic grain size that reflect the properties of the glaciogenic fraction (Figure 5a,b(ii,iii)) [128].

Figure 5. (**a**) Physical and magnetic properties of Upper Klamath Lake non-glacial sediment (NGS) (upper stippled zone in (**b**)) and the calculated glacial sediment end member (GS) from the mixed sediment (lower stippled zone (**b**)). Values in parentheses are the ratio of enrichment of magnetic properties in the GS compared to NGS, e.g., mass normalized MS (χ) of the GS is 8.7 times the value of the NGS. (**b**) Percent of sediment <10 μm (light blue shading) and (**i**) IRM (red line); (**ii**) ARM/χ (blue line); and (**iii**) S-ratios (green line) for the Upper Klamath Lake sediment core [128]. Note the strong agreement between the amount of fine sediment (linked to higher proportions of glacial sediment) and bulk magnetic properties resulting from the concentration of ferrimagnetic minerals in the GS relative to NGS (**a**) in the mixed bulk sediment. All data from Rosenbaum *et al.* [128].

(a)	Median Grain Size (μm)	χ (m^3 kg^{-1})	ARM (Am^2 kg^{-1})	IRM (Am^2 kg^{-1})
Calculated Glacial Sediment	5 μm	5.43×10^{-6} (8.7 x NGS)	4.51×10^{-3} (15.2 x NGS)	9.33×10^{-2} (9.2 x NGS)
Non-Glacial Sediment	24 μm	6.23×10^{-7}	2.96×10^{-4}	1.01×10^{-2}

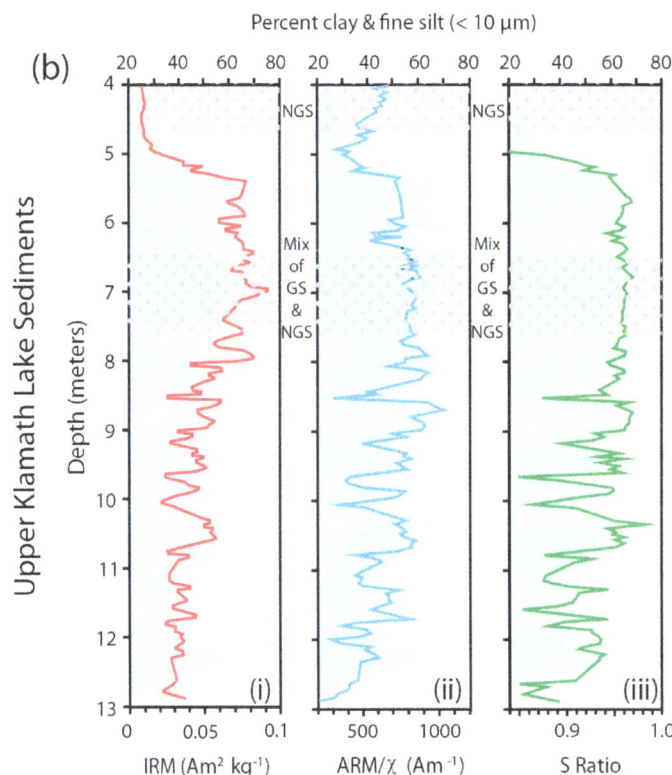

Instead of just a sum of different components, these results suggest that bulk magnetic properties should be considered as a convolved signal of the magnetic heterogeneity of different sediment size fractions (that is influenced by source(s)) weighted by their abundance within the bulk sediment (influenced by transport processes). While bulk magnetic properties are a convolved measure of the two, particle size specific measurements provide the opportunity to quantify the effect of both source and transport. Consideration of both factors has not always been the case in interpretation of bulk sediment magnetic records. For example, in marine sediment cores in the Northern North Atlantic (NNA) basin, source variability is traditionally assessed through measurement of bulk magnetic mineralogy, e.g., s-ratios to determine proportions of magnetically soft and hard components [4,9,22]. In this respect, our assessment of source may be hampered by an ability to easily discriminate between ferrimagnetic dominated sources (e.g., Iceland and Greenland, Figure 4). In practice, NNA records that do not display strong changes in bulk magnetic concentration or the proportion of high coercivity minerals (e.g., hematite) are generally assumed to possess a relatively homogeneous ferrimagnetic dominated source signature. Commonly dominated by PSD grains, these sediments are often assumed to be composed of discrete particles (e.g., [9,22,28]) with many ferrimagnetically dominated bulk magnetic grain size records interpreted in terms of sediment transport. For example, Mrs/Ms and the ratio of the volume normalized susceptibility of ARM to volume susceptibility (κ_{ARM}/κ) frequently used as proxies for variations in bottom current paleo-strength [26–28,62]. Recent data showing that ferrimagnetic grain size in the NNA can vary strongly with source (Figure 4) means that these previously transport related interpretations could instead be significantly influenced by source variations [32], especially in those records surrounding Iceland.

4.3. Isolation of Specific Components and/or Processes

If one magnetic component or process dominates the record then interpretation can be relatively simple (e.g., Figure 5). However, if the signal of interest resides in a fraction that is not abundant within the bulk sediment, and/or it is not reflected by a dominant magnetic component, then its signature can be overwhelmed by other fractions and/or components. If different components associate with different sediment fractions then isolation and measurement of specific grain size ranges can permit the targeting of specific processes and potentially amplification of non-dominant signals from the bulk magnetic record. For example, only the silt and sand fractions can discriminate Iceland and Greenland terrestrial sediments in Figure 4. As NNA sediment cores are often dominated by relatively fine silts and clays [27,129–131], magnetic grain size based source discrimination afforded by the silt and sand fractions would be diluted in the bulk magnetic record by the magnetic properties of the magnetically fine clay fraction (Figure 4). Isolation and measurement of magnetic properties of the silt and sand fractions provides the opportunity to view source variation free from the clay fraction which can vary in abundance depending on the efficiency of sediment transport.

The association of different discrete and included magnetic grains with different sediment grain sizes potentially has implications for interpretation of the processes of post depositional alteration. In sediments that are affected by reductive diagenesis, it is usually assumed that the finest magnetic grain sizes undergo greatest alteration due to their greater surface/volume ratios. Selective dissolution of these (often assumed discrete) grains is usually given as the explanation for decreased magnetic concentration and the coarsening of average bulk magnetic grain size [2,43,45,132]. In a study of Senegalese shelf sediments affected by diagenesis, Razik et al. [12] examined the magnetic properties of 16 sediment size fractions. While all size fractions experienced some degree of alteration, the finest fractions showed a greater degree of change relative to the coarser fractions. This they suggest is explained by the greater surface area and reactive potential of the finest fractions, lower surface to volume ratios in the coarser sediments, and the potential protection afforded by magnetic grains being included within poly-crystalline clasts [12]. In cores less affected by diagenesis, the magnetic properties of different grain size ranges varied strongly with sediment grain size suggesting different source components could be isolated from the bulk sediment [12]. The 10–63 μm fraction is characterized by relatively coarse magnetic grain sizes and lower S-ratios consistent with hematite rich eolian fractions sourced from the Sahara and Sahel, the ferrimagnetically rich <10 μm fraction possesses high concentrations of relatively fine magnetite grains interpreted to reflect delivery of fluvial sediments sourced from the Senegal drainage basin, while the magnetically weaker and coarser sands (63–500 μm) reflected the magnetic properties of marine carbonates [12]. While ferrimagnetic concentration is higher in the fine silt and clay fractions (<10 μm), the greater abundance of the 10–63 μm fraction makes silt an important contributor to bulk magnetic properties [12]. Variances in coercivity of these three different components could potentially be unmixed (see Section 3) to reflect their contribution to bulk magnetic properties. The added advantage of particle size specific measurements in this case is that the magnetic components can be interpreted in terms of their sedimentology and process; the hematite rich component occupies the modal grain size for eolian transport of Saharan and Sahel dust while the fine ferrimagnets in the <10 μm fraction are consistent with fluvial transport of igneous and/or pedogenic magnetite through the Senegal river [12]. Separation also allows the magnetically weaker hematite rich eolian fractions to be studied separately from the ferrimagnetically rich finer fluvial fractions that dominate bulk magnetic properties [12].

In addition to amplification of source and/or process, isolation of specific particle size ranges can also be used to negate the unwanted effect of certain sources of magnetic minerals. Bacterial magnetosomes are a common component in lake and marine sediments [11,31,39,40] that can often dominate bulk magnetic properties and overprint detrital fingerprints (e.g., [83]). These discrete SD grain size particles often reside in the finest fractions during separation [11,31], such that fractionation can isolate them from the detrital signature that might reside the coarser sediment fractions. Figure 6 shows this exemplified in Bassenthwaite Lake in Northwest England. Bassenthwaite Lake has two major sediment inflows, the River Derwent and Newlands Beck. The 31–63 μm fraction of suspended sediments originating from these two sources can be discriminated using IRM acquisition ratios and the MDF of IRM (Figure 6, [11]). The magnetic properties of the bulk sediment from the top 10 cm of four cores recovered from the lake are strongly influenced by that of the clay (<2 μm) fraction which are interpreted to be dominated by bacterial magnetosomes (Figure 6; [11]). When the coarse silt (and medium silt; 8–31 μm (not shown)) fraction in these core top sediments is isolated and measured its properties differ significantly from the bulk sediment and the clay fraction, falling on a mixing line between the coarse silt fraction of two inflow endmembers and suggests Newlands Beck forms the principal contemporary sediment source to Bassenthwaite Lake [11]. Linear unmixing of the bulk magnetic signal might have difficulty in correctly identifying the SD component as an additional magnetic source endmember. Particle size specific measurement reduces this problem to a two end member solution making for an easier assessment of sediment provenance.

In summary, particle size magnetic properties allows an assessment of how magnetic grains are carried in the bulk sediment sample. This methodology allows magnetic properties to be interpreted in a sedimentological context, allows an assessment of the influence of different processes on the bulk magnetic properties, and provides a pathway for amplification and negation of signatures that are potentially difficult to isolate and/or identify within the bulk measurements. However, like all analyses, they are not without their caveats and limitations, most of which pertain to the design and the implementation of their methodology.

Figure 6. Magnetic fingerprints for Bassenthwaite Lake core top sediments and catchment sources. Note how the bulk lake sediments (blue circles) have similar magnetic properties to the clay fraction (black diamonds) while the coarse silt fraction (orange crosses) differs significantly from these two falling on a mixing line between Newlands Beck (green diamonds) and River Derwent (red diamonds) suspended sediments. Isolation of the coarse silt fraction permits comparison with catchment inflows while the bulk magnetic properties are strongly influenced by bacterial magnetosomes that reside in the clay fraction. All data from Hatfield and Maher [11].

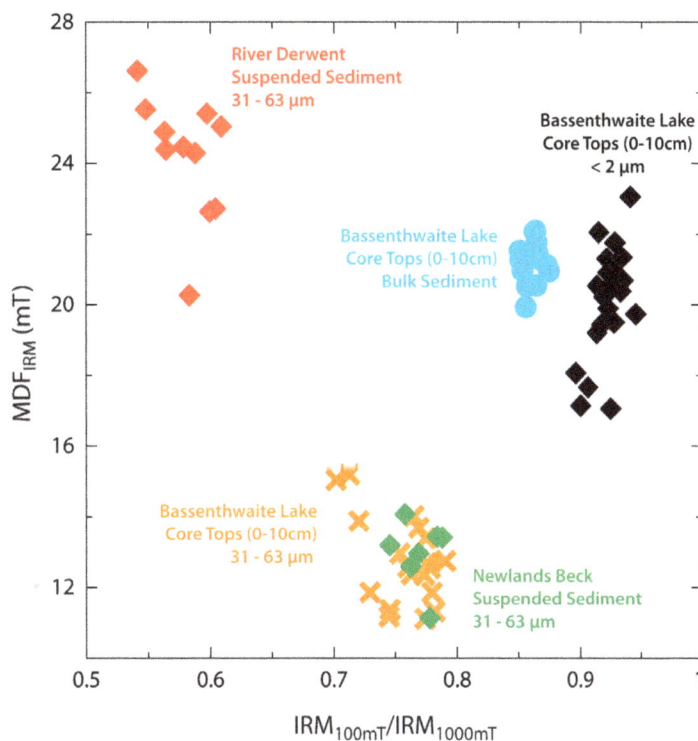

5. Designing a Particle Size-Specific Study

5.1. Number of Fractions

Choosing the optimal number and size range of fractions to separate from the bulk sediment is often the first question when designing any particle size specific study. Selection usually depends on several considerations including the research question(s) posed, the nature of the bulk sediment, the number of potential processes and sources, the time available, and the size of the bulk sample. A large number of separated fractions reduces the volume of material in each fraction and increases the time taken to make separations. This is an important consideration if initial bulk samples are small, or its properties are too weak to be reliably measured. While the current generation of Superconducting Quantum Interface Device (SQUID) magnetometers and Vibrating Sample Magnetometers (VSM) can measure remanence and infield magnetic properties of small sediment samples relatively reliably, if this material is dominated by organic or other weakly magnetically material, as can be the case in many Holocene age lake sediment sequences (e.g., [133–135]), then sample volume becomes particularly important. Resources can also often dictate the number of fractions that can reasonably be generated from bulk samples. For example, separation of 16 fractions from 10 bulk samples quickly generates 160 samples for measurement [12] that might require greater time and material commitments and potentially generates a tradeoff between the number of fractions and the number of samples that can be measured.

Balancing these considerations with the opportunities afforded by particle size specific measurement is therefore one of the central questions in any particle size study. Sometimes, a prior knowledge of process can be exploited to target specific particle size ranges, e.g., the modal grain size of Saharan dust (20–40 µm) [12], or the variation of magnetic properties in glaciogenic silt [10,32]. Other times, the focus on certain grain size ranges pertains to specific research questions, e.g., focusing on the finest fractions to understand the effect of pedogenesis and bacterial magnetosomes [31,122] or potentially understanding how diagenesis might affect different fractions [12]. Utilizing sedimentological information in this way is useful for targeting fractions that might most efficiently reveal and address the scientific research question(s) posed.

When *a-priori* knowledge of how magnetic properties may vary with physical grain size is not available, a sedimentological and/or bulk magnetic property pilot study may provide useful information for targeting fractions of interest. Grain size analysis of the bulk sediment can help identify the fractions most abundant and therefore those which might potentially contribute most to the bulk magnetic signal. While bulk magnetic measurements cannot directly identify the particle size dependence in the sample, they can provide an assessment of the average magnetic grain size and mineralogy and can potentially identify the presence of multiple components which might possess strong particle size dependence. For example, populations of SD sized magnetite that may be related to authigenic bacterial magnetosomes [50,51,112–114] are often identified through IRM acquisition and FORC analyses. With a relatively narrow coercivity range, and characterized as either magnetically hard (BH) or soft (BS) bacterial components (e.g., Figure 2, [49–52]), they often associate with the finest sediment fractions [11,31]. With knowledge of the abundance of the fine fractions, an experiment can be designed to examine how these components might affect the bulk magnetic record. Equally, if the coarser silt and sand fractions form a large part of the sample, then these fractions might also warrant more attention in the fractionation program. If no sample information is available, nor is the opportunity to perform a pilot study, then simple informed decisions about the nature of the sample can be the most useful approach. For example, in samples dominated by a relatively fine granulometry (e.g., soils, loess and eolian deposits, and suspended, lake, and marine sediments), separation into sand (>63 µm), silt (2–63 µm), and clay (<2 µm) fractions can be a very useful and informative first step. If silt turns out to be the dominant granulometric and/or magnetic fraction, it may then be advantageous to further separate this fraction into smaller grain size bins (e.g., [11,17,18,31,32]) as the silt fraction has often shown to be important for carrying a large proportion of the detrital bulk magnetic signal [3,10,11,32,128]. In the literature, the number of generated ranges from two or three up to 16, though commonly around 5–6 fractions are chosen with a focus on the clay, fine silt, and silt fractions [10–12,136,137]. However, ultimately the choice of the number of fractions is a balance between resources and study aims.

5.2. Sample Preparation

Several methods have been outlined in the literature to achieve fractionation of sediment samples [115,138–141]. The simplest and most frequently used involve a combination of sieving of the larger sediment fractions and settling/pipetting of the finer fractions according to Stokes Law. Samples are often pre-treated with a dispersal agent (e.g., sodium hexametaphosphate, tetra-sodiumdiphosphate) to promote disaggregation of flocs in the fine silt and clay fractions. It is important to note the concentration and volume of any added dispersants as they ultimately reside in and contribute to the weight of the finest sediment fraction after settling. Wet sieving is most suited for separation of the sand and coarse silt fractions and is achieved by washing the dispersed sample through a sieve of pre-determined size. For relatively fine sediments, a 63 μm sieve can be used to obtain a sand fraction and, if required, the coarser silt fractions can be obtained using sieves with smaller openings down to about 20 μm (e.g., [32]). Settling is most suited to fractionate samples smaller than those that can be sieved, e.g., fine silts and clays. By making a few assumptions, we can exploit the different settling velocity speeds of differently sized particles according to Stokes' law.

$$V_t = \frac{gd^2(\rho_p - \rho_m)}{18\mu} \tag{1}$$

where V_t is the settling velocity (m·s⁻¹), g is the acceleration due to gravity (m·s⁻²), d is the particle diameter (m), ρ_p and ρ_m are the density of the particle and the medium, respectively (kg·m⁻³), and μ is the viscosity of the fluid (kg·m·s⁻¹).

Assuming all other factors remain equal, settling velocity is largely determined by density which is dependent upon mass and ultimately particle size. Manipulation of the height over which the sediment falls ultimately dictates the duration of the fractionation experiment. Table 1 lists durations and fall heights for some common grain size fractions assuming an average density of quartz (2.65 g·cm⁻³). The finest sediment fraction can be recovered from suspension by using a centrifuge and whether they are sieved or settled different fractions should then be labeled and either freeze dried (preferable) or air dried at room temperature (~25 °C) prior to preparation for measurement.

Table 1. Calculated settling times using Equation (1) to calculate fall velocity for a range of particle sizes over six fall heights. Timings are calculated using the density of quartz 2650 kg·m⁻³ and the density (998 kg·m⁻³) and viscosity (1.002 × 10⁻³ kg·m⁻¹·s⁻¹) of water at 20 °C. "h" = hour(s), "m" = minute(s), "s" = second(s). Micrometer sizes (μm) are given as an approximation of phi (φ) size.

Fall Height	31 μm (5 φ)	16 μm (6 φ)	8 μm (7 φ)	4 μm (8 φ)	2 μm (9 φ)
5 cm	54 s	3 m 37 s	14 m 30 s	58 m 59 s	3 h 52 m
10 cm	1 m 49 s	7 m 15 s	29 m 00 s	1 h 56 m	7 h 44 m
15 cm	2 m 43 s	10 m 52 s	43 m 30 s	2 h 54 m	11 h 36 m
20 cm	3 m 37 s	14 m 30 s	57 m 59 s	3 h 52 m	15 h 28 m
25 cm	4 m 32 s	18 m 07 s	1 h 12 m	4 h 50 m	19 h 20 m
30 cm	5 m 26 s	21 m 45 s	1 h 27 m	5 h 48 m	23 h 12 m

5.3. Magnetic Measurements

Dried sample fractions should be weighed so the granulometric contribution of different particle sizes to the bulk sample can be calculated. Sample measurement cubes, cylinders, capsules, and straws vary between equipment but all packed material should be immobilized and weighed prior to measurement to permit calculation and comparison of magnetic properties on a mass specific basis. To facilitate direct comparison with the bulk record, it is recommended that the bulk magnetic properties of the sample are either measured prior to separation or some bulk sample is retained to permit bulk measurement alongside the measurement of its fractions. This allows comparison of individual fractions to the bulk sediment and examination of which fraction, or combination of fractions, might be important for driving bulk magnetic properties. This also allows for a test of linear additivity as the sum of the different fractions weighted by their abundance should equal the measurement of the bulk sediment. The suite of magnetic measurements employed will largely be dictated by the specific requirements of the investigation, however, measurements of MS, ARM and/or hysteresis measurements, and IRM at several fields, and calculation of their intra-parametric ratios, should be considered a minimum to characterize the concentration, grain-size, and mineralogy of the magnetic grain assemblage.

5.4. Limitations of Particle Size Specific Measurements

The discriminatory power of particle size specific measurements inherently relies on different components associating with different size fractions. Isolation of these different fractions then provides a pathway to isolate different magnetic fingerprints. However, if multiple components reside in the same fraction (e.g., diagenetic products and detrital sources) difficulty may exist in discriminating their separate fingerprints using simple particle size specific measurements. If their coercivity spectra differ, the techniques outlined in Section 3 [49–52,114] might be a useful tool to identify the presence of multiple components within the same sedimentary fraction. Making these high resolution measurements on sediment particles size fractions can be increasingly resource intensive, however, if they can identify and link different components within the fraction with their sedimentological abundance in the bulk sediment, they can potentially provide a better estimate and understanding of how the bulk magnetic record is constructed.

In addition to the resource and time constraints that fractionation imparts on data generation several other factors should be noted. Natural sediments are rarely dominated by spherical grains; instead, they often range from close to spherical grains to needle-like rods and platy clays. This can affect how different particles fractionate into different grain size ranges. For example, platy clays have slower settling velocities than spherical particles and a rod shaped particle that could pass through a sieve along one axis might not be able to if it is rotated 90°. While non-spherical induced drag can enrich the finer fractions in coarser sediments, variations in mineralogical density can enrich the coarse fractions in fine particles (Figure 7). For most settling studies, the average density of quartz (2650 kg·m^{-3}) is often assumed to generate a settling velocity [11,12,17,18,31,32], however, natural samples often differ from this average value resulting in variations in settling velocity for a given particle size [140,141]. Of considerable importance for magnetic studies is that the density of magnetite (5170 kg·m^{-3}) is almost twice that of quartz. While this is not as important for wet sieving of the coarser fractions where grain size binning is purely determined by size, it is especially important for the calculation of fall velocity as discrete magnetite will settle through the sediment column almost twice as fast as similar sized quartz grains. This can result in a finer magnetic population relative to its surrounding sediment matrix which becomes important when comparisons are made between sediment and magnetic grain sizes. For example, quartz dominated silt separated at 7 φ (~32 μm) could contain coarse PSD size discrete magnetite grains at the PSD/MD boundary (~20 μm) within a fraction that purely by size should theoretically be dominated by MD size grains.

Consistency in the methodology used can also be a factor that might affect sediment separation. While the use of multiple personnel to assist in sample fractionation certainly speeds up the separation process, different operators are likely to introduce their own bias through slight differences in technique. Similarly, someone learning these techniques may refine and improve their methodology through practice meaning that after several cycles of repetition, later samples may deviate slightly from the first batch of samples processed. While these biases are harder to account for, they should be considered especially if systematic offsets are apparent in the dataset. Ultimately, these assumptions and nuances mean that there is often an observed distribution of particle sizes outside of the targeted range when the particle size distribution of fractionated samples is analyzed (Figure 7, [140,141]). In the experiment, in Figure 7, samples were only settled once, while in practice separation of each fraction is usually repeated at least twice for each fraction [11,12] to reduce contamination by fines in the coarser fractions. When this methodology is applied, the percentage of material in the finer tail is greatly reduced (e.g., [141]) though coarse and fine particles still exist outside the target range. While it is almost impossible to prevent contamination of fines in the coarser fractions—and *vice versa*—ensuring the same techniques are applied to all samples can help minimize separation variances and permit comparison of similar size distributions.

Figure 7. Laser granulometry measurements of the particle size distribution of separated sediment fractions. Note that the mode occurs within the target range but it accompanied by large tails overlapping into different particle size bins. The coarser fractions are more positively skewed being enriched in fines, while the fines are more negatively skewed being enriched in coarser grains. Figure redrawn from Walden and Slattery [140].

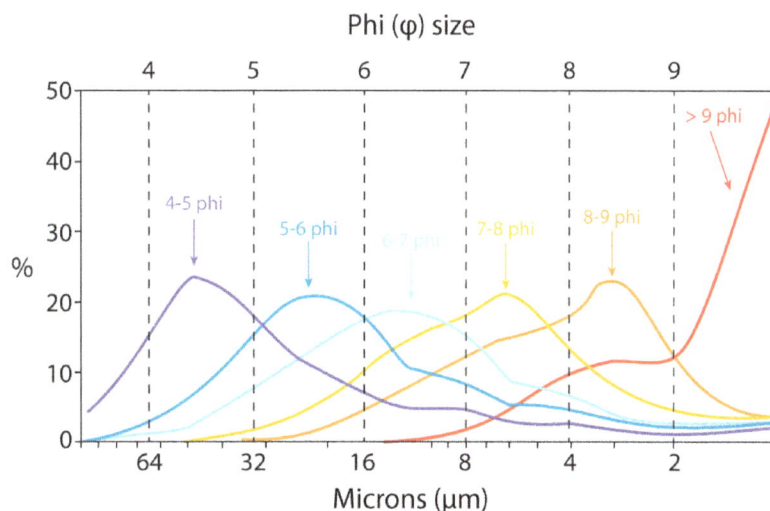

6. Summary and Future Directions

Measurement of particle size-specific magnetic properties demonstrates great potential for understanding the drivers of the bulk magnetic record. This methodology allows a determination of how magnetic minerals and magnetic grain sizes are carried in the bulk sediment, which sediment grain sizes they associate with, how sedimentology affects bulk magnetic properties, to what extent the simple relationship between sediment and magnetic grain size can be assumed, and what processes and/or sources likely affect variances in the bulk magnetic record. Through isolation of specific fractions, certain components can also be amplified and/or cleaned from the study. However, compared to bulk measurements, their generation is time consuming and while not sample destructive they are sample intrusive. This may be a major barrier in studies that utilize magnetic properties because of their rapid, non-destructive nature. As a compromise, an appropriate methodology might be to employ a pilot study to examine the degree of particle size dependence in a few representative samples and examine if bulk magnetic properties can be better understood and interpreted from these pilot data. This approach has been successfully applied in a range of environments, for example, to understand the deposition of silt-size pollution derived magnetic spherules [3,13–15], the generation of ultra-fine magnetite through pedogenesis [3,31,67,77], concentration of heavy magnetic minerals as placer deposits in zones of erosion [123,142], and to understand variations in source and sediments (Figures 4 and 5, [10,32,128]). Alternatively, the pilot study might reveal that particle size specific magnetic measurements are necessary to target reveal certain aspects of the record (e.g., Figure 6) and in these cases they often reveal information that could not otherwise have been gained through measurement of bulk magnetic properties.

Although particle size specific measurements are becoming more frequently employed in the literature, their number is still small in relation to the generation of bulk magnetic records. From this standpoint, particle size specific measurement is very much still in its infancy, despite being used for over 30 years (e.g., [30,69,121]), and the more studies that make these types of measurements the more questions they often produce. Why magnetic fragments partition with different fractions is an important but not fully understood question. In some cases, it appears relatively simple and is related to the presence of discrete magnetic particles, in other situations the hosting of magnetic particles as inclusions may make interpretation more complex (e.g., [32]). How and why magnetic material is hosted within polycrystalline clasts and how they are subsequently weathered into a range of discrete and/or included magnetic particles is also poorly understood. This may be related to host rock type, cooling rate, and/or the nature of the weathering but is yet to be fully investigated. Some rock types may be more preconditioned to release discrete magnetic grains during weathering, while others may be better equipped to retain the non-magnetic matrix magnetic grains that are contained within. This has important implications for the interpretation of bulk magnetic properties and how we view the relationship between sediment and magnetic grain size.

Aside from the tentative suggestion that 3 μm might be the optimal boundary to separate pedogenic SP grains [31], there are at present no criteria that might assist with distinguishing the optimal number of sediment fractions and ranges for separation from the bulk sediment sample. Most sediment size separations are based on additional information and sometimes just scientific instinct. The development of a protocol to best target, select, and isolate different components within the bulk signal would be a useful step and might highlight certain fractions that appear important for hosting bulk magnetic properties. A better protocol to assist separations, and even define the most optimal ranges to use, will not only assist in making the separation process more efficient, but will aid comparison of the same size fractions between studies. With an increased number of particle size-specific records using common fractionation strategies, some of the more fundamental magnetic questions can begin to be addressed rather than just exploiting particle size-specific measurements to answer specific research questions. These broader impacts have not only implications for the drivers of the bulk environmental magnetic record but also for how magnetic recording occurs in sediment paleomagnetic records which demand interpretation of the undisturbed bulk magnetic sediment.

Acknowledgments

Robert G. Hatfield would like to acknowledge support from the National Science Foundation through a Paleo Perspectives on Climate Change Grant ARC-0902751 to Joseph S. Stoner and Anders E. Carlson and an Ocean Drilling Program Grant OCE-1260671 to Robert G. Hatfield and Joseph S. Stoner. Robert G. Hatfield would also like to thank Barbara Maher, Maria Cioppa, and Joseph Stoner for training, guidance, conversations, and support during investigations into the particle size dependence of magnetic properties and three reviewers who greatly improved this manuscript.

Conflicts of Interest

The author declares no conflict of interest.

References

1. Thompson, R.; Battarbee, R.W.; O'Sullivan, P.E.; Oldfield, F. Magnetic susceptibility of lake sediments. *Limnol. Oceanogr.* **1975**, *20*, 687–698.

2. Thompson, R.; Oldfield, F. *Environmental Magnetism*; Allen & Unwin: London, UK, 1986.

3. Maher, B.A. Characterisation of soils by mineral magnetic measurements. *Phys. Earth Planet. Inter.* **1986**, *42*, 76–92.

4. Robinson, S.G. The late Pleistocene palaeoclimatic record of North Atlantic deep-sea sediments revealed by mineral-magnetic measurements. *Phys. Earth Planet. Inter.* **1986**, *42*, 22–47.

5. Maher, B.A.; Thompson, R. Mineral magnetic record of the Chinese loess and paleosols. *Geology* **1991**, *19*, 3–6.

6. Maher, B.A.; Thompson, R. Paleoclimatic significance of the mineral magnetic record of the Chinese loess and paleosols. *Quat. Res.* **1992**, *37*, 155–170.

7. Grousset, F.E.; Labeyrie, L.; Sinko, J.A.; Cremer, M.; Bond, G.; Duprat, J.; Cortijo, E.; Huon, S. Patterns of ice-rafted detritus in the glacial North Atlantic (40–55° N). *Paleoceanography* **1993**, *8*, 175–192.

8. Rasmussen, T.L.; van Weering, T.C.E.; Labeyrie, L. Climatic instability, ice sheets and ocean dynamics at high northern latitudes during the last glacial period (58–10 KA BP). *Quat. Sci. Rev.* **1997**, *16*, 71–80.

9. Kissel, C.; Laj, C.; Labeyrie, L.; Dokken, T.; Voelker, A.; Blamart, D. Rapid climatic variations during marine isotopic stage 3: Magnetic analysis of sediments from Nordic Seas and North Atlantic. *Earth Planet. Sci. Lett.* **1999**, *171*, 489–502.

10. Rosenbaum, J.G.; Reynolds, R.L. Basis for paleoenvironmental interpretation of magnetic properties of sediment from upper Klamath lake (Oregon): Effects of weathering and mineralogical sorting. *J. Paleolimnol.* **2004**, *31*, 253–265.

11. Hatfield, R.G.; Maher, B.A. Suspended sediment characterization and tracing using a magnetic fingerprinting technique: Bassenthwaite Lake, Cumbria, UK. *Holocene* **2008**, *18*, 105–115.

12. Razik, S.; Dekkers, M.J.; von Dobeneck, T. How environmental magnetism can enhance the interpretational value of grain-size analysis: A time-slice study on sediment export to the NW African margin in Heinrich Stadial 1 and Mid Holocene. *Palaeogeogr. Palaeoclimatol. Palaecol.* **2014**, *406*, 33–48.

13. Kapička, A.; Petrovský, E.; Ustjakb, S.; Macháčková, K. Proxy mapping of fly-ash pollution of soils around a coal-burning power plant: A case study in the Czech Republic. *J. Geochem. Explor.* **1999**, *66*, 291–297.

14. Hoffmann, V.; Knab, M.; Appel, E. Magnetic susceptibility mapping of roadside pollution. *J. Geochem. Explor.* **1999**, *66*, 313–326.

15. Hansard, R.; Maher, B.A.; Kinnersley, R. Biomagnetic monitoring of industry-derived particulate pollution. *Environ. Pollut.* **2011**, *159*, 1673–1681.

16. Walling, D.E.; Peart, M.R.; Oldfield, F.; Thompson, R. Suspended sediment sources identified by magnetic measurements. *Nature* **1979**, *281*, 110–113.

17. Hatfield, R.G.; Maher, B.A. Holocene sediment dynamics in an upland temperate catchment: Climatic and land-use impacts in the English Lake District. *Holocene* **2009**, *19*, 427–438.

18. Hatfield, R.G.; Maher, B.A.; Pates, J.M.; Barker, P.A. Sediment dynamics in an upland temperate catchment: Changing sediment sources, rates, and deposition. *J. Paleolimnol.* **2008**, *40*, 1143–1158.

19. Maher, B.A.; Thompson, R.; Zhou, L.P. Spatial and temporal reconstructions of changes in the Asian palaeomonsoon: A new mineral magnetic approach. *Earth Planet. Sci. Lett.* **1994**, *125*, 461–471.

20. Maher, B.A.; Hu, M.; Roberts, H.M.; Wintle, A.G. Holocene loess accumulation and soil development at the western edge of the Chinese Loess Plateau: Implications for magnetic proxies of paleorainfall. *Quat. Sci. Rev.* **2002**, *22*, 445–451.

21. Geiss, C.E.; Egli, R.; Zanner, C.W. Direct estimates of pedogenic magnetite as a tool to reconstruct past climates from buried soils. *J. Geophys. Res.* **2008**, *113*, 1–15.

22. Stoner, J.S.; Channell, J.E.T.; Hillaire-Marcel, C. The magnetic signature of rapidly deposited detrital layers from the Deep Labrador Sea: Relationship to North Atlantic Heinrich layers. *Paleoceanography* **1996**, *11*, 309–325.

23. Watkins, S.J.; Maher, B.A. Magnetic characterisation of present-day deep-sea sediments and sources in the North Atlantic. *Earth Planet. Sci. Lett.* **2003**, *214*, 379–394.

24. Bloemendal, J.; de Menocal, P. Evidence for a change in the periodicity of tropical climate cycles at 2.4 Myr from whole-core magnetic susceptibility measurements. *Nature* **1989**, *342*, 897–900.

25. Stoner, J.S.; Channell, J.E.T.; Hillaire-Marcel, C. Magnetic properties of deep—Sea sediments off southwest Greenland: Evidence for major differences between the last two deglaciations. *Geology* **1995**, *23*, 241–244.

26. Kissel, C.; Laj, C.; Mulder, T.; Wandres, C.; Cremer, M. The magnetic fraction: A tracer of deep water circulation in the North Atlantic. *Earth Planet. Sci. Lett.* **2009**, *288*, 444–454.

27. Snowball, I.; Moros, M. Saw-tooth pattern of North Atlantic current speed during Dansgaard-Oeschger cycles revealed by the magnetic grain size of Reykjanes Ridge sediments at 59° N. *Paleoceanography* **2003**, *18*, 1026–1037.

28. Ballini, M.; Kissel, C.; Colin, C.; Richter, T. Deep-water mass source and dynamic associated with rapid climatic variations during the last glacial stage in the North Atlantic: A multiproxy investigation of the detrital fraction of deep-sea sediments. *Geochem. Geophys. Geosyst.* **2006**, *7*, Q02N01.

29. Heller, F.; Liu, T.S. Palaeoclimatic and sedimentary history from magnetic susceptibility of loess in China. *Geophys. Res. Lett.* **1986**, *13*, 1169–1172.

30. Oldfield, F.; Maher, B.A.; Donoghue, J.; Pierce, J. Particle-size related mineral magnetic source sediment linkages in the Rhode River catchment, Maryland, USA. *J. Geol. Soc.* **1985**, *142*, 1035–1046.

31. Oldfield, F.; Hao, Q.; Bloemendal, J.; Gibbs-Eggar, Z.; Patil, S.; Guo, Z. Links between bulk sediment particle size and magnetic grain size: General observations and implications for Chinese loess studies. *Sedimentology* **2009**, *56*, 2091–2106.

32. Hatfield, R.G.; Stoner, J.S.; Carlson, A.E.; Reyes, A.V.; Housen, B. Source as a controlling factor on the quality and interpretation of sediment magnetic records from the northern North Atlantic. *Earth Planet. Sci. Lett.* **2013**, *368*, 69–77.

33. Canfield, D.E.; Berner, R.A. Dissolution and pyritization of magnetite in anoxic marine sediments. *Geochim. Cosmochim. Acta* **1987**, *51*, 645–659.

34. Roberts, A.P.; Turner, G.M. Diagenetic formation of ferrimagnetic iron sulphide minerals in rapidly deposited marine sediments, South Island, New Zealand. *Earth Planet. Sci. Lett.* **1993**, *115*, 257–273.

35. Passier, H.F.; Dekkers, M.J. Iron oxide formation in the active oxidation front above sapropel S1 in the eastern Mediterranean Sea as derived from low-temperature magnetism. *Geophys. J. Int.* **2002**, *150*, 230–240.

36. Larrasoaña, J.C.; Roberts, A.P.; Musgrave, R.J.; Gràcia, E.; Piñero, P.; Vega, M.; Martínez-Ruiz, F. Diagenetic formation of greigite and pyrrhotite in gas hydrate marine sedimentary systems. *Earth Planet. Sci. Lett.* **2007**, *261*, 350–366.

37. Rowan, C.J.; Roberts, A.P.; Broadbent, T. Reductive diagenesis, magnetite dissolution, greigite growth and paleomagnetic smoothing in marine sediments: A new view. *Earth Planet. Sci. Lett.* **2009**, *277*, 223–235.

38. Maher, B.A.; Taylor, R.M. Formation of ultrafine-grained magnetite in soils. *Nature* **1988**, *336*, 368–370.

39. Snowball, I.F. Bacterial magnetite and the magnetic properties of sediments in a Swedish lake. *Earth Planet. Sci. Lett.* **1994**, *126*, 129–142.

40. Yamazaki, T. Magnetostatic interactions in deep-sea sediments inferred from first-order reversal curve diagrams: Implications for relative paleointensity normalization. *Geochem. Geophys. Geosyst.* **2008**, *9*, Q02005, doi:10.1029/2007GC001797.

41. Robinson, S.G.; Maslin, M.A.; McCave, I.N. Magnetic susceptibility variations in Upper Pleistocene deep-sea sediments of the NE Atlantic: Implications for ice rafting and paleocirculation at the last glacial maximum. *Paleoceanography* **1995**, *10*, 221–250.

42. Zhang, W.; Yu, L.; Lu, M.; Hutchinson, S.; Feng, H. Magnetic approach to normalizing heavy metal concentrations for particle size effects in intertidal sediments in the Yangtze Estuary, China. *Environ. Pollut.* **2007**, *147*, 238–244.

43. Karlin, R.; Levi, S. Diagenesis of magnetic minerals in recent haemipelagic sediments. *Nature* **1983**, *303*, 327–330.

44. Karlin, R.; Levi, S. Geochemical and sedimentological control of the magnetic properties of hemipelagic sediments. *J. Geophys. Res.* **1985**, *90*, 10373–10392.

45. Robinson, S.G.; Sahota, J.T.S.; Oldfield, F. Early diagenesis in North Atlantic abyssal plain sediments chatracterized by rock-magnetic and geochemical indicies. *Mar. Geol.* **2000**, *163*, 77–107.

46. Russell, M.A.; Walling, D.E.; Hodgkinson, R.A. Suspended sediment sources in two small lowland agricultural catchemnts in the UK. *J. Hydrol.* **2001**, *252*, 1–24.

47. Robertson, D.J.; France, D.E. Discrimination of remanence-carrying minerals in mixtures using isothermal remanent magnetization acquisition curves. *Phys. Earth Planet. Inter.* **1994**, *82*, 223–234.

48. Heslop, D.; Dekkers, M.J.; Kruiver, P.P.; van Oorschot, I.H.M. Analysis of isothermal remanent magnetization acquisition curves using the expectation—Maximization algorithm. *Geophys. J. Int.* **2002**, *148*, 58–64.

49. Egli, R. Analysis of the field dependence of remanent magnetization curves. *J. Geophys. Res.* **2003**, *108*, 2081, doi:10.1029/2002JB002023.

50. Egli, R. Characterization of individual rock magnetic components by analysis of remanence curves. 1. Unmixing natural sediments. *Stud. Geophys. Geod.* **2004**, *48*, 391–446.

51. Egli, R. Characterization of individual rock magnetic components by analysis of remanence curves. 2. Fundamental properties of coercivity distributions. *Phys. Chem. Earth* **2004**, *29*, 851–867.

52. Egli, R. Characterization of individual rock magnetic components by analysis of remanence curves. 3. Bacterial magnetite and natural processes in lakes. *Phys. Chem. Earth* **2004**, *29*, 869–884.

53. Lascu, I.; Banerjee, S.K.; Berquo, T.S. Quantifying the concentration of ferrimagnetic particles in sediments using rock magnetic methods. *Geochem. Geophys. Geosyst.* **2010**, *11*, Q08Z19, doi:10.1029/2010GC003182.

54. Heslop, D.; Roberts, A.P. Estimating best fit binary mixing lines in the Day plot. *J. Geophys. Res.* **2012**, *117*, B01101, doi:10.1029/2011JB008787.

55. Heslop, D.; Roberts, A.P. A method for unmixing magnetic hysteresis loops. *J. Geophys. Res.* **2012**, *117*, B03103, doi:10.1029/2011JB008859.

56. Evans, M.E.; Heller, F. *Environmental Magnetism: Principles and Applications of Enviromagnetics*; Academic Press: London, UK, 2003.

57. Dekkers, M.J. Magnetic proxy parameters. In *Encyclopedia of Geomagnetism and Paleomagnetism*; Gubbins, D., Herrero-Bervera, E., Eds.; Springer: Amsterdam, The Netherlands, 2007.

58. Dunlop, D.J.; Ozdemir, O. *Rock Magnetism. Fundamentals and Frontiers*; Cambridge University Press: Cambridge, UK, 2001.

59. Kletetschka, G.; Wasilewski, P.J. Grain size limit for SD hematite. *Phys. Earth Planet. Inter.* **2002**, *129*, 173–179.

60. Thompson, R. Modelling magnetization data using SIMPLEX. *Phys. Earth Planet. Inter.* **1986**, *42*, 113–127.

61. Gee, J.; Kent, D.V. Calibration of magnetic granulometric trends in oceanic basalts. *Earth Planet. Sci. Lett.* **1999**, *170*, 377–390.

62. Stanford, J.D.; Rohling, E.J.; Hunter, S.E.; Roberts, A.P.; Rasmussen, S.O.; Bard, E.; McManus, J.; Fairbanks, R.G. Timing of meltwater pulse 1a and climate responses to meltwater injections. *Paleoceanography* **2006**, *21*, PA4103.

63. Maher, B.A.; Thompson, R. *Quaterary Climates, Environments and Magnetism*; Cambridge University Press: Cambridge, UK, 1999.

64. Butler, R.F. *Paleomagnetism: Magnetic Domains to Geologic Terranes*; Blackwell: London, UK, 1992.

65. Zhou, W.; van der Voo, R.; Peacor, D.R.; Zhang, Y. Vatiable Ti-content and grain size of titanomagnetite as a function of cooling rate in very young MORB. *Earth Planet. Sci. Lett.* **2000**, *179*, 9–20.

66. Maher, B.A. Magnetic properties of some synthetic sub-micron magnetites. *Geophys. J. R. Astron. Soc.* **1988**, *94*, 83–96.

67. Maher, B.A. Magnetic properties of modern soils and loessic paleosols: Implications for paleoclimate. *Palaeogeogr. Palaeoclimatol. Palaeocol.* **1998**, *137*, 25–54.

68. Grimley, D.A.; Arruda, N.K.; Bramstedt, M.W. Using magnetic susceptibility to facilitate more rapid, reproducible and precise delineation of hydric soils in the midwestern USA. *Catena* **2004**, *58*, 183–213.

69. Yu, L.; Oldfield, F. Quantitative sediment source ascription using magnetic measurements in a reservoir-catchment system near Nijar, S.E. Spain. *Earth Surf. Process. Landf.* **1993**, *18*, 441–454.

70. Caitcheon, G.G. Applying environmental magnetism to sediment tracing. In *Tracers in Hydrology*; International Association of Hydrological Sciences Publication No. 215; Peters, N.E., Hoehn, E., Leibundgut, C., Tase, N., Walling, D.E., Eds.; IAHS Press: Wallingford, UK, 1993; pp. 285–292.

71. Roberts, A.P. Magnetic characteristics of sedimentary greigite (Fe$_3$S$_4$). *Earth Planet. Sci. Lett.* **1995**, *134*, 227–236.

72. Rowan, C.J.; Roberts, A.P. Magnetite dissolution, diachronous greigite formation, and secondary magnetizations from pyrite oxidation: Unravelling complex magnetizations in Neogene marine sediments from New Zealand. *Earth Planet. Sci. Lett.* **2006**, *241*, 119–137.

73. Roberts, A.P.; Chang, L.; Rowan, C.J.; Horng, C.-S.; Florindo, F. Magnetic properties of sedimentary greigite (Fe$_3$S$_4$): An update. *Rev. Geophys.* **2011**, *49*, RG1002, doi: 10.1029/2010RG000336.

74. Kent, D.V.; Lowrie, W. Origin of magnetic instability in sediment cores from the Central North Pacific. *J. Geophys. Res.* **1974**, *79*, 2987–3000.

75. Passier, H.F.; de Lange, G.J.; Dekkers, M.J. Rock-magnetic properties and geochemistry of the active oxidation front and the youngest sapropel in the eastern Mediterranean. *Geophys. J. Int.* **2001**, *145*, 604–614.

76. Dearing, J.A.; Dann, R.J.L.; Hay, K.; Lees, J.A.; Loveland, P.J.; Maher, B.A.; O'Grady, K. Frequency-dependant susceptibility measurements of environmental materials. *Geophys. J. Int.* **1996**, *124*, 228–240.

77. Zheng, H.; Oldfield, F.; Yu, L.; Shaw, J.; An, Z. The magnetic properties of particle-sized samples from the Luo Chuan loess section: Evidence for pedogenesis. *Phys. Earth Planet. Inter.* **1991**, *68*, 250–258.

78. Blakemore, R. Magnetotactic bacteria. *Science* **1975**, *190*, 377–379.

79. Petersen, N.; von Dobeneck, T.; Vali, H. Fossil bacterial magnetite in deep-sea sediments from the South Atlantic Ocean. *Nature* **1986**, *320*, 611–615.

80. Kopp, R.E.; Kirschvink, J.L. The identification and biogeochemical interpretation of fossilized magnetotactic bacteria. *Earth Sci. Rev.* **2008**, *86*, 42–61.

81. Kirschvink, J.L.; Chang, S.R. Ultrafine-grained magnetite in deep-sea sediments: Possible bacterial magnetofossils. *Geology* **1984**, *12*, 559–562.

82. Paasche, Ø.; Løvlie, R.; Dahl, S.O.; Bakke, J.; Nesje, A. Bacterial magnetite in lake sediments: Late glacial to Holocene climate and sedimentary changes in northern Norway. *Earth Planet. Sci. Lett.* **2004**, *223*, 319–333.

83. Shen, Z.; Bloemendal, J.; Mauz, B.; Chiverrall, R.C.; Dearing, J.A.; Lang, A.; Liu, Q. Holocene environmental reconstruction of sediment-source linkages at Crummock Water, English Lake District, based on magnetic measurements. *Holocene* **2008**, *18*, 129–140.

84. Egli, R. VARIFORC: An optimized protocol for calculating non-regular first-order reversal curve (FORC) diagrams. *Glob. Planet. Chang.* **2013**, *110*, 302–320.

85. Harrison, R.J.; Feinberg, J.M. FORCinel: An improved algorithm for calculating first-order reversal curve distributions using locally weighted regression smoothing. *Geochem. Geophys. Geosyst.* **2008**, *9*, Q05016.

86. Roberts, A.P.; Pike, C.R.; Verosub, K.L. First-order reversal curve diagrams: A new tool for characterizing the magnetic properties of natural samples. *J. Geophys. Res.* **2000**, *105*, 28461–28475.

87. Chen, A.P.; Egli, R.; Moskowitz, B.M. First-order reversal curve (FORC) diagrams of natural and cultured biogenic magnetic particles. *J. Geophys. Res.* **2007**, *112*, B08S90.

88. Hanesch, M.; Scholger, R. The influence of soil type on the magnetic susceptibility measured throughout soil profiles. *Geophys. J. Int.* **2005**, *161*, 50–56.

89. Mitchell, R.; Maher, B.A.; Kinnersley, R. Rates of particulate pollution deposition onto leaf surfaces: Temporal and inter-species analyses. *Environ. Pollut.* **2010**, *158*, 1472–1478.

90. Hanesch, M.; Scholger, R. Mapping of heavy metal loadings in soils by means of magnetic susceptibility measurements. *J. Environ. Geol.* **2002**, *42*, 857–870.

91. Shi, R.; Cioppa, M. Magnetic survey of topsoils in Windsor-Essex County, Canada. *J. Appl. Geophys.* **2006**, *60*, 201–212.

92. Day, R.; Fuller, M.; Schmidt, V.A. Hysteresis properties of titanomagnetites: Grain size and composition dependence. *Phys. Earth Planet. Inter.* **1977**, *13*, 260–267.

93. Dunlop, D.J. Theory and application of the Day plot (Mrs/Ms *versus* Hcr/Hc): 1. Theoretical curves and tests using titanomagnetite data. *J. Geophys. Res.* **2002**, *107*, doi:10.1029/2001JB000486.

94. Dunlop, D.J. Theory and application of the Day plot (Mrs/Ms *versus* Hcr/Hc): 2. Application to data for rocks, sediments, and soils. *J. Geophys. Res.* **2002**, *107*, doi:10.1029/2001JB000487.

95. Oldfield, F. Toward the discrimination of fine-grained ferrimagnets by magnetic measurements in lake and near-shore marine sediments. *J. Geophys. Res.* **1994**, *99*, 9045–9050.

96. Oldfield, F. Sources of fine-grained magnetic minerals in sediments: A problem revisited. *Holocene* **2007**, *17*, 1265–1271.

97. Van der Post, K.D.; Oldfield, F.; Haworth, E.Y.; Crooks, P.R.J.; Appleby, P.G. A record of accelerated erosion in the recent sediments of Blelham Tarn in the English Lake District. *J. Paleolimnol.* **1997**, *18*, 103–120.

98. Peters, C.; Dekkers, M.J. Selected room temperature magnetic parameters as a function of mineralogy, concentration and grain size. *Phys. Chem. Earth* **2003**, *28*, 659–667.

99. Lees, J.A. Mineral magnetic properties of mixtures of environmental and synthetic materials: Linear additivity and interaction effects. *Geophys. J. Int.* **1997**, *131*, 335–346.

100. Lees, J.A. Evaluating magnetic parameters for use in source identification, classification and modelling of natural and environmental materials. In *Environmental Magnetism: A Practical Guide*; Technical Guide No. 6; Oldfield, F., Walden, J., Smith, J., Eds.; Quaternary Research Association: London, UK, 1999; pp. 113–138.

101. Lascu, I.; McLauchlan, K.K.; Myrbo, A.; Leavitt, P.R.; Banerjee, S.K. Sediment-magnetic evidence for last millennium drought conditions at the prairie—Forest ecotone of northern United States. *Palaeogeogr. Palaeoclimatol. Palaeocol.* **2012**, *337*, 99–107.

102. Lees, J.A. Modelling the Magnetic Properties of Natural and Environmental Materials. Ph.D. Thesis, Coventry University, Coventry, UK, 1994.

103. Rowan, J.S.; Goodwill, P.; Franks, S.W. Uncertainty estimation in fingerprinting suspended sediment sources. In *Tracers in Geomorphology*; Foster, I.D.L., Ed.; Wiley: Chichester, UK, 2000; pp. 279–290.

104. Stockhausen, H. Some new aspects for the modelling of isothermal remanent magnetization acquisition curves by cumulative log Gaussian functions. *Geophys. Res. Lett.* **1998**, *25*, 2217–2220.

105. Kruiver, P.P.; Dekkers, M.J.; Heslop, D. Quantification of magnetic coercivity components by the analysis of acquisition curves of isothermal remanent magnetization. *Earth Planet. Sci. Lett.* **2001**, *189*, 269–276.

106. Heslop, D.; McIntosh, G.; Dekkers, M.J. Using time and temperature dependant Preisach models to investigate the limitations of modelling isothermal remanent magnetization acquisition curves with cumulative log Gaussian functions. *Geophys. J. Int.* **2004**, *157*, 55–63.

107. Spassov, S.; Egli, R.; Heller, F.; Nourgaliev, D.K.; Hannam, J. Magnetic quantification of urban pollution sources in atmospheric particulate matter. *Geophys. J. Int.* **2004**, *159*, 555–564.

108. Sagnotti, L.; Macri, P.; Egli, R.; Mondino, M. Magnetic properties of atmospheric particulate matter from automatic air sampler stations in Latium (Italy): Toward a definition of magnetic fingerprints for natural and anthropogenic PM10 sources. *J. Geophys. Res.* **2006**, *111*, B12S22, doi:10.1029/2006JB004508.

109. Yamazaki, T. Paleoposition of Intertropical Convergence Zone in the eastern Pacific inferred from glacial-interglacial changes in terrigenous and biogenic magnetic mineral fractions. *Geology* **2012**, *40*, 151–154.

110. Chang, L.; Winklhofer, M.; Roberts, A.P.; Heslop, D.; Florindo, F.; Dekkers, M.J.; Krijgsman, W.; Kodama, K.; Yamamoto, Y. Low-temperature magnetic properties of pelagic carbonates: Oxidation of biogenic magnetite and identification of magnetosome chains. *J. Geophys. Res. Sol. Earth* **2013**, *118*, 6049–6065.

111. Jackson, M.; Solheid, P. On the quantitative analysis and evaluation of magnetic hysteresis data. *Geochem. Geophys. Geosyst.* **2010**, *11*, Q04Z15, doi:10.1029/2009GC002932.

112. Pike, C.R.; Roberts, A.P.; Verosub, K.L. Characterizing interactions in fine magnetic particle systems using first order reversal curves. *J. Appl. Phys.* **1999**, *85*, 6660–6667.

113. Roberts, A.P.; Pike, C.R.; Verosub, K.L. First-order reversal curve diagrams and thermal relaxation effects in magnetic particles. *Geophys. J. Int.* **2001**, *145*, 721–730.

114. Egli, R.; Chen, A.P.; Winklhofer, M.; Kodama, K.P.; Horng, C.-S. Detection of noninteracting single domain particles using first-order reversal curve diagrams. *Geochem. Geophys. Geosyst.* **2010**, *11*, Q01Z11, doi:10.1029/2009GC002916.

115. Carter-Stiglitz, B.; Moskowitz, B.; Jackson, M. Unmixing magnetic assemblages and the magnetic behavior of bimodal mixtures. *J. Geophys. Res.* **2001**, *106*, 26397–26412.

116. Yu, Y.; Dunlop, D.J.; Özdemir, Ö. Partial anhysteretic remanent magnetization in magnetite, 1: Additivity. *J. Geophys. Res.* **2002**, *107*, B001249, doi:10.1029/2001JB001249.

117. Channell, J.E.T.; Hodell, D.A.; Lehman, B. Relative geomagnetic paleointensity and $\delta^{18}O$ at ODP Site 983 (Gardar Drift, North Atlantic) since 350 ka. *Earth Planet. Sci. Lett.* **1997**, *153*, 103–118.

118. Channell, J.E.T. Geomagnetic paleointensity and directional secular variation at Ocean Drilling Program (ODP) Site 984 (Bjorn Drift) since 500 ka: Comparisons with ODP Site 983 (Gardar Drift). *J. Geophys. Res.* **1999**, *104*, 22937–22951.

119. Evans, H.F.; Channell, J.E.T.; Stoner, J.S.; Hillaire-Marcel, C.; Wright, J.D.; Neitzke, L.C.; Mountain, G.S. Paleointensity-assisted chronostratigraphy of detrital layers on the Eirik Drift (North Atlantic) since marine isotope stage 11. *Geochem. Geophys. Geosyst.* **2007**, *8*, Q11007, doi:10.1029/2007GC001720.

120. Mazaud, A.; Channell, J.E.T.; Stoner, J.S. Relative paleointensity and environmental magnetism since 1.2 Ma at IODP site U1305 (Eirik Drift, NW Atlantic). *Earth Planet. Sci. Lett.* **2012**, *357–358*, 137–144.

121. Oldfield, F.; Yu, L. The influence of particle size variations on the magnetic properties of sediments from the north-eastern Irish Sea. *Sedimentology* **1994**, *41*, 1093–1108.

122. Hao, Q.; Oldfield, F.; Bloemendal, J.; Guo, Z. Particle size separation and evidence for pedogenesis in samples from the Chinese Loess Plateau spanning the last 22 Ma. *Geology* **2008**, *36*, 727–730.

123. Hatfield, R.G.; Cioppa, M.T.; Trenhaile, A.S. Sediment sorting and beach erosion along a coastal foreland: Magnetic measurements in Point Pelee National Park, Ontario, Canada. *Sediment. Geol.* **2010**, *231*, 63–73.

124. Jakobsson, S.P. Chemistry and distribution pattern of recent basaltic rocks in Iceland. *Lithos* **1972**, *5*, 365–386.

125. Andrews, J.T.; Hardardóttir, J.; Stoner, J.S.; Principato, S.M. Holocene sediment magnetic properties along a transect from Ísafjardardjúp to Djúpáll, Northwest Iceland. *Arct. Antarct. Alp. Res.* **2008**, *40*, 1–15.

126. Willigers, B.J.A.; Krogstad, E.J.; Wijbrans, J.R. Comparison of thermochron- ometers in a slowly cooled granulite Terrain: Nagssugtoqidian Orogen, West Greenland. *J. Petrol.* **2001**, *42*, 1729–1749.

127. Willigers, B.J.A.; van Gool, J.A.M.; Wijbrans, R.; Krogstad, J.; Mezger, K. Posttectonic cooling of the Nagssugtoqidian orogen and a comparison of contrasting cooling histories in Precambrian and Phanerozoic orogens. *J. Geol.* **2002**, *110*, 503–517.

128. Rosenbaum, J.G.; Reynolds, R.L.; Colman, S.M. Fingerprinting of glacial silt in lake sediments yields continuous records of alpine glaciation (35–15 ka), western USA. *Quat. Res.* **2012**, *78*, 333–340.

129. McCave, I.N.; Manighetti, B.; Robinson, S.G. Sortable silt and fine sediment size/composition slicing: Parameters for palaeocurrent speed and palaeoceanography. *Paleoceanography* **1995**, *10*, 593–610.

130. Fagel, N.; Hillaire-Marcel, C.; Humblet, M.; Brasseur, R.; Weis, D.; Stevenson, R. Nd and Pb isotope signatures of the clay-size fraction of Labrador Sea sediments during the Holocene: Implications for the inception of the modern deep circulation pattern. *Paleoceanography* **2004**, *19*, PA3002, doi:10.1029/2003PA000993.

131. Praetorius, S.K.; McManus, J.F.; Oppo, D.W.; Curry, W.B. Episodic reductions in bottom-water currents since the last ice age. *Nat. Geosci.* **2008**, *1*, 449–452.

132. Liu, J.; Zhu, R.; Roberts, A.P.; Li, S.; Chang, J.-H. High-resolution analysis of early diagenetic effects on magnetic minerals in post-middle-Holocene continental shelf sediments from the Korea Strait. *J. Geophys. Res.* **2004**, *109*, B03103.

133. Li, Y.-X.; Yu, Z.; Kodama, K.P.; Moeller, R.E. A 14,000-year environmental change history revealed by mineral magnetic data from White Lake, New Jersey, USA. *Earth Planet. Sci. Lett.* **2006**, *246*, 27–40.

134. Abbott, M.B.; Edwards, M.E.; Finney, B.P. A 40,000-yr record of environmental change from Burial Lake in Northwest Alaska. *Quat. Res.* **2010**, *74*, 156–165.

135. Balascio, N.L.; Bradley, R.S. Evaluating Holocene climate change in northern Norway using sediment records from two contrasting lake systems. *J. Paleolimnol.* **2012**, *48*, 259–273.

136. Wang, Y.; Yu, Z.; Li, G.; Oguchi, T.; He, H.; Shen, H. Discrimination in magnetic properties of different-sized sediments from the Changjiang and Huanghe Estuaries of China and its implication for provenance of sediment on the shelf. *Mar. Geol.* **2009**, *260*, 121–129.

137. Wang, Y.; Dong, H.; Li, G.; Zhang, W.; Oguchi, T.; Bao, M.; Jiang, H.; Bishop, M.E. Magnetic properties of muddy sediments on the northeastern continental shelves of China: Implications for provenance and transportation. *Mar. Geol.* **2010**, *274*, 107–119.

138. Bush, D.C., Jenkins, R.E.; McCaleb, S.B. Separation of swelling clay minerals by a centrifugal method. *Clays Clay Miner.* **1966**, *14*, 407–418.

139. Smith, J.P. Mineral Magnetic Studies on Two Shropshire-Cheshire Meres. Ph.D. Thesis, University of Liverpool, Liverpool, UK, 1985.

140. Walden, J.; Slattery, M.C. Verification of a simple gravity technique for separation of particle size fractions suitable for mineral magnetic analyses. *Earth Surf. Process. Landf.* **1993**, *18*, 829–833.

141. Clifton, J.; McDonald, P.; Plater, A.; Oldfield, F. An investigation into the efficiency of particle size separation using Stokes' law. *Earth Surf. Process. Landf.* **1999**, *24*, 725–730.

142. Gallaway, E.; Trenhaile, A.S.; Cioppa, M.T.; Hatfield, R.G. Magnetic mineral transport and sorting in the swash-zone: Northern Lake Erie, Canada. *Sedimentology* **2012**, *59*, 1718–1734.

3

Review of Biohydrometallurgical Metals Extraction from Polymetallic Mineral Resources

Helen R. Watling

CSIRO Mineral Resources Flagship, PO Box 7229, Karawara, WA 6152, Australia;
E-Mail: Helen.Watling@csiro.au

Academic Editor: Karen Hudson-Edwards

Abstract: This review has as its underlying premise the need to become proficient in delivering a suite of element or metal products from polymetallic ores to avoid the predicted exhaustion of key metals in demand in technological societies. Many technologies, proven or still to be developed, will assist in meeting the demands of the next generation for trace and rare metals, potentially including the broader application of biohydrometallurgy for the extraction of multiple metals from low-grade and complex ores. Developed biotechnologies that could be applied are briefly reviewed and some of the difficulties to be overcome highlighted. Examples of the bioleaching of polymetallic mineral resources using different combinations of those technologies are described for polymetallic sulfide concentrates, low-grade sulfide and oxidised ores. Three areas for further research are: (i) the development of sophisticated continuous vat bioreactors with additional controls; (ii) *in situ* and in stope bioleaching and the need to solve problems associated with microbial activity in that scenario; and (iii) the exploitation of sulfur-oxidising microorganisms that, under specific anaerobic leaching conditions, reduce and solubilise refractory iron(III) or manganese(IV) compounds containing multiple elements. Finally, with the successful applications of stirred tank bioleaching to a polymetallic tailings dump and heap bioleaching to a polymetallic black schist ore, there is no reason why those proven technologies should not be more widely applied.

Keywords: polymetallic; concentrates; tailings; black shales; bioleaching; mine wastes; sulfide ores; oxidised ores

1. Introduction

Currently, there are strong drivers to become more efficient in metal extraction because, for some metals, discoveries of new high-grade deposits or large low-grade deposits are too few to match the predicted growing demand. Cases in point are copper, for which there is already a supply deficit that is likely to continue and zinc, for which the surplus of recent years (Figure 1) is likely to become a deficit. With 16 planned zinc mine closures in the period 2013–2017 and a predicted increase in consumption from 13.5 to 20.5 Mt by 2025 [1], the zinc deficit could increase to as much as 30% of zinc production. These two metals with a long history are seemingly indispensable to the well being of developed nations. Other "indispensable" metals include:

- Ta, In, Ru, Ga, Ge and Pd used in electrical and electronic equipment,
- Ga, Te, In and Ge in photovoltaic cells,
- Co, Li and rare earths in batteries, and
- Pt, Pd and rare earths in catalysts [2].

Conclusions based on the analysis of an extensive data set for the use of seven metals (Ag, Ni, Pb, Cr, Zn, Cu and Fe) in 49 countries [3], were that: (i) per capita metal use is more than 10 times the global average in developed countries; (ii) countries that use more than the average of any metal do so for all metals; and (iii) as wealth and technology increase in developing countries, there will be strong demand for all industrial metals. The growth of the mobile phone industry provides a good example. Mobile phones have become an important means of communication worldwide. Sales rose from about 100 million in 1997 to nearly a billion in 2009, more than six billion phones in the period [4]. The raw materials required for their production are mined around the world and, while the amounts per mobile range from small to trace, the amounts required for a billion phones (annual production) measure in tonnes: 15,000 Cu, 3000 Fe/Al, 2000 Ni, 1000 Sn, 500 Ag, <100 Au, ~15 Pd, ~4 Ta and ~2 In [4].

Based on known reserves, and consumption and disposal at current rates, it was estimated [5] that the elements Zn, Ga, Ge, As, Rh, Ag, In, Sn, Sb, Hf and Au will be exhausted within 50 years, and that Ni, Cu, Cd, Tl and U will be exhausted within 100 years. Estimates of the same order for metals depletion within 50 years (Cu, Pb, Mn, Ag, Sn, Zn) and within 100 years (Fe, Ni, U) were obtained independently [6] and it was suggested that the platinum group metals would be exhausted in 150 years. Estimates such as these highlight the need for new discoveries, more efficient element extraction from known reserves, the processing of mine waste or tailings and the recycling of industrial and urban metal-rich waste The focus of this review is on the extraction of metals from minerals, specifically polymetallic ores, waste or tailings, using proven biotechnologies and highlighting some innovative potential variants.

The declining grades of Australian ores (Figure 2) is representative of declining grades in other mining regions [7]. However, greatly improved process efficiencies have made it possible to extract metals economically from ores of much lower grade than those processed historically. Thus, part of the apparent decline can be attributed to the processing of lower-grade ores, when these are accounted in statistical analyses [8,9]. Nevertheless, in the absence of new discoveries of sufficiently large deposits of high-grade ores containing the targeted elements, there is a need to process ores of generally lower grade than was the case in the 20th century.

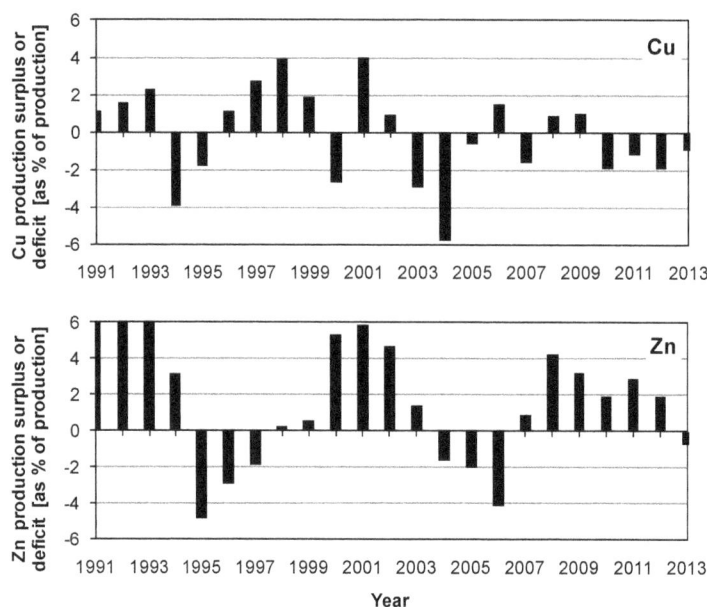

Figure 1. Supply/demand relationships for copper and zinc; annual surpluses and deficits are shown as percentages of total annual production (data from [10–12]).

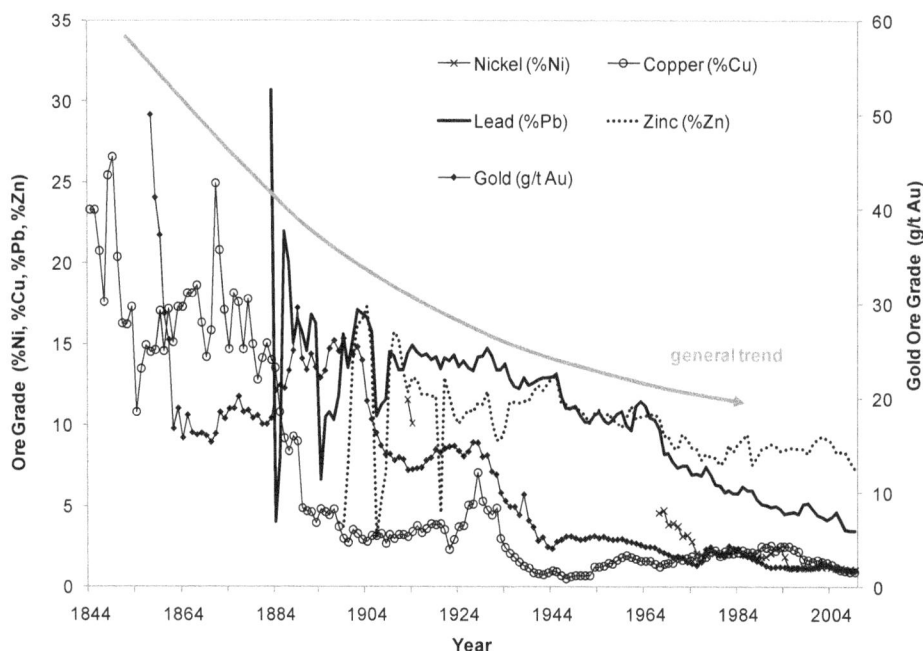

Figure 2. Declining grades of Australian base and precious metal ores. Reproduced with permission from Prior *et al.* [7].

In respect of the greater mineralogical complexity of polymetallic materials, there is a need to develop efficient technologies that recover a suite of metals rather than a single metal, from resources for which the considerable costs of resource definition, mining and ore processing, and waste management are unavoidable, should a decision to construct and operate a commercial plant be undertaken. "Ancillary" metals co-extracted, separated and purified may make the processing of some complex ores economic where the production of a single metal does not. At the same time, alternative sources of metals should be sought and tested. Historically, thousands of tonnes of metal values lie in waste ores and tailings abandoned at mine sites because today's more efficient metal extraction technologies were not available at the time.

In this review the roles and contributions of microorganisms in mineral bioprocessing are briefly reviewed, and bioleaching and biooxidation technologies trialled at pilot scale or demonstration scale, or implemented commercially are described. Examples are taken from publicly-available literature on the bioprocessing of polymetallic mineral resources including ores, concentrates and tailings. Ores are sulfidic or lateritic, or low- or high-organic content (shales or schists).

There are many applications in which microorganisms mobilise or immobilise elements in the environment using innovative biochemical processes. However, those processes and applications of bioleaching, biooxidation and bioreduction not targeting metals extraction have been excluded in order to keep the review to manageable size. In addition, the review does not include data on metal recovery from manufactured products, sludges, electronic waste, *etc.*; topics for which there are already substantial bodies of literature, including some informative reviews [13–18].

2. The Chemistry and Microbiology of Mineral Dissolution

The deliberate exploitation of microorganisms in the process of extracting metals (mainly copper) from ores under acidic conditions has an extremely long history, even though the roles of those microorganisms in both extraction and the generation of acid rock drainage (ARD) were not originally recognised. Some thirty years prior to the "discovery" of *Thiobacillus ferrooxidans*, now *Acidithiobacillus (At.) ferrooxidans* [19], a possible biological method for the "economic utilization of low-grade zinc sulphide ores" was proposed [20]. Broader recognition of bacterial roles followed from studies of ARD associated with bituminous coal [21–23] and copper mine waste dumps [24–26].

Laboratory-scale studies on the extraction of metals from sulfide minerals followed, initially on the application of *At. ferrooxidans* and/or *At. thiooxidans*, but expanding to include newly discovered microorganisms that also enhanced the dissolution rates of many sulfide minerals under acidic conditions [27–30]. Not surprisingly, the main focus of these studies was on microbial iron(II) oxidation and reduced inorganic sulfur compound (RISC) oxidation, the two key microbial capabilities for metals extraction. At the same time, the possible exploitation of heterotrophic microorganisms in the leaching of oxidised silicate ores was tested [31–33] and the important roles of heterotrophic bacteria in bioleaching were studied [34,35].

The main functions of microorganisms useful in bioleaching were readily elucidated because the microorganisms accelerated already understood chemical reactions. From a processing point of view, the microbial functions were considered to be catalytic. However, biohydrometallurgists worldwide have pursued fundamental research on the physiology of microbiological functions and increasingly developed novel approaches to their exploitation either for metals extraction for commercial purposes or remediation of acid rock drainage (ARD). Not surprisingly, with the development of sophisticated experimental and analytical tools, including the application of deoxyribonucleic acid (DNA)-based techniques, a wealth of knowledge about the activities of microorganisms present in bioleaching or ARD environments has been described and discussed in numerous research articles and some recent reviews [36–39]. Only brief summaries of the leaching chemistry and microbial characteristics that assist mineral dissolution are presented in this review.

2.1. Mineral Dissolution in Acidic Environments

Acid leaching of ores or sediments involves the dissolution of minerals. Some minerals dissolve congruently (e.g., calcite, reaction (1)), in which case the resulting soluble species have the same stoichiometry as the source material. Other minerals dissolve incongruently (e.g., biotite, reaction (2)) in reactions that generate soluble species not representative of the source mineral stoichiometry and one or more secondary insoluble minerals.

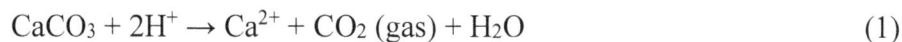

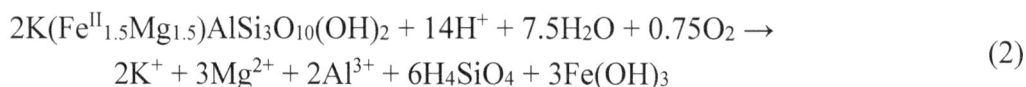

$$CaCO_3 + 2H^+ \rightarrow Ca^{2+} + CO_2 \text{ (gas)} + H_2O \tag{1}$$

$$2K(Fe^{II}_{1.5}Mg_{1.5})AlSi_3O_{10}(OH)_2 + 14H^+ + 7.5H_2O + 0.75O_2 \rightarrow$$
$$2K^+ + 3Mg^{2+} + 2Al^{3+} + 6H_4SiO_4 + 3Fe(OH)_3 \tag{2}$$

The important role of iron chemistry in bioleaching environments should not be forgotten because it is inextricably linked with ore mineralogy, mineral dissolution and solution acidity. Ferric ions in bioleaching solutions containing sulfate anions (from sulfuric acid) and monovalent cations (Na^+, K^+ from the dissolution of carbonate or silicate minerals) readily form insoluble iron(III) hydroxides, oxides, sulfates or hydroxysulfates, or mixtures of them, depending on the composition and acidity (pH) of the solutions. A guide to the conditions of formation of the iron(III) precipitates most often detected in bioleached residues [40] is as follows:

Ferrihydrite (approximate formula $5Fe^{III}_2O_3 \cdot 9H_2O$): Formation is favoured in solutions of pH > 5. Other elements can be adsorbed from solution onto ferrihydrite [41,42]. Ferrihydrite is a poorly crystalline compound that in solution pH 2–5, transforms to goethite (reaction (3)) [43,44].

Goethite ($Fe^{III}OOH$): Formation is favoured from solutions of pH > 4 with low sulfate concentrations [43,45].

Schwertmannite (approximate formula $Fe^{III}_8O_8(OH)_6SO_4$): Formation is favoured in solutions of pH 3–4 containing moderate to high ferric sulfate but low monovalent cation concentrations [46]. Various elements can be adsorbed from solution onto schwertmannite [47–49]. Schwertmannite is poorly crystalline and unstable and transforms to either goethite (reaction (4)) [45,50] or jarosite (reaction (5)) [51,52].

Jarosite ((Na, K, H_3O^+, NH_4^+)$Fe^{III}_3(SO_4)_2(OH)_6$): Formation is favoured in solutions of pH 1.7–2.3 containing monovalent cations, moderate to high ferric ion and sulfate concentrations. Conditions such as these prevail in most laboratory-scale batch bioleaching tests because the microbial culture medium contains potassium and ammonium salts and the leaching is conducted at solution pH 1.5–2 [46,53].

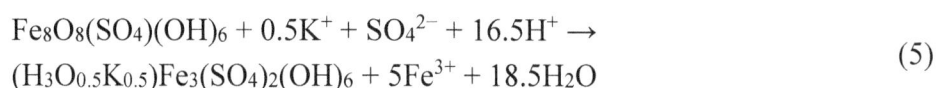

$$Fe_2O_3 \cdot 9H_2O + H_2O \rightarrow 2FeOOH + 9H_2O \tag{3}$$

$$Fe_8O_8(SO_4)(OH)_6 + 2H_2O \rightarrow 8FeOOH + H_2SO_4 \tag{4}$$

$$Fe_8O_8(SO_4)(OH)_6 + 0.5K^+ + SO_4^{2-} + 16.5H^+ \rightarrow$$
$$(H_3O_{0.5}K_{0.5})Fe_3(SO_4)_2(OH)_6 + 5Fe^{3+} + 18.5H_2O \tag{5}$$

2.2. Bio-Generation of Inorganic Acids

Mineral structures can be weakened through the action of microbially-generated inorganic acids such as nitric and nitrous acids, sulfuric and sulfurous acids, and carbonic acid [54,55]. Sulfuric and sulfurous acids are produced by *Acidithiobacillus* species as well as *Thiothrix, Beggiatoa* [56] and some fungi [57]. Nitric and nitrous acids are produced by ammonia and nitrite oxidising organisms, heterotrophic nitrifying organisms and some fungi [58–60]. Carbonic acid is the end product of energy metabolism when carbon dioxide reacts with water but it is a weak acid and unlikely to contribute greatly to mineral dissolution with the exception of carbonate minerals [55].

Sulfuric acid is most often the acid responsible for low-pH leaching environments and is produced by the oxidation of RISCs such as sulfur (reaction (6)). Naturally occurring RISCs are present wherever sulfide minerals are exposed to air and moisture [61]. In the oxidation reaction, the RISC is the electron donor and oxygen is the electron acceptor. Theoretically the amount of energy obtained by microorganisms during RISC (bio)oxidation to sulfate is much greater than when iron(II) is (bio)oxidised [62].

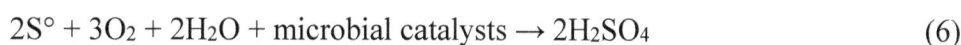

$$2S° + 3O_2 + 2H_2O + \text{microbial catalysts} \rightarrow 2H_2SO_4 \tag{6}$$

2.3. Bio-Generation of Organic Acids and Chelating Agents

All microorganisms can excrete organic acids especially when growth is unbalanced [54]. Lactic acid bacteria and acetic acid bacteria are well known and mineral weathering by fungi (and algae) largely occurs through organic acid production [63].

Organic acids, such as oxalic, citric, gluconic, malic and succinic acids, together with amino acids, nucleic acids and uronic acids, can dissolve minerals via salt formation and complexation reactions [55,64,65]. Organic acids adhere to mineral surfaces and extract nutrient elements from mineral particles by electron transfer. Oxygen links in minerals are broken and ions present in solution are chelated via carboxyl and hydroxyl functional groups [66,67]. Polyfunctional acids like oxalic acid enhance the dissolution of silicates by creating an imbalance between cation and anion concentrations in solution [68–72] but they can also protect calcareous rocks through the formation of, for example, calcium oxalate films [73,74]. Sterflinger [75] collated fungal species reported to colonise building stones (sandstone, marble and granite) and, where known, listed the acids produced by them. In many studies on the bioleaching of oxidised ores such as nickel laterites, the aim was to exploit organic acid production by selected fungi such as Aspergillus species to extract nickel and cobalt [33,76,77].

Siderophores are a group of organic compounds produced by microorganisms to obtain iron in circumstances of low iron availability [78]. Microbial siderophores contain carbonyl structures and have a strong affinity for iron(III) and manganese(III), which they can chelate and transport into cells [79–81]. In specific studies, siderophores have been shown to promote the dissolution of manganese oxides and hematite in the presence of organic acids [82,83]. Siderophores and organic acids act synergistically, the microorganism and/or organic acid interacting with the mineral surface and extracting iron or manganese from the mineral, and the siderophore chelating the iron in solution, thus reducing the free iron concentration and driving the dissolution reaction [82].

2.4. Biodegradation of Organo-Metallic Compounds

Deposits of interest in this review include the black schists and shales, for which there are considerable data on ore geochemistry and ore genesis. Black schists in Finland encountered as interlayers in mica schists contain 1%–2% of non-carbonate carbon and those associated with serpentinite-quartz rock-skarn assemblages contain, on average, 7% non-carbonate carbon [84,85]. The Chimiari shale of Pakistan contains 18% carbon, part of which is carbonate (not quantified) [86]. The black shale horizon of the Kupferschiefer black shales, Poland, contains between 5%–14% organic matter of marine origin (type II kerogen) [87,88], including a range of metallo-porphyrins and metallo-porphyrin-derivatives that contained one or more of Ni, Pb, Co, Cu, Mg, Zn, V, Al, Cr), as well as organo-metallic compounds that contained Sn, Te, W, Pt or Zr [89].

Studies of the bacterial diversity of black shales are an important part of understanding how deposits were formed but fewer studies detailing the diversity of culturable species have been found. Two strains of heterotrophic bacteria, *Bacillus (B.) cereus* and *B. amyloliqueficiens*, were isolated and used to extract metals (at pH 7) from the organometallic component of Kupferschiefer black shale ore [90]. Recoveries after 24–28 days of leaching were Cu (2.5%), Ni (9.3%) and Zn (<0.01%). Isolates of heterotrophic bacteria that degraded organo-metallic components included *Pseudomonas, Acinetobacter, Aeromonas, Brevibacillus, Microbacterium* and *Bacillus* species [91–93]. Most of the isolates were able to utilise simple organic compounds, such as acids or sugars, one isolate was able to degrade phenanthrene (an aromatic hydrocarbon) and several isolates could degrade synthetic metallo-porphyrins [94]. All strains could grow on black shale ore as the sole energy and carbon source, resulting in a slightly increased dissolved organic carbon concentration (14–16 $mg \cdot L^{-1}$) compared with the control (10 $g \cdot L^{-1}$) after 30 days.

DNA-based and other microbiological methods were used to enumerate and describe microbial communities to a depth of 1500 m in a deep borehole through sulfidic black schist [95]. The high microbial diversity comprised communities influenced by depth and differing mineralogical strata. Diverse bacterial communities similarly influenced by depth and sample mineralogy were also reported for a group of black shales from China [96]. Proteobacteria, actinobacteria and firmicutes were more dominant than other phyla and the communities changed progressively with the degree of black shale weathering. The roles of microorganisms in utilising and oxidising sedimentary organic matter were summarised [97], from which it was concluded that black shales supported aerobic heterotrophs, anaerobic heterotrophs such as sulfate reducing bacteria (SRB) and fermentative bacteria, chemoautotrophs (iron(II)- and sulfur oxidising bacteria) and methanogenic bacteria.

In the context of metals extraction, biodegradation of the organic matter in black schist and shale deposits is considered important because some of the values may be bound to organic matter. For example, metallo-porphyrins, which bind elements such as vanadium, nickel, molybdenum and rhenium, are considered to be the compounds most resistant to degradation. In a study using various defined mixed cultures containing species of *Bacillus, Streptomyces, Burkholderia* and/or *Pseudomonas* bacteria and a natural consortium of indigenous bacteria, it was shown that the degradation of synthetic Cu-, V-, Ni- and Fe-octaethylporphyrin compounds was slow [98]. Maximum extractions from the synthetic metallo-porphyrins were 80%, 72% and 4% in 8 weeks, while extractions from shale organic matter were 32%, 81% and 12% in 8 weeks, for Cu, Ni and V, respectively, under the test conditions. However, bacteria isolated from Kupferschiefer ore could grow in salts medium containing synthetic copper and cobalt metallo-porphyrins as the sole source of energy and carbon [99]. In those experiments the increase in metal extraction was accompanied by a decrease in dissolved organic carbon in the medium. Similarly, the same indigenous bacterial cultures grew in salts medium with ground black shale ore as sole energy and carbon source [94,100,101]. SEM or TEM visualization of black shale surfaces after 30 days of leaching revealed the surface corrosion caused by bacterial action and the accumulations of organic matter in cells [89,101].

2.5. Bio-Participation in Redox Reactions

Some bacteria and archaea are able to oxidise reduced species of manganese(II), iron(II), cobalt(II), copper(I), arsenic (AsO_2^-) or selenium (SeO_4^{2-} or SeO_3^{2-}) and others can reduce manganese(IV), iron(III), cobalt(III), arsenic (AsO_4^{2-}) or selenium (SeO_4^{2-} or SeO_3^{2-}), obtaining energy from the reactions [102]. Well known examples are *At. ferrooxidans* and *Leptospirillum (L.) ferrooxidans*, both of which can obtain all of their energy for growth from the oxidation of iron(II) to iron(III). Less well known examples are *Stibiobacter senarmontii*, which obtains energy for growth from the oxidation of antimony(III) to antimony(V) [103] and *Pseudomonas (Pm.) arsenitoxidans*, which obtains energy from the oxidation of arsenic(III) to arsenic(V) [104]. In their comprehensive review on redox reactions of iron in acidic environments, Johnson *et al.* [105] discussed the diverse metabolic characteristics of acidophiles that catalyse iron redox transformations at low pH and the mechanisms employed by acidophiles engaged in iron oxidation and reduction, and described examples of iron cycling in acidic environments, including the degradation of iron(III) compounds under microaerobic or anaerobic conditions. The manganese(IV) mineral asbolane can be solubilised, releasing cobalt, and chromium is released from chromite mineral as the less toxic chromium(III) species [106].

2.5.1. Fe(II) and RISC Biooxidation in Oxygenated Environments

The dissolution of sulfidic minerals, the most widely applied biotechnology for the extraction of metals, requires the presence of an oxidant. Typically, the oxidant is ferric ions and the reactions take place in an oxidising environment. The sulfide moiety in most mineral sulfides is oxidised to sulfur, releasing the metal ion and, for iron-containing mineral sulfides such as chalcopyrite (reaction (7)), ferrous ions. Among sulfide minerals, pyrite is an exception, because oxidation yields sulfate rather than sulfur (reaction (8)). While ferric ions may be released directly during the dissolution of some gangue minerals, for example chamosite ($Fe^{II}_3MgFe^{III}$)$_5AlSi_3AlO_{10}(OH)_8$, the iron-rich end-member of the chlorite group of minerals, or nontronite $Na_{0.3}Fe^{III}_2(Si,Al)_4O_{10}(OH)_2 \cdot nH_2O$), a swelling clay found in some ores, it is mainly regenerated from ferrous ion oxidation by acidophilic microorganisms (reaction (9)). Many acidophiles can also oxidise the elemental sulfur or soluble polythionates formed in bioleaching environments to sulfate (reaction (6)).

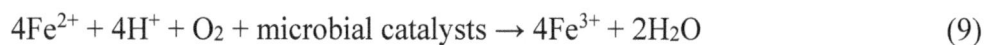

$$CuFeS_2 + 2Fe_2(SO_4)_3 \rightarrow 5FeSO_4 + CuSO_4 + 2S^0 \tag{7}$$

$$FeS_2 + 7Fe_2(SO_4)_3 + 8H_2O \rightarrow 15FeSO_4 + 8H_2SO_4 \tag{8}$$

$$4Fe^{2+} + 4H^+ + O_2 + \text{microbial catalysts} \rightarrow 4Fe^{3+} + 2H_2O \tag{9}$$

The oxidation of ferrous ions and sulfur are the key microbial functions exploited in managed acidic bioleaching processes for the extraction of metals from sulfide minerals. The advantage of the sulfide-chemical-microbial interactions is that the acidophilic microorganisms gain energy for growth from both iron(II) and sulfur oxidation while catalysing the breakdown of the sulfide mineral structure.

Most of the microorganisms utilise carbon dioxide from the air as a carbon source and obtain phosphorus, nitrogen, potassium and micronutrients from the ore, thus minimising the need for, and costs of, microbial maintenance beyond that of providing a suitably acidic environment and a supply of air.

Iron(II)- and sulfur-oxidising acidophiles are relatively few in number (refer to [39] for an overview of biodiversity in acid environments and [40] for species found in heap or tank leaching bioreactors). More are being discovered/described each year using microbiological and molecular techniques for species identification and characterisation [38,107–109] and physico-chemical methods for monitoring substrate oxidation and growth under bioleaching conditions [110,111].

2.5.2. RISC Biooxidation in Anoxic or Oxygen-Limited Environments

Many acidophiles that oxidise iron(II) and RISCs in the presence of oxygen, also reduce iron(III) in the absence of oxygen (reaction (10)). For example, *At. ferrooxidans* uses RISCs as the electron donor and iron(III) as the electron acceptor [112] and also reduces molybdenum(VI) and copper(II) using elemental sulfur as an electron donor [113,114]. The most effective conditions of iron(III) reduction by *Sulfobacillus (S.) thermosulfidooxidans*, *S. acidophilus* and *Acidimicrobium ferrooxidans* were reported to be during mixotrophic or heterotrophic growth in oxygen-limited environments [115]. *At. ferrooxidans* growing on sulfur under aerobic conditions could, in the same reactor operated anaerobically, couple sulfur oxidation with iron(III) reduction and accelerate the dissolution of goethite in mildly acidic medium [116].

$$CuFeS_2 + 2Fe_2(SO_4)_3 \rightarrow 5FeSO_4 + CuSO_4 + 2S^0 \qquad (10)$$

2.5.3. RISC Bioreduction in Anoxic or Oxygen-limited Environments

In addition, heterotrophic acidophiles unable to oxidise iron(II), may nevertheless reduce iron(III) by coupling the reaction with the utilisation of organic compounds. For example, *Acidicaldus organivorus* grown on glucose (the electron donor) under anaerobic conditions used ferric ion as electron acceptor in an acid producing reaction (reaction (11)) [117]. *Acidiphilium* SJH reduced iron(III) in insoluble compounds such as amorphous ferric hydroxide ($Fe^{III}(OH)_3$), akaganeite (β-$Fe^{III}O(OH,Cl)$), goethite or jarosite [115]. In the case of schwertmannite, the reaction produced hydroxyl ions, resulting in raised pH (reaction (12)) [118]. Iron(III) mineral dissolution occurs indirectly because the bioreduction of soluble ferric ion under oxygen-limited conditions causes the solubility equilibrium between the iron(III) mineral phase and iron(III) in solution to become unbalanced and further mineral to dissolve [119].

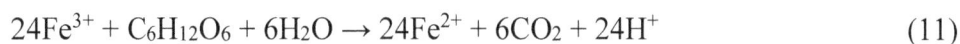

$$24Fe^{3+} + C_6H_{12}O_6 + 6H_2O \rightarrow 24Fe^{2+} + 6CO_2 + 24H^+ \qquad (11)$$

$$3Fe_8O_8(OH)_6(SO_4) + C_6H_{12}O_6 + 6H_2O \rightarrow 24Fe^{2+} + CO_2 + 3SO_4^{2-} + 42OH^- \qquad (12)$$

In both modes of operation, either iron(III) reduction coupled with sulfur oxidation (reaction (10)) or iron(III) reduction coupled with utilisation of an organic compound (reaction (11)), there is a need to supply the electron donor, which adds to the cost of a process. Hallberg *et al.* [116] estimated that glycerol, a relatively low-cost, organic by-product of bio-diesel fuel production, was nevertheless more expensive than sulfur and that the use of sulfur had the additional advantage of lowering the amount of acid required to maintain solution acidity during mineral dissolution. They also noted that, in an "open" bioleaching system, acidophiles other than those that produce acid would colonise the reactor and compete for the supply of organic compound, adding to the cost of maintaining the necessary acidity for mineral dissolution.

SRB are ubiquitous in anoxic environments and play important roles in the carbon and sulfur cycles [120]. SRB may be beneficial in removing sulfate from wastewater but may cause problems through the production of sulfide which can be toxic and corrosive. An example of the use of SRB in the mining industry is a 500 m³ gas-lift reactor to treat zinc sulfate-containing process water at a zinc smelter [121]. In their review, Johnson and Hallberg [35] noted that only four genera of acidophilic archaea are known to grow anaerobically by the reduction of elemental sulfur: *Acidianus*, *Stygiolobus*, *Sulfurisphaera* and *Thermoplasma*, and that the majority of acid-tolerant or acidophilic SRB are sensitive to mild acidity (no growth at pH < 5). The strong sensitivity to acid would preclude their use in typical bioleaching environments.

2.6. Microbial Growth under Element Stress

A key challenge to microbial growth is solution chemistry. High concentrations of cations and anions build up in recycled process solutions during leaching. Thus, depending on the mineral concentrate and processing conditions, concentrations in stirred tank leachates may be up to (g·L⁻¹): Zn 65, Fe 60, Cu 35, Ni 25, As 20, Co 5, Mg <1, and SO_4^{2-} 145. Solution ionic strengths up to 8.5 M (estimated assuming ideal conditions) may exist and are especially influenced by iron(III) and sulfate ion concentrations and different concentrations exist in primary, secondary and tertiary tanks [122]. Concentrations are markedly different in heaps or dumps because of gangue mineral dissolution, pH gradients with depth and the condition of unsaturated leaching, which results in spatial variations in effective contact between particles of the target minerals and the percolating solution. Thus recycled heap solutions may contain (g·L⁻¹): Fe 25, Al 25, Zn 23, Mg 10, Cu 6, Ni 5, As 8, Co <1 and SO_4^{2-} 130. Solution ionic strengths up to 7.6 M have been estimated for heap process solutions, reflecting the high Fe^{3+}, Al^{3+} and SO_4^{2-} concentrations.

Several accounts of microbial adaptation to arsenic emerged as a result of the commercialisation of biooxidation plants for the treatment of gold-containing arsenopyrite concentrates. A mixed culture in a 40 wt % arsenopyrite slurry became adapted to growth in a pH 0.5 solution containing 27 g·L⁻¹ As and 90 g·L⁻¹ Fe [123]. That culture was shown to oxidise both iron(II) and sulfur in a pH 0.9 solution, and to oxidise iron(II) in a pH 2.3 solution. Evidence of adaptation to arsenic was also obtained during the development of the BIOX™ process. After two years of continuous pilot-scale operation, the required retention time was reduced from 12 to 3.5 days in solutions of 13 g·L⁻¹ As [124]. Similarly, during the development of the Bacox process [125], a moderately thermophilic culture was adapted to 25 g·L⁻¹ As, much higher than the 6 g·L⁻¹ subsequently present in plant process solution.

Descriptions of the piloting and demonstration plants for copper, nickel, cobalt and zinc are lacking in direct evidence of microbial adaptation. In a cobalt process, the microorganisms grew in process solution with >5 g·L⁻¹ Co [126] and mineral oxidation was three times faster in continuous reactors than in batch reactors [127]. In a nickel process, the microorganisms grew in process solution with up to 23 $g \cdot L^{-1}$ Ni and 38 $g \cdot L^{-1}$ Fe [128], and in a high-temperature copper process, microorganisms grew in process solution with up to 36 $g \cdot L^{-1}$ Cu and could be adapted to 45 $g \cdot L^{-1}$ Cu [129]. Given that these concentrations are much higher than those encountered in most metalliferous environments, it can be inferred that the microorganisms have adapted to the extreme environments encountered in tank leaching. The report that active bacterial strains were present in mine water with up to 12 $g \cdot L^{-1}$ U_3O_8, [130] is also considered to be indirect evidence of adaptation.

Anions in process water also affect the growth and activity of microorganisms. While it might be expected that the highest concentrations would occur in the more intense tank leaching processes, this is not always the case. Sulfate concentrations can be up to 145 or 130 $g \cdot L^{-1}$ in tanks and heaps, respectively. In laboratory studies, cell replication was halved when a mixed mesophilic culture was "adapted" to 40 $g \cdot L^{-1}$ SO_4^{2-} [131]. Thus microbial adaptation can be inferred from the success of bioleaching in process waters of high sulfate concentrations.

Chloride in process water inhibits iron(II) oxidation more so than sulfur oxidation and microbial growth [132–134]. Although examples of the use of salt-tolerant cultures for ore leaching in process solutions with up to 115 $g \cdot L^{-1}$ total dissolved solids [135] or of the use of a salt-tolerant strain of *Leptospirillum* sp. in heap leaching [136] have been reported, details of the organisms, conditions and duration of adaptation and the limits of sulfur- or iron(II)-oxidation were not reported for either study. However, a mixed population of acidophiles, dominated by an *L. ferriphilum* strain, grew under extreme conditions ($g \cdot L^{-1}$): 0.50 Cd^{2+}, 3.75 Cu^{2+}, 0.2 Pb^{2+}, 92 Zn^{2+}, 6.4 Na^+, 5.5 Cl^-, 154 SO_4^{2-} and total dissolved solids 393.8 (ionic strength 7.47 M) extracting up to 78% Cu and 70% Zn from a polymetallic concentrate [137].

Fluoride in process water presents a greater challenge than chloride. Concentrations greater than 0.5 $g \cdot L^{-1} \cdot F$ are considered to be problematic to microbial growth [129]. However, if fluoride release to solution is accompanied by strong aluminium or iron solubilisation from gangue minerals, then the effects on the microorganisms may be substantially mitigated, due to the formation of complexes [138–142].

The impacts of nitrate ions on microorganisms inhabiting heaps have received less attention. Studies using *At. ferrooxidans* as the test organism showed that 6 $g \cdot L^{-1}$ $NaNO_3$ inhibited iron(II) oxidation by 40%; in the same study iron(II) oxidation was arrested by 8 $g \cdot L^{-1}$ [143]. Nitrate was more inhibitory to iron(II) oxidation than sulfur oxidation by *At. ferrooxidans* [143,144]. However, a recent comparison of the effects of nitrate on substrate utilisation by bacteria and archaea capable of both iron(II)- and sulfur-oxidation showed that mesophilic and moderately thermophilic bacteria adapted to the presence of nitrate and resumed iron(II) oxidation, but iron(II) oxidation by the archaea was suppressed for the duration of the 12-week experiment [111].

2.7. Technology Developments

Different bioleaching technologies have been developed and/or refined during the last 65 years. They include some novel designs tested at laboratory and pilot scale but not finding wide industry acceptance: the flood-drain bioreactor, the aerated trough bioreactor, the airlift bioreactor and the rotating-drum bioreactor.

The flood-drain bioreactor [145,146] comprises a lined container with a perforated-pipe solution distribution system into which is placed ore to a depth of several metres. The reactor is divided into sections. Intermittently, for 1–2 min, bacterial culture is pumped from below to fluidise one section of the bed. This practice uniformly wets and inoculates the ore, at the same time destroying any anoxic zones and washing out secondary precipitates or fines that might plug the bed. Air (O_2, CO_2) is drawn down into the bed as the fluid drains between periods of fluidisation. Sequential fluidisation and draining of sections, together with controlled flow rate and solution management, achieved a degree of size separation and variation in residence time, as well as allowing the washing of the product solids in counter-current mode of operation.

The aerated trough reactor, developed for bio-assisted coal depyritisation [147,148] was a long, rectangular tank with V-shaped base along which was a perforated pipe for air sparging. The trough was divided into sections with solid baffles each having a small hole at the slurry level for flow between sections. Each section had an independent aeration system and a drain for high-density solids removal (the pyrite-enriched fraction, subsequently the feed for the bacterial feed to the reactor).

Airlift reactors have been used relatively frequently for laboratory studies on bacterial leaching of soils, sediments and sludges [149,150] but less frequently for bioleaching and metals extraction from ores or concentrates [151–155]. One of the reasons is that the rate of metal extraction decreases rapidly when the solids loading is greater than 20% [156], which, at the time, was one of the key improvements that was being sought in bioreactor design.

The rotating drum bioreactor (Biorotor) [157,158] is a baffled cylindrical barrel with an aeration inlet (CO_2 and O_2) at one end and an air outlet at the other end. The cylinder sits on rollers and during rotation the baffles lift solids and then discharge them to fall through the solution, maximising mixing, eliminating "dead" zones and minimising shear stresses on bacterial cells. Tests carried out with 30% solids suspension of pyrite yielded solubilisation rates an order of magnitude greater than thought possible [158]. The rotating bioreactor has attracted recent attention with the development of a modified continuous reactor capable of processing a 40% solids suspension [159].

Dumps, heaps, *in situ* and in stope leaching, vat leaching, and continuous stirred tank leaching are the main technologies applied commercially to the bio extraction of metals from minerals.

Dump, *in-stope* and *in-situ* leaching have been applied to copper [160,161] and uranium run-of-mine (ROM) ores [162,163]. Heap leach and bioleach technology for the extraction of Cu from oxidised ores and secondary copper sulfides have been applied most widely in Chile [164]. Heap or dump leaching of copper ores (Figure 3) has spread worldwide [165] and the technology has been adapted and applied to other commodities, for example gold [166], nickel [167], uranium [168] and zinc [169].

The Geocoat® technology is a means of treating concentrates with heap bioleach technology [170]. It combines the advantages of concentrating valuable metals into much smaller bulk and conducting the biooxidation/bioleaching unit process under heap leaching conditions, at lower cost than by stirred-tank processing. Most development tests for the Geocoat® process were conducted with a single target element, such as copper, gold or zinc [171–173]. In the case of refractory gold concentrates or platinum group element (PGE) extraction, the leached de-sulfidised residues would be washed off the support rocks and re-processed for the extraction of the metal. For base metals, soluble metals were recovered from the pregnant leach solutions (e.g., laboratory-scale tests for zinc and copper concentrates [174–176]).

Figure 3. Run-of-mine dump bioleach at Escondida, Chile showing (**a**) surface with irrigation lines and (**b**) covers to retain heat and moisture. Photographs by D.W. Shiers, reproduced with permission.

The use of vat leaching for readily-leached oxidised ores has been largely superseded by dumps or heaps. Domic [164] described the introduction of vat leaching percolation (followed by direct electrowinning of copper) in 1915 at Chuquicamata and in 1928 at Potrerillos, and noted the first solvent extraction-electrowinning trials in Chile (1969–1970) using vat leach solutions from the Exótica (now Mina Sur) mine, and the commercialisation of Solvent extraction—electro winning (SX-EW) for the treatment of solutions from the vat leach of Mina Sur oxide ores in 1987. Currently, the only vat leaching operation for copper ore is located at Mantos Blancos mine in Chile (installed in 1961) (Figure 4). However, vat-leaching is well placed for a revival and wider application with the development of continuous vat technologies [177–179] within which conditions can be controlled and additives can be introduced to enhance extraction.

Agitated tank leaching was developed for the treatment of concentrates but has not found widespread commercial application except in the treatment of refractory gold ores [165,180,181]. Nevertheless, base-metals processes have been developed to pilot and demonstration scale for the bioleaching of copper (Figure 5), nickel and zinc sulfide concentrates using mesophiles, moderate thermophiles and thermophiles [128,182–186] and a process for the extraction of Co from pyrite concentrate has been commercialised [187].

A modification of the stirred tank technology is the separation of the leaching and metals recovery process from the biological regeneration of ferric ions; these are considered to be two-stage processes, termed "indirect bioleaching" or "effects separation" processes with which it is possible to optimise the leaching and biological processes independently and thus maximise metals recovery [188,189]. Another modification for the stirred tank technology is the operation of aerobic reactors followed by anaerobic reactors to effect acid leaching (air plus carbon dioxide sparge) and then bioreduction under anaerobic conditions (nitrogen plus carbon dioxide sparge) [190].

Figure 4. Vat leaching of oxidised copper ore at Mantos Blancos, Chile. Photograph by D.W. Shiers, reproduced with permission.

Figure 5. Bioleaching reactors at the Peñoles Mintek-Bactech demonstration plant at Monterrey, Mexico. Photograph by P.C. Miller, reproduced with permission.

Thus, in summary, the technologies required for the bio-processing of more complex, polymetallic ores and other materials have been developed, piloted and/or demonstrated. In addition, the advances briefly described above have been underpinned by concomitant progress in the separation and purification of many of the elements in the periodic table using solvent extraction or ion-exchange technologies [191–194]. Given the technology developments and the remarkable resilience and adaptability of acidophilic microorganisms, it is not surprising that numerous studies on the application of biohydrometallurgy to many different metal-rich materials from primary mineral resources, secondary mining products and numerous "manufactured" resources have been described. Subsequent discussion is therefore focused on compiling public-domain data on polymetallic bioleaching processes for the recovery or removal of metals from mineral resources, mine wastes and metalliferous sediments, for the purposes of either the production of metals or the remediation of contaminated mine sites.

3. Polymetallic Sulfide Concentrates

Studies using (nominally) pure mineral sulfide concentrates and well known biomining microorganisms as individual strains or prepared mixed cultures comprise the vast majority of fundamental investigations on bioleaching [195–198]. While describing the extraction of only one element, these studies instil confidence that microorganisms are capable of enhancing the extraction of many elements that occur as, or are associated with, sulfide minerals.

The much smaller body of literature on the bioleaching of polymetallic concentrates are, for the most part, conducted at laboratory scale with only a few being developed to pilot scale and integrated with down-stream purification and separation of the precipitates or metallic products. In most laboratory studies, the bioleaching of sulfide minerals or elements associated with them was investigated using known species of iron(II)- and sulfur-oxidising microorganisms in reactors operated at pH 1.5–2 and temperatures in the range of 25–55 °C; the extractions of elements known to leach readily (Cu, Zn, Ni, Co, As) were reported [199–204]. The extractions of more refractory metals not necessarily associated with sulfide minerals such as Mo, Re, V and U were reported in only a few studies [205,206].

Among the studies using polymetallic concentrates are examples of:

- direct bioleaching in continuous stirred tanks [182,207–209],
- two-stage indirect bioleaching, in which ferric ion oxidation of sulfides and ferric ion regeneration by microorganisms are effected in separately optimised bioreactors [137,210–212], and
- combinations of bioleaching and chemical leaching [213,214].

The purposes of the studies varied but included:

- extraction of metals;
- removal of unwanted contaminating elements from "dirty" concentrates [215–218];
- stabilisation and/or immobilisation of unwanted waste products, facilitating residue storage, disposal or subsequent treatment [214,219–221]; or
- descriptions of the microorganisms involved at various stages of processing [222–226].

In some studies, methods for the separation and recovery of a variety of metal products were either tested or suggested in proposed flow sheets. Solvent extraction—electro winning (SX-EW) comprises a well developed suite of technologies applied to most hydrometallurgical processes in which base metals are to be recovered from solution and, as such, does not receive particular attention in the examples cited. However, intermediate products like hydroxide or sulfide precipitates or cementation are also possible. In a number of studies note was made that precious metals reporting to the leached residues could be recovered using cyanide or more aggressive reagents [176,213,214,219,227].

One of the research streams in the Bioshale FP6 project [228] was directed towards process developments for the copper- and silver-enriched black shale ore and concentrate from Lubin mine and concentrator, Poland. The copper concentrator also produced a middling shale fraction that had similar properties to the shale ore but caused problems in the flotation plant. The team showed that both the concentrate and the middlings could be leached using acidophilic iron(II)- and sulfur-oxidising microorganisms in a range of temperatures from 30 to 78 °C [213,229–231]. However, in respect of the tank bioleaching tests [213] it was concluded that the grades for nickel, copper and zinc in the concentrate were too low for this process option to be economic.

4. Low Grade Sulfide Ores

Low-grade sulfide ores are those from which it is not economical to prepare a concentrate, or those of complex mineralogy that precludes upgrading by gravity or flotation concentration processes. Being of low grade, economic processing necessitates the application of low-cost technology. Currently, the technologies of choice are heap or dump leaching, both of which have been commercialised on very large scales for the leaching of copper sulfide ores [30,164] but historically, *in situ* and in stope bioleaching have received considerable attention for the extraction of copper [232,233] and uranium [162,234]. Investigations of the amenabilities of different ores generally involve laboratory-scale test work, small-medium or large-scale columns or cribs to simulate the heap environment, and larger test heaps. It is not surprising, therefore, that many laboratory-scale studies using pulverised ore are reported as part of the preliminary assessment of different ores and the growth of microorganisms, and that far fewer studies address the larger-scale issues. For the purposes of this review, low-grade sulfide ore types have been grouped as "low" organic carbon content and "high" organic carbon content.

4.1. Ores with Low Organic Content

There are relatively few studies of the extraction of more than one element from low-grade, low organic-carbon-content ores using bioleaching and most studies have been focused on copper or nickel. Typical of many ore bioleaching studies, Groudeva and Groudev [235] applied a mixed culture of the then-known iron(II)-oxidising, sulfur-oxidising bacteria and heterotrophic bacteria to the extraction of copper and zinc from a pyritic ore that contained chalcopyrite, sphalerite and galena. A depth profile of bacterial presence and activity was obtained from the 2 m columns. In that study, cell numbers, iron(II) oxidation and carbon dioxide fixation decreased with depth and the amounts of copper and zinc extracted after 300 days also decreased with depth. Maximum extraction of both metals in the top section of the column was 91%–92%.

Marine hydrothermal polymetallic sulfide materials may contain high concentrations of Cu, Pb, Zn, Au, Ag, In, Ge, Bi and Se, and are more like polymetallic concentrates than ores. For example, the test material used by Korehi and Schippers [236] was mainly composed of chalcopyrite (38.5% Cu). Copper was monitored during bioleaching at three temperatures using known biomining microorganisms. Extractions correlated with temperature, 56%, 42% and 32% after 28 days at 70, 50 and 30 °C, respectively.

The unprecedented rise in the price of nickel in the period up to 2006 together with encouraging results from bench-scale testwork prompted pilot- and demonstration-scale trials of nickel sulfide ores in heaps and agitated tanks in Australia, Finland, China and South Africa [237–240]. Watling *et al.* [241] undertook retrospective studies aimed at understanding the nickel and copper leaching chemistry in a copper-nickel test heap in Western Australia [237]. In that heap, 91% Ni was recovered in one year but copper extraction lagged behind (50% recovery). Examination of leached heap and column materials showed that chalcopyrite had been oxidised and the copper mobilised, but that copper had precipitated from solution as covellite or possibly chalcopyrite at greater depth in the heap or column and only subsequently redissolved when the solution chemistry changed [242–244].

Another extensive study of the bioleaching of nickel sulfide in a pentlandite-pyrrhotite ore was conducted under conditions selected to minimise magnesium solubilisation (pH > 3) from Mg-rich silicate minerals while maintaining efficient nickel extraction [245–247]. Subsequently, the bioleaching of six Ni-Co-Cu-containing ores from different geographical regions were compared using stirred tanks and solutions of pH 2–5 [248]. The results indicated that, by lowering the solution acidity from pH 2 to pH 3, nickel and cobalt extractions were the same within statistical error but acid consumption by the gangue minerals was significantly reduced. As a pre-inoculation strategy, a test heap of high-magnesium Ni-Co-Cu sulfide ore was leached for 80 days with dilute acid to remove the readily leached magnesium [249]. This strategy, while not minimising overall acid consumption, did generate conditions conducive to microbial activity throughout the ore bed with the result that 84.6% Ni and 75% Co was extracted after 350 days.

In addition to "standard" testwork, research topics of strong interest included:

- *in-situ* or in-stope bioleaching [250–255];
- strategies to deal with gangue mineral acid consumption while maintaining conditions conducive to microbial growth [245–248,256];
- comparisons of different reactor designs or modes of operation [250,257–259]; and
- development of hybrid processes in which the base metals were bioleached and precious metals or lead were extracted subsequently from the leached residues in a secondary process [235,260–262].

As far as is known, none of the above studies resulted in commercial developments, although flowsheets were described for two of the proposed processes [261,263].

4.2. Sulfidic Schists and Shales with High Organic Content

The main driver for the development of technologies to extract metals from black shales and schists is the uranium content. It is estimated that the black shales around the Black Sea collectively contain the largest known uranium resource in the European Union [264]. However, in general, black shales occur in many parts of the world and contain many base and precious metals (Table 1). Estimated averages for trace element contents in black shales and coals have been collated [265]. The elements V, Mn, Co, Ni, Cu, Zn, Y, Zr, Mo, Ag, Cd, Ce, Re, Au, Hg, Th and U, were identified as target metals for possible production from black shales. V, Ni, Mo, Re and U dominate the more-difficult-to-extract porphyry-bound metals in shales but Cu, Fe, Se, Ag, Cd and Bi typically occur as discrete mineral phases.

4.2.1. Microbiological Aspects

Fundamental multi-element laboratory-scale studies were conducted on shales from a variety of regions. The studies fall into two groups, those that employed autotrophic iron(II)- and sulfur-oxidising acidophiles and those that employed organic-acid-producing heterotrophs. As was the case with the bioleaching studies on low-carbon complex sulfide ores (Section 4.1), many of the laboratory studies on black shales and schists provided valuable data on the different ore types and the kinetics and extent of metals extraction from them under varied but controlled conditions. More than one element was monitored in most studies, mainly those elements that occurred as discrete mineral sulfides and the associated precious metals, because technologies are available for their recovery (Section 2.4).

Bioleaching with autotrophic acidophiles presupposes that the target material will supply the required iron(II) and reduced inorganic sulfur species to support microbial growth [266,267] or that these substrates can be added economically [268–270]. Many metalliferous black shales and schists contain discrete pyrite grains within the organic matrix, thus meeting that particular requirement [87,266,267,271,272] (Figure 6). In two studies on the Alum shale ore, bioleaching was undertaken in tanks or columns inoculated with mixed mesophilic and moderately thermophilic cultures enriched from acidic pools at an auto-heating coal mine [266,267]. For the tank bioleaching of ground ore, maximum extractions after 28–30 days at 55 °C were for Cu (72%–80%), Zn (96%–97%), Ni (46%–50%) and Co (82%–83%). For the columns using −25 mm crushed quarter drill core, maximum extractions at 50 °C and pH 1.6, were Co (>90%), Ni and Cu (65%) and Zn (70%), after 102 days. Acid consumption was lower than anticipated from the shale carbonate content.

Table 1. Some base metal districts with black shale and schist deposits potentially suited to bioleaching [86,87,272–279].

Deposit/Region	Enriched Elements
Talvivaara, Finland	Ni, Cu, Co, Zn, Mn, U, Ag, As,
Kainuu, Finland	Ni, Cu, Co, Zn, V, Mo
Alum shale, Sweden	Ni, Cu, Co, Zn, Pb, V, Mo, U, Cr, Mn, Ba
Viken, Sweden	Ni, V, Mo, U
Kupferschiefer, Poland	Cu, Ag, Zn, Ni, Co, Pb, Mo, V, U, As, Se, Cd, Bi, Tl, Re, PGE
Kamenec, Czech Republic	Ni, Cu, Zn, Mo, Cr, PGE
Hromnice, Czech Republic	Ni, Cu, Zn, V, Mo, Au
Pyrenées, France	Zn, Pb, P, Ge, Cd
Eastern Pyrenées, France	Ni, Cu, Co, Zn, Pb, Au, W, Sb
Dauphiné Basin, France	Ni, Cu, Pb, U, Ba
Selwyn Basin, YT, Canada	Ni, Zn, Pb, Mo, Ag, Cd, In, Sb, Se, As, Au, Tl, Re, PGE
Athabaska region, AB, Canada	Ni, Cu, Co, Zn, V, Mo, U, Ag, Au, Li, Cd, REE
Kimberley, BC, Canada	Cu, Zn, Pb, Sn, Ag, Sb, Cd, Bi
Red Dog, AK, USA	Zn, Pb, Ag, Se, Ba
Carlin region, NV, USA	Ni, Zn, V, Mo, Se, Au, Ag, As
Mina Aguilar, Argentina	Zn, Pb, Ag
Rajasthan, India	Zn, Pb, Ag
KPK Region, Pakistan	Cu, Zn, V, Mn, U, Ti
Zunyi, China	Ni, Zn, Mo, Au, Se, As, PGE
Changba, China	Zn, Pb
Western Yunnan, China	Cu, Zn, Pb, Tl, Cd, Ag, As
Okcheon, South Korea	Ni, Cu, Zn, V, Mo, U, Ba, Cr, Y, Au, PGE
Mt Isa, Australia	Zn, Pb, Ag
Gauteng, RSA	Ni, Cu, Co, Zn, Pb, V, Mo, Cr, Au
Konkola, Zambia	Cu, Co, U, Pd, Re

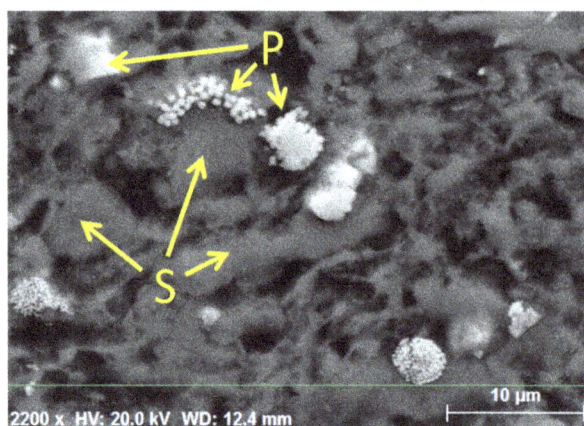

Figure 6. SEM visualization of a polished block of Alum shale from Sweden showing silicate minerals (S) and pyrite (P) embedded within the porous organic matrix.

In another study [280], columns containing polymetallic sulfide black-schist ore high in pyrrhotite were inoculated with a mixed moderately thermophilic culture or thermophilic microorganisms and operated at 47 °C or 68 °C, respectively, for approximately a year. Maximum extractions in inoculated columns were 39%, 34%, 17%, 19% at 47 °C and 69%, 64%, 11% and 32% at 68 °C for Ni, Zn, Cu and Co, respectively. An interesting feature of the tests was the repacking of columns mid-way through the leach, which resulted in an immediate but short-lived increase in the extraction rates of metals in the 47 °C columns. The enhanced extraction rates were of longer duration in the 68 °C columns and were attributed to the disturbance of well-established solution channels and resultant effective contact of the lixiviant with fresh ore particles after column repacking. The possibility that insoluble iron(III) compounds deposited on particle surfaces during percolation leaching of the black schist ore could be removed was tested using *At. ferrooxidans* or *S. thermosulfidooxidans* during autotrophic growth on sulfur under anaerobic conditions [281]. A strategy of alternating periods of aeration and anoxic conditions during leaching resulted in much greater nickel, copper and zinc extraction.

Among those studies in which biogenic organic acids were utilised, key parameters were (i) the variety of microorganisms that could be exploited; (ii) the predominant acids produced by them and (iii) the best carbon sources to be used should larger scale bioleaching reactors be commissioned.

Heterotrophic bacteria, *Pm. fluorescens*, *Shewanella (Sh.) putrefaciens* and *Pm. stutzeri*, were applied to the extraction of uranium from shale tailings at the Ranstad uranium mine, Germany (now closed) [282]. The focus of that study was on metals mobilisation from the tailings rather than economic metals extraction, and the species selected for the tests were known ligand-producing bacteria. The authors noted that ligands may increase the mobility of micronutrients such as Cr, Co, Cu, Mn, Fe, Mo, Ni, V or Zn and possibly lanthanide and actinide elements at pH ~7. Bacteria isolated from the black shale included strains of *Pseudomonas*, *Acinetobacter*, *Bacillus* and *Microbacterium* spp. All were able to grow in salts medium containing black shale as sole energy and carbon source.

Among the heterotrophic fungi that have been applied to uranium and/or metals extraction from black shales are *Aspergillus (A.) niger*, *Penicillium (Pc.) notatum*, *Ganoderma (G.) lucidum* or native strains of *A. flavus*, *Curvularia clavata* and *Cladosporium oxysporum* [283–287]. Fungal leaching is generally slower than bacterial leaching. For example, native fungal strains supplied with sucrose mobilised up to 71% U in 10 days [283], compared with 80% extraction using *At. ferrooxidans* supplied with ferrous ions [268]. However, based on their own results, Anjum *et al.* [269,285,288,289] concluded that fungi may be a better choice when the polymetallic ore (concentrate or tailings) has high organic-carbon content.

When leaching with heterotrophs, questions arise about what organic carbon source must be supplied for maximum growth and organic acid production, which acids are most effective for the target metals, and whether the supply of an organic carbon source is economic for a proposed process. The studies by Anjum, Bhatti and co-workers [285–288] go some way to providing answers in respect of shale ores. In chemical leaching tests using black shale, citric acid was most efficient for copper extraction, oxalic acid for zinc and cobalt extractions and tartaric acid was the least effective for the three metals [288]. In similar tests using brown shale (shale formed in an oxidising environment), citric acid was best for copper, manganese and aluminium extractions and tartaric acid was also effective for copper, but oxalic acid was the most effective for magnesium extraction from brown shale [287].

While four organic acids, citric, oxalic, tartaric and malic acids, were detected in bioleaching solutions from tests on black shale using *A. niger* provided with different substrates, citric acid was the main acid produced [285]. In the case of *Pc. notatum*, the same four acids were detected but the substrate influenced dominant acid production. With glucose, *Pc. notatum* produced citric acid and Cu and Mg were preferentially extracted (>80%), but with molasses as substrate, tartaric acid was produced and Mn and Cu were extracted (>70%) [286]. *G. lucidum* produced up to 10 times more tartaric acid than the other acids [287].

The substrates typically used in laboratory bioleaching tests are refined organic carbon compounds that would be prohibitively expensive for commercial-scale plants. Thus, the use of less expensive materials has been trialled. In a comparison between glucose, molasses or breadcrumbs as sources of carbon, it was reported that *Pc. notatum* produced very little acid from the breadcrumbs [286]. When glucose, molasses, sawdust or cottonseed cake were compared, acid production by *G. lucidum* (mainly tartaric acid) was low from either sawdust or cottonseed cake [287]. In addition to glucose and molasses, the alternative substrates tested by Anjum *et al.* [285] were mango peel, seedcake and rice bran. Citric and malic acids, the two main acids produced by *A. niger*, decreased in concentration in bioleaching solutions in the order mango peel > seedcake > rice bran but maximum biomass production was obtained using acidified mango peel as the organic carbon source (compared with glucose and molasses). Sjoberg *et al.* [290] also focused on the cost of refined organic carbon compounds. They selected aspen wood shavings in a low-water-content system. In that study, overall extraction was very low (1.7% U in 56 days) but the authors showed that the minimum effort and cost process could mobilise uranium from shale.

4.2.2. Prospects for Commercialisation

It is not surprising that black shales have been studied extensively with a view to multi-metal extraction and some deposits are being considered for commercialisation. A key parameter in the decision-making process is the size of the deposit because the concentrations of the target elements tend to be low. Jowitt and Keays [276] reported that there had been significant research in three of the shale-hosted nickel regions, either because of their size or because of the high metal grades; they are the Talvivaara deposit in Finland, a number of nickel-molybdenum deposits in southern China (e.g., Zunyi and Zhijin, Guizhou province; Dayong and Cili, Hunan, Dexhe, Yunnan, Duchang, Jiangxi and Lizhe, Zhejiang [291]), and deposits in the Selwyn Basin, Yukon, Canada. Copper-enriched polymetallic black-shale ores and their concentrates have been a focus of two large, international projects supported by the European Union, the BioMinE project [292] and the BioShale FP6 project [228]. It should be noted that, while grades or deposit size might preclude the development of economic processes for today's industry, smaller, lower-grade deposits might become economic in the future [276], particularly in view of the success of the Talvivaara heap bioleaching process [238] and the pilot-scale studies on Kupferschiefer concentrates and ores [228].

The considerable body of research on Finnish ore deposits dates back more than 25 years but the main deposit of interest in the context of bioleaching is that at Talvivaara. The Talvivaara ore body is a large black schist deposit (>400 Mt) containing disseminated sulfide minerals, pyrrhotite (~15%), pentlandite/violarite, sphalerite, chalcopyrite and alabandite, and carbon as graphite (~10%) [293].

Initial studies involved batch bioleaching tests in which copper, zinc, nickel and cobalt extractions were monitored, bioleaching conditions were optimised, and the effects of temperature, pH and oxidation-reduction potential (ORP) on extraction were examined [229,294,295]. In column leaching studies, a microbial culture enriched from mine water was active over a wide temperature range in bench scale columns, different microbial populations evolved at different temperatures, and pH 2 was optimal for nickel, zinc and cobalt extractions [296,297], but copper extraction was minimal in the time frame of the experiment. The benefit of supplementing bioleaching tests inoculated with an iron(II)- and sulfur-oxidising mixed culture with additional ferrous ions was tested and the altered mineralogy that evolved during bioleaching investigated [298–300].

Two major studies were undertaken as part of the Bioshale FP6 European project, specifically, a pilot-scale column bioleaching trial on agglomerated ore [301] and a tank bioleaching trial using flotation concentrate [213] (see Section 3). Complementary laboratory-scale experimental and theoretical studies included the application of bioflotation to increase the amount of concentrate obtainable and/or its grade, and modelling—simulation of heat transfer in heap bioleaching using data from both the pilot crib operated by the Geological Survey of Finland and the pilot heap conducted at Talvivaara mine [228].

The large-scale column tests [301] comprised two columns charged with ore of different particle size ranges. In addition to metal-extraction kinetics, the evolution of the microbial community was monitored as a function of time. Similarly, Halinen et al. [302] monitored microbial community diversity in the demonstration heaps at the Talvivaara mine site, showing that microbial diversity decreased with time and differing community profiles evolved with changes in temperature and location in the heap. From these column and heap studies it was jointly concluded that heterogeneity in heaps (depth, temperature, acidity, changes induced during leaching) will favour different suites of microorganisms and that overall efficient metal extraction depends on the presence of a sufficient diversity of microorganisms.

Together, these multi-element studies contributed directly and indirectly to the commercialisation of the Talvivaara heap bioleaching project [238]. As a result, interest in developing Scandinavian ore bodies in the Outokumpu region, the Viken and Storsjon deposits, Alberta Canada and other regions has increased markedly [272,273,275,303–305].

While large areas of the Kupferschiefer black shale contain only average concentrations of base and precious metals, some deposits are enriched to ore grades and often exceed 4 wt % Cu, mainly as chalcocite, bornite and chalcopyrite enriched with silver. In the Mansfeld district (Germany), near-surface zones were exploited for silver and copper from mediaeval times. Pyrite (enriched with nickel and cobalt), marcasite, galena (enriched with silver), sphalerite and native Ag occur with the black shale deposits and many other elements in the Kupferschiefer shale were reported (Table 1) [87,306].

The body of research within the BioShale FP6 European project on processing of the "Kupferschiefer" black shale follows the historical trend of focusing on copper and silver extraction [153,307–309] but with broad research objectives [228]: to assess bioprocessing methods and complementary processing routes for hydrometallurgical recovery of metals, undertake a risk assessment regarding waste management and a techno-economic evaluation of new processes, and to characterise microbial communities engaged in leaching.

In direct bioleaching tests in stirred, aerated 4 L reactors inoculated with *Acidithiobacillus* strains, up to 65% Cu was extracted in 28 days (pH 2, 40 °C), but there was a lag before nickel extraction rose to 65%, and zinc extraction was poor (6.5%) [310]. The absence of lead in solutions was attributed to the oxidation of galena followed by the precipitation of $PbSO_4$ in the residues. When indigenous *Bacillus* strains were used to inoculate flasks or 4 L tanks containing shale ore there was an initial increase in metals (dissolution of soluble components), a lag period before extraction commenced and low extractions in 28 days (for $T = 40$ °C; 3% Cu, 10% Ni and <0.1% Zn) [90]. It was concluded that nickel, and to a lesser extent copper and zinc, were bound as organo-metallic compounds in the shale [90].

Two two-stage bioleaching investigations were conducted. In the first, carbonaceous black shale ore (22 wt % inorganic carbon) was pre-leached with sulfuric acid before the columns were inoculated with mixed mesophilic iron(II)- and sulfur-oxidising acidophiles and irrigated with ferric ion-rich acidic medium [309]. Copper (82%) was recovered from leach solutions using cementation with metallic iron, and silver (51%) was recovered from the leached, neutralised residues using ammonium thiosulfate (pH 10) and subsequently recovered by cementation using Zn. This process was deemed not to be economic because of reagent consumption (ammonium thiosulfate, copper sulfate, and zinc).

In the second two-stage process [308], heterotrophic bacteria isolated from flotation waste [91] were used to inoculate flasks containing medium (pH 7, 30 °C) and ore. The underlying premise of this research was that heterotrophic microorganisms should enhance the degradation of the organic fraction of the ore and thus release bound metals. However, only 1% Cu and 6% Ag were extracted during the first stage of bioleaching using heterotrophs. After 25 days, the treated residues were separated and dried, resuspended in sulfuric acid medium (pH 1.8, $T = 25$ °C) and inoculated with *Acidithiobacillus* strains to oxidise sulfide minerals. At the end of the second stage (60 days of bioleaching), 72% Cu was extracted but only 9% Ag. In 35-day tests using the autotrophs, extractions were 41% Cu and 11% Ag. However it is unclear to what extent the increased extraction of copper in the two-stage process was due to the extended period of leaching or how much might be attributed to the effects of the heterotrophs, for example on mineral surfaces, given they would not be expected to function during the acidic bioleach. Overall, these results do not present convincing evidence of the benefit of a two-stage process.

The research undertaken in the BioShale project and reported extensively, finds a natural extension within the newly-funded BioMOre project, an objective of which is "to extract metal from deep mineralised zone in Europe (Poland and Germany) by coupling solution mining and bioleaching technologies". This project will build on the knowledge from both the BioShale FP6 project, that copper and other metals can be efficiently extracted from black shale "Kupferschiefer" deposits, and the EU-funded ProMine Fp7 project, within which it was demonstrated that huge potential reserves in copper and other metals occur at depths greater than 1500 m in Central Europe [311]. One of the proposed research topics is the application of biotechnology to in-stope leaching of polymetallic black shale at the Lubin mine, Poland. Research into *in situ* or in stope bioleaching is an important part of the future for mineral processing, even although deposits with the physical and geological requisites may be rare. Underground technologies are perceived as better for the environment (the "invisible mine") and likely to attract community acceptance, and may also be the means of accessing deposits too deep for current economic mining technology. Therefore research to advance any part of the many aspects of successful technology development is to be welcomed.

5. Oxidised Ores

While oxidised ores may contain more than one element of interest, they are not generally considered to be polymetallic and mineral processing operations are often focused on the production of a single element and seldom produce more than two elements. However, that may change as global resources of different elements become diminished.

Metals extraction from oxidised ores using biotechnologies have not been commercialised to date but have been developed to laboratory and pilot scale. Most process developments are chemical rather than biochemical, as evidenced by the numbers of publications found in literature searches using the terms: "laterite AND leach*" (766) against "laterite AND bioleach" (40), "manganese dioxide" AND leach*" (576) against "manganese dioxide AND bioleach*" (15), or "copper silicate AND leach" (7) against "copper silicate AND bioleach" (1). As a further example, in a recent review [312], commercial processes and prospective processes for nickel laterites were described and critically assessed; none of them utilised microorganisms. Similarly, in a review of manganese metallurgy [313], pyrometallurgy and hydrometallurgy (chemical leaching) technologies dominated the discussion. Nevertheless, methods that target the bioleaching of manganese nodules and nickel laterites are topics that continue to inspire the development of novel biotechnologies.

Manganese is the twelfth most abundant element in the earth's crust [314]. Pyrolusite (MnO_2) is the most common manganese mineral but manganese also occurs in carbonate and silicate minerals. Manganeses nodules or polymetallic nodules are rock concretions formed of iron and manganese hydroxides around a core and often contain nickel, cobalt and copper. Nodules of economic interest occur in the north central Pacific Ocean, the Peru Basin in the south east Pacific, and the central north Indian Ocean. The existence of many microorganisms with enzymes capable of oxidising or reducing manganese has been reported [314].

In the future, nickel production from lateritic ores is expected to far exceed that from sulfides because laterites comprise about 70% of land-based nickel reserves [315]. Most of the bioleaching studies on laterites relate to bodies of research on Greek laterites that contain nickel, cobalt and chromium [32,76,316–318], Indian laterites that contain nickel, cobalt and chromium [319–323], New Caledonian laterites that contain nickel, cobalt and manganese [33,324–330] and African laterites that contain nickel, cobalt, manganese and chromium [77,331,332]. In metal extraction, the key difference between copper oxide or sulfide ores and nickel laterite ores is the absence of specific nickel phases in the lateritic ores [333]. As a consequence, in order to extract the nickel, it is necessary to dissolve or alter the host phases, which are mainly silicates and iron oxides. Interest in the application of one or more biotechnologies to the extraction of nickel (and other elements) from laterites has a long history, including some innovative process suggestions, but few have reached pilot stage.

Tested biotechnologies for oxidised ores fall into three groups, those that use: (i) inorganic acid generated by the biooxidation of RISCs; (ii) organic acid generated by heterotrophic metabolism of organic compounds; or (iii) a combination of bioreduction and leaching. Some examples are described for each of the modes, where possible using studies in which more than one element was extracted.

5.1. Leaching with Acidophilic Autotrophs/Inorganic Acid

The use of inorganic acids to dissolve oxidised ores such as manganese oxides or nickel laterites, comprising silicate and iron oxide minerals, is well known. The use of autotrophic sulfur oxidising microorganisms is a means of minimising costs because elemental sulfur is generally a less expensive commodity than sulfuric acid. Thus, in its simplest form, autotrophs such as *At. ferrooxidans* and *At. thiooxidans* and more recently discovered sulfur oxidising species oxidise elemental sulfur and the biogenic acid dissolves the manganese dioxide or laterite minerals.

Using this basic system of acid bioleaching to extract manganese from manganese dioxide, Imai [334] noted that the microorganisms contributed more than just acid, as extractions of manganese were higher in the presence of microorganisms than with chemical acid leaching. It was postulated that intermediate reduced sulfur species, such as hydrogen sulfide or sulfite produced during the biooxidation of sulfur, accelerated the dissolution of manganese dioxide. When ferromanganese crust of similar mineralogy to manganese nodules was leached using *At. ferrooxidans* with elemental sulfur as substrate, nickel and copper were dissolved rapidly and cobalt, manganese and iron dissolved as the pH decreased, due to sulfur biooxidation [335]. The substitution of pyrite for elemental sulfur was effective and bacterial adaptation to the crust material resulted in increased extraction rates. Similar enhanced extraction was recorded when low-grade manganese ores were bioleached with pyrite addition [336] and the substitution of nickel sulfide for elemental sulfur also resulted in enhanced manganese and cobalt extraction from manganese nodules [337].

In a process developed for the dissolution of manganese nodules containing four elements (Cu, Zn, Ni and Co), the acid required for leaching was generated from sulfur biooxidation by *At. ferrooxidans* or *At. thiooxidans* (mesophiles, 30 °C) or *Acidianus brierleyi* (thermophiles, 65 °C) [338]. Many tests aimed at optimising process parameters for metals extraction were included in that study, with a final test conducted under optimised conditions. Distinctive features of the process were the differences in leaching behaviour for the target metals, the strong temperature dependence (accelerated extraction at 65 °C), and the selective leaching of base metals and manganese against negligible extraction of iron. The consumption of elemental sulfur was an unavoidable cost in the process. To mitigate the cost of sulfur, Konishi *et al.* [338] proposed that sulfate could be reduced to hydrogen sulfide using SRB and then, using photosynthetic bacteria, the hydrogen sulfide could be oxidised to elemental sulfur for recycle to the process. Zafiratos *et al.* [190] developed a stirred-tank process for which the techno-economic assessment was favourable. They used a mixed culture of *At. ferrooxidans*-like strains isolated from different mine sites and grown in medium containing elemental sulfur in a two-stage process involving acid leaching of the test ore (20% Mn, 1% Fe) under aerobic and then anaerobic conditions to yield a pH 1–2 solution with manganese (12 $g \cdot L^{-1}$) and iron (0.65 $g \cdot L^{-1}$). Iron was removed in a neutralisation stage and disposed of, while manganese in pH 7–8 solution was recovered by electrolysis.

The strategy of exploiting the sulfur-oxidising capability of some chemolithotrophs to provide acid was also trialled for the extraction of nickel from nickel laterites. In bioleaching tests using mixed cultures of *At. ferrooxidans*, *At. caldus* and *L. ferrooxidans* it was found that elemental sulfur supported bacterial growth, acidification and nickel recovery more strongly than if pyrite was used as the substrate [331,332,339]. Under optimised conditions, 2.6% solids loading, 63 μm ore particle size, initial pH 2 and elemental sulfur as substrate, the nickel yield was 74% in 26 days [332]. Kumari and Natarajan [340] described a micro-aerobic leaching system utilising *At. ferrooxidans* and *At. thiooxidans* to oxidise elemental sulfur, with and without added sucrose (reducing agent), to leach manganese nodules. In inoculated tests without sucrose, extractions were Cu ≥ Ni (50%–60% in 8 days) >> Co > Fe > Mn for both test species. However, bioleaching in the presence of sucrose provided sufficiently reducing conditions at lower pH to reduce manganese and iron(III) oxides and enhance acid dissolution with the concomitant release of the ancillary elements copper, nickel and cobalt.

5.2. Leaching with Acidophilic Heterotrophs/Organic Acid

The use of heterotrophic bacteria or fungi to leach nickel and cobalt from laterites or manganese oxide ores relies upon the bio-production of organic chelating acids (especially carboxylic acids) or other metabolic products from supplied substrates. In their review on the biohydrometallurgy of non sulfide minerals, Jain and Sharma [341] listed bacteria (four genera) and fungi (seven genera) that produced organic acids during fermentation. Oxalic acid (dicarboxylic acid) and citric acid (tricarboxylic acid) are produced by many fungal species, especially *Aspergillus* and *Penicillium* spp., during the metabolism of carbohydrates.

Examples of the application of heterotrophs include the extraction of potassium from leucite (a silicate mineral) [233], extraction of copper and zinc from oxidised copper and lead-zinc ores, respectively [342], bioleaching of aluminium from clays [343], and the extraction of nickel from lateritic materials [31,76,316,319,324,344]. Extractions of 97% Cu, 98% Ni, 86% Co and 91% Mn from manganese nodules were reported by Das *et al.* [345]. In each of those studies *Aspergillus* and/or *Penicillium* were among the genera tested.

The acids have a dual function, to provide protons for mineral corrosion and to chelate soluble metals. Often, as a prelude to bioleaching, chemical leaching tests using the acids were undertaken to assess and compare their efficacy in metals extraction. For example, in comparative tests Bosecker [31] examined the extraction of nickel by fifteen organic acids using laterites from New Caledonia, Australia, Brazil and the Philippines, and concluded that citric acid was effective for nickel extraction from silicates but not for nickel extraction from iron oxides in laterite ores. In contrast, using laterite ores from Western Australia and Indonesia, McKenzie *et al.* [346] reported that citric, tartaric and pyruvic acids increased the yield of soluble nickel for limonite type ores. In a comparative study,

Das *et al.* [347] showed that oxalic acid was more efficient at extracting manganese (66%) than citric acid (40%) from low-grade ferromanganese ore, attributing the difference to the reducing capacity of oxalic acid and its affinity for manganese complexation.

The method is seemingly simple but, compared with traditional bioleaching using chemolithotrophs, the use of heterotrophs presents different challenges and is yet to be commercialised as a stand-alone biotechnology. These challenges have been comprehensively reviewed [77] and include:

- the need to provide large amounts of inexpensive sources of organic carbon for the production of the required organic acids in economic scaled-up processes [32,76,348];
- the need to restrict the use of the carbon source to acid-producing species and to prevent the biomass from inhibiting the metal extraction [32,77];
- the low and variable tolerance exhibited by different fungal species to soluble elements in process solutions and the resultant reduced acid production [76,325,326,349];
- the need to dispose of the biomass [350]; and
- differences in extraction rates and affinities of extracted metals for the minerals in saprolite, nontronite and limonite ores [328].

Nickel and cobalt are the main elements recovered from laterites but in some instances the economics of a process might be improved by extracting additional elements. Zhang and Cheng [313] noted that the ratio of Ni:Mn in laterite ores at Murrin Murrin (Western Australia) is 2:1 and estimated that about 18,000 ton per annum of manganese is discarded to a waste stream.

The paucity of studies using heterotrophic bacteria to generate organic acids suggests that they are less productive. However, Sukla *et al.* [319] found that nickel extraction using *B. circulans* (85% in 20 days) compared favourably with that using *A. niger* (92% in 20 days). A *Bacillus* sp. strain was isolated from seawater and applied to the leaching of marine nodules [351]. In saline growth medium (pH 8.2) containing 1% ground nodule material (50–75 μm particle size range) inoculated with actively growing *Bacillus* sp. culture, extractions of copper, cobalt, nickel and manganese were in the range of 20%–30% for 4-h leach duration. When starch (1%) was included in the growth medium to create reducing conditions, extractions in the range of 80%–85% for cobalt, copper and manganese, and 65% for nickel were achieved in 4 h [351].

5.3. Reductive Bioleaching (Organic Carbon Compound as Electron Donor)

Numerous microorganisms have been shown to reduce iron(III) during the metabolism of carbon compounds, fewer to reduce manganese(IV), uranium(VI), and some other elements in various oxidation states [352–354]. In the case of iron(III) and manganese(IV) reduction, these abilities have been harnessed at laboratory scale for the extraction of metals [314,355].

Manganese nodules, with "average" composition (wt %) 24 Mn, 14 Fe and minor copper, nickel and cobalt, were subjected to systematic study using heterotrophic microorganisms. Thus, similar challenges apply as those identified for the leaching of laterites using heterotrophs. However, because the manganese in nodules is present mainly as manganese dioxide there is the additional need to reduce the manganese(IV) to manganese(II) to render it soluble [314].

Most studies were focused on the extraction of manganese and only a few studies included the extraction of ancillary elements. Lee *et al.* [356] investigated manganese, cobalt and nickel release by manganese-reducing bacteria (not identified). Efficient recovery of metals (75%–80% after 48 h of incubation) was achieved during anaerobic leaching at pH 5–6.5 with glucose as the source of organic carbon and temperatures in a range of 30–45 °C. While not directly comparable with other studies, the leach duration was considerably shorter than those reported for fungal production of organic acids [32,357] or the use of RISC-oxidising autotrophs [335].

The release of nickel and cobalt from crushed ferromanganese nodules correlated directly with MnO_2 reduction and dissolution when treated with a *Bacillus* strain [358]. The reduction of manganese(IV) in manganese nodules released simultaneously the ancillary elements copper, nickel and cobalt but not iron. The co-solubilisation of copper, nickel and cobalt was attributed to those elements being bound in the MnO_2 matrix, and the poor release of iron as due to the reduction of iron(III) being less favoured than the reduction of MnO_2 under the conditions used. A hypothetical anaerobic process utilising manganese(IV)-reducing bacteria was proposed [358] involving the suspension of ground nodules (-2 mm) at optimal solids loading in salts medium containing a specific organic carbon compound that would act as the electron donor as well as supply carbon for the growth of the selected microorganism(s). Ehrlich *et al.* [358] thought that a mixture of *Geobacter metallireducens*, a strict anaerobe, and *Sh. putrefaciens*, a facultative anaerobe that could scavenge any oxygen initially present in the medium, would provide a suitable culture to reduce manganese(IV) and release Mn^{2+} and other cations to solution.

A batch and semi-continuous process under micro-aerobic conditions for the bioleaching of ferromanganese minerals was investigated [350]. With molasses as the organic carbon source for the mixed cultures of heterotrophic bacteria, 95%–100% Mn extraction was achieved in 36–48 h of treatment using a solids loading of 20% (*w/v*) of ore. Subsequent tests at pilot plant scale, conducted under non-sterile conditions provided evidence that a full-scale process was technically feasible, assuming that a suitable strategy for biomass disposal could be developed, the leach liquor could be purified ahead of manganese separation as the carbonate, and the overall cost of the process could be reduced.

Sh. alga is a facultatively anaerobic bacterium that can couple the oxidation of an organic compound (electron donor) to the reduction of iron(III) and other high-oxidation-state elements [359–361]. Konishi *et al.* [362] proposed to exploit the capability of *Sh. alga* to produce ferrous ions that simultaneously affected the dissolution of manganese nodules under anaerobic conditions (pH 7, ambient temperature). Element solubilisation was rapid, with more than 80% for Ni, Co and Mn in 5 h. As a consequence, it was concluded that, as ferromanganese crusts might contain titanium, cerium, zirconium, molybdenum and other elements, further study of the bioleaching behaviour of such elements should be undertaken.

5.4. Reductive Bioleaching (RISC as Electron Donor)

Nickel extraction from laterites is strongly dependent on ore mineralogy. Three generic nickel laterite ore types have been described [363,364]; (i) limonite ores in which the nickel is mainly hosted in goethite and in which cobalt- and nickel-enriched manganese oxides may be abundant; (ii) saprolite ores with dominant magnesium-nickel hydrous silicates; and (iii) low-quartz, clay silicate deposits with dominant nickel-smectites. Goethitic nickel-laterite ores should be the most amenable to iron(III)-reductive bioleaching.

Hallberg *et al.* [116] showed that the dissolution of iron(III)-containing laterite ores under anaerobic conditions using *At. ferrooxidans* in medium supplemented with sulfur resulted in rapid and selective extraction of nickel compared with iron. Johnson *et al.* [106] monitored nickel, cobalt and manganese recoveries of up to 82%, 64% and 116%, respectively, based on pregnant leach solution (PLS) and feed analysis (head grade), in 20 days of leaching (30 °C, pH 1.8). The use of sulfur and *At. ferrooxidans* offered the added advantages that sulfur is generally less costly than a suitable organic compound for use with heterotrophs, can be separated and recycled if not utilised during leaching, and lowered the overall amount of acid required for the proposed process. Further evaluation of the proposed "Ferredox" process, was undertaken [365] with the development of a conceptual flowsheet that incorporated four process components: (i) bioreductive leaching of limonite resulting in a pregnant leach solution (PLS) containing ferrous ion; (ii) extraction of metal values from the PLS; (iii) ferrous ion biooxidation in PLS generating either soluble ferric sulfate or iron(III) oxysulfate precipitates that can be used to regenerate sulfuric acid; and (iv) a bioreductive acid-generation step (ferric ion oxidation of sulfur) and re-cycle to the leaching stage. It was noted that, while the process components had been demonstrated independently in other studies, their integration into a process flowsheet would require further development [365]; six key steps towards that goal were described.

6. Biotechnologies for Polymetallic Mine Waste and Tailings

Waste materials at mines comprise material that must be moved to expose the target mineralisation, material that is intermixed with the target mineralisation and thus of lower grade or otherwise unsuited to the selected mineral processing route, and residues from processing, often flotation tailings. The historical context of the large waste dumps and flotation tailings impoundments that litter the world's metal-rich mining regions was briefly described in a succinct, informative and timely review on "Sustainability in metal mining" [366] where it was noted that, at the time of mining, mine waste would contain metal grades lower than can be extracted economically, waste stockpiles could contain metal grades higher than are generally present in the earth's crust. Not surprisingly, the older the stockpile, the greater that difference, because modern extraction methods have become more efficient, driven by the need to process ores of lower grades (Figure 2). According to Mudd [367], "the extent of waste rock or overburden produced by mining is closely linked to the [increased] use and scale of open cut mining". Shaw *et al.* [368] presented an estimate for global mine waste of "several thousand million tonnes per year" and noted that as demand for metals "is expected to grow for the foreseeable future, it is likely that mine waste generation will follow a similar trend."

Regulatory frameworks for mining industries, with regard to mining practices, water use and its contamination, and mine waste management, have been implemented in industrialised countries from about the 1970s and more recently in developing countries. The scarcity of new, high-grade resources, the need for regulatory compliance and the rising costs of environmental remediation have together created a "developing interest in the recovery of resources from mine waste" [368].

6.1. Waste and Tailings from Sulfide Ores

Tailings and waste from sulfidic deposits are more problematic than those from oxidised ores because the residual sulfide grains in the material are exposed to oxidising conditions. The numerous reports in the literature on acid mine drainage and the need for its prevention or remediation are testament to the reactivity of, and subsequent rapid metal mobilisation from, disturbed sulfidic waste ores or tailings. Some post-2000 examples of the application of developed bioleaching technologies (see Section 2) to polymetallic mine waste and tailings for the extraction of base and precious metals are summarised. Among those studies there are examples of various biotechnologies:

- the application of direct bioleaching/biooxidation using autotrophic acidophiles in batch reactors, stirred tanks, vats or heaps, the latter with or without the ground tailings being supported on host rocks [369–375];
- the application of heterotrophic microorganisms/organic acids to treat sulfidic tailings comparing direct and indirect bioleaching [376];
- studies directed towards a fundamental understanding of ARD and the development of strategies to mitigate its impact [282,377–381];
- studies driven by anticipated values contained in mine waste and tailings dumps that could support dump remediation or removal [88,382–386]; or
- modifications of developed technologies on a case-by-case basis, to accommodate peculiarities of the tailings arising from their mineralogy, prior treatment and/or storage, or as means of accelerating metals extraction [370,387–389].

The Kasese project (Uganda), is the only application of base-metal stirred tank biotechnology to have been commercialised thus far. According to Morin and d'Hugues [187], it is "the first industrial installation incorporating bioleaching into a sophisticated hydrometallurgical flowsheet allowing the selective extractions of various metals." As background to the project, 16 million tons of copper were produced from the Kilembe deposit, Uganda, in the period 1956–1982. The flotation tailings stockpile near Kasese (900,000 t; 80% pyrite containing 1.38% Co) became dispersed by heavy rainfall, produced acid, and was a threat to a national park, a large lake and the local communities. After 10 years of process development at the Bureau de Recherches Géologiques et Minières (BRGM), France, bioleach tanks on site were inoculated in 1998 (Figure 7) and the process of cobalt (zinc, copper and nickel) production and tailings remediation was initiated [187]. Since commissioning, process improvements were undertaken by the plant operators to overcome unforeseen technical difficulties. The Kasese Cobalt Company Limited closed the plant in 2013 through a lack of raw materials but it may be re-opened by Tibet Hima Mining to process cobalt from the Kilembe mines [390].

BacTech Mining Corporation is raising funds for a pilot plant at Cobalt, Ontario, to process tailings from the Castle Mine 50 km to the west of Cobalt, with the potential to expand capacity and process other tailings from nearby locations (Figure 8) [391,392]. BacTech intends to apply its proprietary bioleaching technology to the remediation of arsenic-rich tailings (potentially 16–18 million tonnes) that also contain recoverable silver, cobalt and nickel. In this process iron(III) and arsenic(V) will be precipitated as ferric arsenate (scorodite) for safe disposal [392]. The successful installation of a biohydrometallurgical process at Cobalt, Ontario, will benefit the environment and local communities and will provide confidence in the technology, leading to other cost-effective tailings-remediation projects.

Figure 7. Bioleaching plant (**a**) for the processing of pyritic tailings (**b**) containing cobalt, copper, nickel and zinc. Photographs from D. Morin, Bureau de Recherches Géologiques et Minières (BRGM), reprinted with permission.

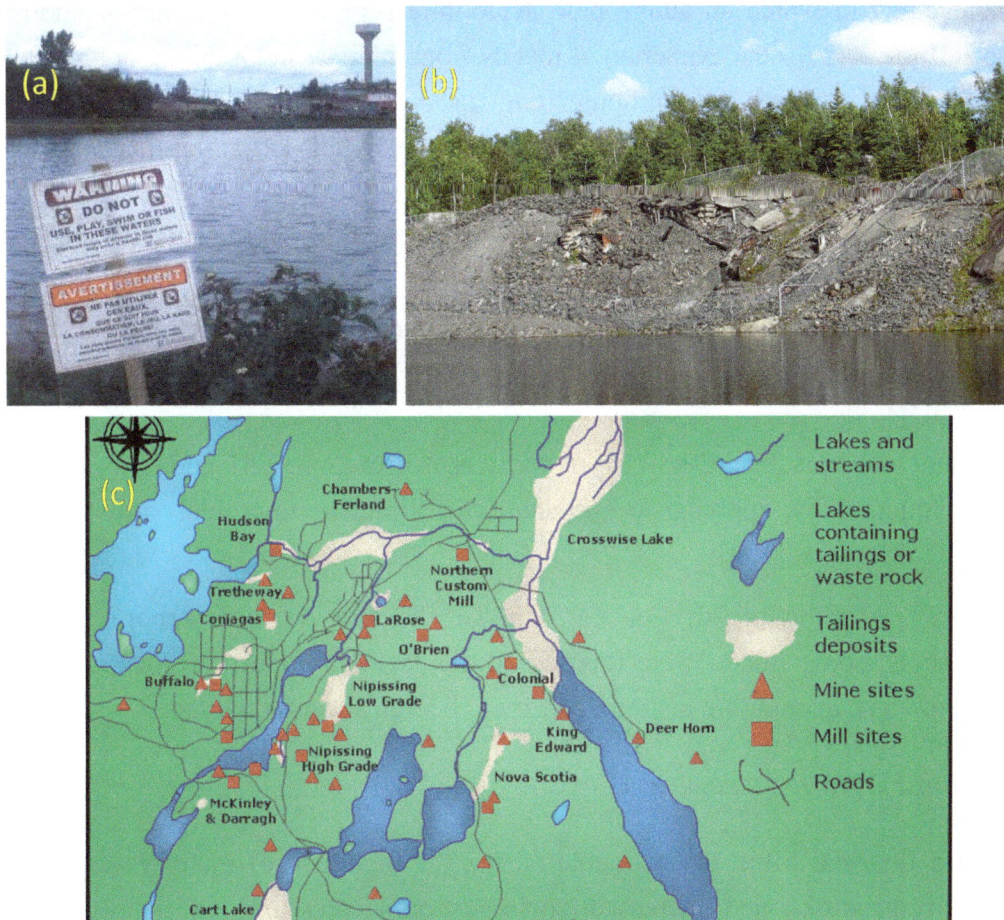

Figure 8. Water (**a**) and tailings (**b**) severely contaminated with arsenic; (**c**) Regional map of old workings in the region near Cobalt, Ontario, Canada with extensive tailings deposits indicated. Pictures from P.C. Miller, BacTech Environmental, reproduced with permission.

In a third tailings project relying on values from cobalt (copper and gold) production to offset the cost of remediation, a process for the *in-situ* bio-oxidation (*i*-BO™) of refractory tailings dumps

(4 million tonnes) at the Peko Mine, Tennant Creek, Australia was developed [393,394]. The strategy was to accelerate the natural weathering of the tailings by scarifying the tailings surface, introducing large numbers of chemolithotrophic bacteria, and allowing the ingress of oxygen and carbon dioxide from air. Periodic irrigation and collection of drainage facilitated base metals extraction. It was planned that leached residues would be returned to the open-cut mine after gold extraction. The bacterial activity also cleaned up the magnetite, making it another commercial product. Unfortunately, low commodity prices forced the mine into care and maintenance (2010) before the full potential of the technology could be realised.

6.2. Waste and Tailings from Oxidised Ores

Fewer studies on the application of biohydrometallurgy to metals extraction from oxidised waste or tailings and no commercial plants were found during this review. Among those studies were examples of:

- bioleaching in batch reactors, stirred tanks or columns, including one column leaching experiment in which fungi were applied as inoculum [395];
- direct bioleaching in which an autotrophic acidophile supplied with sulfur or another RISC produced sulfuric acid for the extraction of metals from mine waste or tailings [344,396–400];
- direct bioleaching in which heterotrophic fungi produced organic acids when supplied with sources of organic carbon and the organic acids effected metals extraction from oxidised mine waste or tailings [344,395,397–405];
- a two stage process, in which biological acid production and chemical leaching were separated [399];
- optimisation of acid production and metals extraction and improved economics of the process, by trialling low-cost sources of carbon for heterotrophs [401,402] or pre-treating the mine waste [344,403,405];
- reductive (anaerobic) leaching using *At. ferrooxidans*/sulfur [344,405]; and
- enhanced extraction by applying a surfactant to stimulate microbial growth but reduce carbon-source utilisation [404].

Extractions of two or more of Ni, Co, Zn, Mn, Cr, Fe and As were reported for different experimental conditions. Some studies were focused on the economic return of metals extraction but others had the goal of clean up and safe disposal of a more benign waste or tailings.

6.3. Towards Responsible Waste Management

Whether or not a financial offset is obtainable during the treatment of mine wastes and tailings, stricter regulatory frameworks and increased socio-economic pressures in many countries are compelling the mining industry to move in the direction of more sustainable and cleaner metal production. Mining activity impacts on soils, water and air, by accelerating naturally slow biogeochemical processes through materials disturbance, comminution and exposure of reactive material surfaces to the atmosphere. Dold [366] notes that "according to the United States Environmental Protection Agency (EPA), water contamination from mining is one of the top three ecological-security threats in the world." The mining and processing of ores may occur on a relatively short time frame but the wastes generated may react for decades to centuries.

The few examples of bioprocesses for the extraction of metals from polymetallic waste materials summarised above apply known methods for the reduction of potentially hazardous components. Not surprisingly, most applications were applied to sulfidic materials that may generate acid over long periods. Some recent microbiological approaches to securing mine wastes and recovering metals from mine waters are highlighted in a review published in this journal ([406] and references therein). The review covers a number of chemical methods for remediation of mine water, some of which have been commercialised but relevant to this review are the bioprocesses, including:

- understanding systems that display "natural attenuation" of contaminated water with a view to exploiting them;
- "bioshrouding"—the formation of biofilms around pyrite grains by heterotrophic microorganisms to limit the attachment of pyrite-oxidising acidithiobacilli;
- "ecological engineering" to reverse the reactions that generate ARD—inoculating freshly-deposited tailings with acidophilic algae and heterotrophic bacteria, sustaining the growth of iron(III)- and sulfate-reducing bacteria within the submerged tailings
- iron oxidation and removal from AMD—in a continuous three-stage modular bioreactor comprising (i) an iron(II)-oxidation reactor inoculated with "*Ferrovum myxofaciens*"; (ii) a schwertmannite precipitation reactor and (iii) a packed bed polishing unit and
- selective precipitation of copper and zinc and other elements as sulfides from mine waters using acidophilic SRB
- using neutral or acidic SRB to convert sulfate to elemental sulfur for removal;

Knowledge gained from research into both the processing of ores for metals extraction and the remediation of acid mine/rock drainage systems, are together invaluable for mine waste management. The processes developed from them for application in the mining industry, while challenging, should "improve the economic outcome of an ore deposit on the long-term perspective by extracting the complete metal content and prevent the uncontrolled release of contaminants to the environment" [366].

7. Summary

Diminishing ore grades and estimates that many valued elements may become exhausted in 50 years (a short time in process development terms) *versus* increased demand for many elements in the periodic table are the key drivers for:

- a longer-term view of what constitutes a mineral resource and a broader approach to what elements can be extracted, co-extracted, and/or separated from resultant process solutions;
- the development of efficient new or modified technologies for the extraction of a suite of elements, rather than the most lucrative element, from complex polymetallic ores, concentrates, mine wastes and tailings, thereby gaining as much benefit as possible from the considerable costs of mining, moving and pre-treating large tonnages of materials;
- discarding waste and tailings that are benign, thus lowering the costs of environmental remediation during mine closure; and

- concomitant development of new separation and purification technologies for the recovery of all elements of interest in the multiple-product mines of the future.

Not all of these challenges will be met using bioprocessing, but the claim may reasonably be made that the key technologies for acid-soluble oxidic or lateritic mineral resources and oxidisable sulfidic mineral resources are robust and proven.

- Heap and dump leaching/bioleaching can be a secondary process or a stand-alone technology applied across a wide range of climatic conditions in terrain that is complex and/or remote. Its flexibility makes it suited to deposits from small to very large and each of the unit processes can be very basic (the use of cementation or precipitation to recover metal products).
- The widespread adoption of stirred tanks for the processing of refractory gold ores and concentrates is one of the success stories of mining biotechnology. Stirred tank processes have been piloted and/or demonstrated for the processing of base metals and in one instance commercialised for cobalt production from tailings.
- Both technologies have benefited from improved SW-EW for the separation and recovery of an increased number of elements through the development of novel reagents.

In my view, the main technology that would benefit from systematic study and further development are vat (bio)reactors. Vats (saturated reactors) need to be redesigned to make them "user friendly" on mine sites where large amounts of ores are processed on a daily basis. Continuous reactors are essential but other design advances should include features that make conditions more conducive to chemical-, electrical- and/or microbial-assisted leaching under controlled but variable conditions. A key advantage that could be exploited is that ores can be crushed to finer particle sizes than is possible in heaps, thereby liberating a greater proportion of the target mineral grains and increasing extraction rates as a result.

With respect to the processing of polymetallic sulfidic concentrates, there is a wealth of knowledge available from studies on "pure" concentrates to build upon. Direct bioleaching, indirect bioleaching with the separation of microbial growth and ferric iron generation from chemical leaching, and hybrid processes comprised of chemical leaching and bioleaching have been developed for polymetallic concentrates using stirred tanks, and in some cases integrated flowsheets described. Considerable research on the polymetallic Kupferschiefer shale was undertaken as part of the Bioshale FP6 project but it was concluded that (at the present time) the grades for nickel, copper and zinc were too low for the process to be economic.

Sulfide ores, with or without high organic content, are also low grade and unlikely to be processed economically in stirred tanks (other than peculiarly rich gold-silver ores). Testwork has therefore mainly been focused on heap, dump, *in situ* or in stope bioleaching, strategies to mitigate acid consumption while maintaining conditions conducive to microbial growth, and some hybrid processes for multi-element processing. The main commercial development is the Talvivaara heap leach yielding nickel, cobalt, copper, zinc and possibly uranium and manganese products, but of greater benefit to biomining would be the interest generated among other mining companies with polymetallic deposits of similar mineralogy. Research into *in situ* or in stope bioleaching are an important part of the future for mineral processing because they may be the means of accessing deposits too deep for economical extraction by current underground mining technology and they are of interest also from the points of view of environmental and community acceptance. Therefore, it is noteworthy that a key objective of the recently initiated, European-funded, BioMOre project is to extract metal from deep mineralised zones by coupling solution mining (*in situ* or in stope processing) with bioleaching.

The bioleaching of oxidised ores such as laterites and manganese oxide ores have a long history but as yet no commercial processes have been developed. Three modes of leaching are employed, depending upon ore mineralogy: (i) the production or acid for direct bioleaching processes; and/or (ii) the production of acid and complexing agents; and (iii) the conduct of bioleaching under conditions amenable to redox reactions.

The need to provide low-cost carbon sources for those microorganisms that produce organic acids, the requirement to control growth conditions to maximise the production of an effective organic acid, the fact that organic acids are not as strong as inorganic acids, and the need in some cases to dispose of large amounts of biomass, are challenges that have yet to be overcome in the development of a large-scale, integrated process. Nevertheless, some innovative processes have been trialled.

One such process exploited sulfur biooxidation of elemental sulfur as a means of generating the required acid for the dissolution of oxidised ores such as manganese nodules or nickel laterites. There are two perceived benefits: (i) the cost of elemental sulfur is often lower than the cost of equivalent amounts of sulfuric acid; and (ii) other metabolic products from the microorganisms may enhance leaching. The further benefit of changing leaching conditions from aerobic to anaerobic introduced the possibility of reducing iron(III) compounds such as goethite or manganese(IV) compounds such as manganese dioxide. In the case of laterites, where nickel and other elements are held in the goethite crystal matrix, this strategy releases them to solution; in the case of manganese nodules, reductive leaching causes solubilisation of the MnO_2 and any ancillary elements. This process exploits the natural ability of some autotrophs to couple iron(III) reduction with sulfur oxidation under anaerobic conditions and is worthy of further research.

The processing of mine waste, overburden and tailings can serve two purposes, (i) the extraction of values from discarded materials perhaps quite high in concentration because historic processes were inefficient and (ii) a means of cleaning up a mine environment as part of closure. Processes in which mine remediation is supported by the extraction of values are a bonus. All of the above technologies have been trialled separately or in combination. A successful application of stirred tank technology to the clean up tailings from a copper mine, supported by the production of cobalt from the pyritic tailings, is the best example of a multiple-product (Co, Zn, Cu and Ni) stirred-tank base-metal leaching process. It is planned to apply a similar stirred-tank process to the remediation of extensive arsenic-contaminated tailings, leading to the safe disposal of benign leached residues. Many of the technologies and their variations have been trialled for the remediation of oxidised ore wastes but none are known to have been implemented thus far. More generally, new bioprocesses to secure reactive tailings and to remove metals from ARD are being developed.

Acknowledgments

The author thanks D.B. Johnson, P.C. Miller, D.H.R. Morin, T. Prior, D.E. Ralph and D.W. Shiers for helpful discussions, review of the manuscript, and/or permission to use illustrative material. The financial support of the Australian Government through the CSIRO Mineral Resources Flagship is gratefully acknowledged.

Conflicts of Interest

The author declares no conflict of interest.

References

1. Zinc Commodity Update. Available online: www.hdrsalva.com/market-news/zinc-commodity-update (accessed on 22 October 2014).
2. Critical Metals for Future Sustainable Technologies and Their Recycling Potential. Available online: www.unep.fr/scp/publications/details.asp?id=DTI/1202/PA (accessed on 22 October 2014).
3. Graedel, T.E.; Cao, J. Metal spectra as indicators of development. *Proc. Natl. Acad. Sci. USA* **2010**, *107*, 20905–20910.
4. Reller, A.; Bublies, T.; Staudinger, T.; Oswald, I.; Meissner, S.; Allen, M. The mobile phone: Powerful communicator and potential metal dissipator. *GAIA* **2009**, *18*, 127–135.
5. Dodson, J.R.; Hunt, A.J.; Parker, H.L.; Yang, Y.; Clark, J.H. Element sustainability: Towards the total recovery of scarce metals. *Chem. Eng. Process.* **2012**, *51*, 69–78.
6. Backman, C.M. Global supply and demand of metals in the future. *J. Toxicol. Environ. Health* **2008**, *71*, 1244–1253.
7. Prior, T.; Giurco, D.; Mudd, G.; Mason, L.; Behrisch, J. Resource depletion, peak minerals and the implications for sustainable resource management. *Glob. Environ. Chang.* **2012**, *22*, 577–587.
8. West, J. Decreasing metal ore grades: Are they really being driven by the depletion of high-grade deposits? *J. Ind. Ecol.* **2011**, *15*, 165–168.
9. Crowson, P. Some observations on copper yields and ore grades. *Resour. Policy* **2012**, *37*, 59–72.
10. USGS Minerals Yearbooks 1994–2011 and Mineral Commodity Summaries 2012–2014. Available online: http://minerals.usgs.gov/minerals/pubs/commodity/ (accessed on 22 October 2014).
11. World Copper Factbook 2013. Available online: http://www.icsg.org/index.php/component/jdownloads/finish/170/1188 (accessed on 22 October 2014).
12. The Outlook for Metals Markets. World Bank: Washington, DC, USA, 2006. Available online: http://siteresources.worldbank.org/INTOGMC/Resources/outlook_for_metals_market.pdf (accessed on 22 October 2014).
13. Cui, J.; Zhang, L. Metallurgical recovery of metals from electronic waste: A review. *J. Hazard. Mater.* **2008**, *158*, 228–256.
14. Pathak, A.; Dastidar, M.G.; Sreekrishnan, T.R. Bioleaching of heavy metals from sewage-sludge: A review. *J. Environ. Manag.* **2009**, *90*, 2343–2353.
15. Lee, J.; Pandey, B.D. Bio-processing of solid wastes and secondary resources for metal extraction—A review. *Waste Manag.* **2012**, *32*, 3–18.
16. Krebs, W.; Brombacher, C.; Bosshard, P.P.; Bachofen, R.; Brandl, H. Microbial recovery of metals from solids. *FEMS Microbiol. Rev.* **1997**, *20*, 605–617.
17. Hoque, M.E.; Philip, O.J. Biotechnological recovery of heavy metals from secondary sources—An overview. *Mater. Sci. Eng. C* **2011**, *31*, 57–66.

18. Asghari, I.; Mousavi, I.M.; Amiri, F.; Tavassoli, S. Bioleaching of spent refinery catalysts. *J. Ind. Eng. Chem.* **2013**, *19*, 1069–1081.

19. Kelly, D.P.; Wood, A.P. Reclassification of some species of *Thiobacillus* to the newly designated genera *Acidithiobacillus* gen. nov., *Halothiobacillus* gen. nov. and *Thermithiobacillus.* gen. nov. *Int. J. Syst. Evol. Microbiol.* **2000**, *50*, 511–516.

20. Rudolfs, W.; Helbronner, A. Oxidation of zinc sulfide by microorganisms. *Soil Sci.* **1922**, *14*, 459–464.

21. Colmer, A.R.; Temple, K.L.; Hinkle, H.E. An iron-oxidizing bacterium from the acid mine drainage of some bituminous coal mines. *J. Bacteriol.* **1950**, *59*, 317–328.

22. Temple, K.L.; Colmer, A.R. The autotrophic oxidation of iron by a new bacterium: *Thiobacillus ferrooxidans. J. Bacteriol.* **1951**, *62*, 605–611.

23. Leathen, W.W.; Braley, S.S., Sr.; McIntyre, L.D. The role of bacteria in the formation of acid from sulfuritic constituents associated with bituminous coal: II. Ferrous iron oxidizing bacteria. *Appl. Microbiol.* **1953**, *1*, 65–68.

24. Bryner, L.C.; Beck, J.V.; Davis, D.B.; Wilson, D.G. Microorganisms in leaching sulfide minerals. *Ind. Eng. Chem.* **1954**, *46*, 2587–2592.

25. Bryner, L.C.; Anderson, R. Microorganisms in leaching sulfide minerals. *Ind. Eng. Chem.* **1957**, *49*, 1721–1724.

26. Bryner, L.C.; Jameson, A.K. Microorganisms in leaching sulfide minerals. *Appl. Microbiol.* **1958**, *6*, 281–287.

27. Rawlings, D.E. Mesophilic autotrophic bioleaching bacteria: Description, physiology and role. In *Biomining: Theory, Microbes and Industrial Processes*; Rawlings, D.E., Ed.; Springer-Verlag: Berlin, Germany, 1997; pp. 229–245.

28. Norris, P.R. Thermophiles and bioleaching. In *Biomining: Theory, Microbes and Industrial Processes*; Rawlings, D.E., Ed.; Springer-Verlag: Berlin, Germany, 1997; pp. 247–258.

29. Hallberg, K.B.; Johnson, D.B. Biodiversity of acidophilic prokaryotes. *Adv. Appl. Microbiol.* **2001**, *49*, 37–84.

30. Watling, H.R. The bioleaching of sulphide minerals with emphasis on copper sulphides—A review. *Hydrometallurgy* **2006**, *84*, 81–108.

31. Bosecker, K. Leaching of lateritic nickel ores with heterotrophic microorganisms. In *Fundamental and Applied Biohydrometallurgy*; Lawrence, R.W., Branion, R.M.R., Ebner, H.G., Eds.; Elsevier: Amsterdam, The Netherlands, 1986; pp. 367–382.

32. Tzeferis, P.G. Leaching of a low-grade hematitic laterite ore using fungi and biologically produced acid metabolites. *Int. J. Miner. Process.* **1994**, *42*, 267–283.

33. Valix, M.; Usai, F.; Malik, R. Fungal bio-leaching of low-grade laterite ores. *Miner. Eng.* **2001**, *14*, 197–203.

34. Johnson, D.B.; Roberto, F.F. Heterotrophic acidophiles and their roles in the bioleaching of sulfide minerals. In *Biomining: Theory, Microbes and Industrial Processes*; Rawlings, D.E., Ed.; Springer-Verlag: Berlin, Germany, 1997; pp. 259–279.

35. Johnson, D.B.; Hallberg, K.B. Carbon, iron and sulfur metabolism in acidophilic micro-organisms. *Adv. Microb. Physiol.* **2008**, *54*, 201–255.

36. Ehrlich, H.L. Beginnings of rational bioleaching and highlights in the development of biohydrometallurgy: A brief history. *Eur. J. Miner. Process. Environ. Prot.* **2004**, *4*, 102–112.

37. Hallberg, K.B.; Johnson, D.B. Novel acidophiles isolated from moderately acidic mine drainage waters. *Hydrometallurgy* **2003**, *71*, 139–148.

38. Jerez, C.A. The use of genomics, proteomics and other OMICS technologies for the global understanding of biomining organisms. *Hydrometallurgy* **2008**, *94*, 162–169.

39. Johnson, D.B. Extremophiles: Acidic environments. In *Encyclopedia of Microbiology*; Schaechter, M., Ed.; Elsevier: Oxford, UK, 2009; pp. 107–126.

40. Watling, H.R.; Watkin, E.L.J.; Ralph, D.E. The resilience and versatility of acidophiles that contribute to the bio-assisted extraction of metals from mineral sulfides. *Environ. Technol.* **2010**, *31*, 915–933.

41. Schultz, M.F.; Benjamin, M.M.; Ferguson, J.F. Adsorption and desorption of metals on ferrihydrite: Reversibility of the reaction and sorption properties of the regenerated solid. *Environ. Sci. Technol.* **1987**, *21*, 863–869.

42. Fuller, C.C.; Davis, J.A.; Waychunas, G.A. Surface chemistry of ferrihydrite: Part 2. Kinetics of arsenate adsorption and coprecipitation. *Geochim. Cosmochim. Acta* **1993**, *57*, 2271–2282.

43. Cudennec, Y.; Lecerf, A. The transformation of ferrihydrite into goethite or hematite, revisited. *J. Solid State Chem.* **2006**, *179*, 716–722.

44. Liu, H.; Li, P.; Zhu, M.; Wei, Y.; Sun, Y. Fe(II)-induced transformation from ferrihydrite to lepidocrocite and goethite. *J. Solid State Chem.* **2007**, *180*, 2121–2128.

45. Bigham, J.M.; Nordstrom, D.K. Iron and aluminium hydroxysulfates from acidic sulfate waters. *Rev. Mineral. Geochem.* **2000**, *40*, 351–403.

46. Gramp, J.P.; Jones, F.S.; Bigham, J.M.; Tuovinen, O.H. Monovalent cation concentrations determine the types of Fe(III) hydroxysulfate precipitates formed in bioleach solutions. *Hydrometallurgy* **2008**, *94*, 29–33.

47. Carlson, L.; Bigham, J.M.; Schwertmann, U.; Kyek, A.; Wagner, F. Scavenging of As from acid mine drainage by schwertmannite and ferrihydrite: A comparison with synthetic analogues. *Environ. Sci. Technol.* **2002**, *36*, 1712–1719.

48. Fukushi, K.; Sato, T.; Yanase, N.; Minati, J.; Yamada, H. Arsenic sorption on schwertmannite. *Am. Mineral.* **2004**, *89*, 1728–1734.

49. Regenspurg, S.; Peiffer, S. Arsenate and chromate incorporation in schwertmannite. *Appl. Geochem.* **2005**, *20*, 1226–1239.

50. Acero, P.; Ayora, C.; Torrento, C.; Nieto, J.-M. The behavior of trace elements during schwertmannite precipitation and subsequent transformation into goethite and jarosite. *Geochim. Cosmochim. Acta* **2006**, *70*, 4130–4139.

51. Wang, H.; Bigham, J.M.; Tuovinen, O.H. Formation of schwertmannite and its transformation to jarosite in the presence of acidophilic iron-oxidizing microorganisms. *Mater. Sci. Eng.* **2006**, *26*, 588–592.

52. Sanchez-España, J.; Yusta, I.; López, G.A. Schwertmannite to jarosite conversion in the water column of an acidic mine pit lake. *Mineral. Mag.* **2012**, *76*, 2659–2682.

53. Daoud, J.; Karamanev, D. Formation of jarosite during Fe^{2+} oxidation by *Acidithiobacillus ferrooxidans*. *Miner. Eng.* **2006**, *19*, 960–967.

54. Sand, W. Microbial mechanisms. In *Microbially Influenced Corrosion of Materials*; Heitz, E., Flemming, H.-C., Sand, W., Eds.; Springer: Heidelberg, Germany, 1996; pp. 15–25.

55. Sand, W.; Jozsa, P.-G.; Manasch, R. Weathering microbiology. In *Environmental Microbiology*; Britton, G., Ed.; Wiley: New York, NY, USA, 2002; Volume 6, pp. 3364–3375.

56. Williams, T.M.; Unz, R.F. Filamentous sulfur bacteria of activated sludge: Characterization of *Thiothrix*, *Beggiatoa*, and *Eikelboom* Type 021N strains. *Appl. Environ. Microbiol.* **1985**, *49*, 887–898.

57. Grayston, S.J.; Nevell, W.; Wainwright, M. Sulphur oxidation by fungi. *Trans. Br. Mycol. Soc.* **1986**, *87*, 193–198.

58. Lang, E.; Jagnow, G. Fungi of a forest soil nitrifying at low pH values. *FEMS Microbiol. Lett.* **1986**, *38*, 257–265.

59. Mansch, R.; Bock, E. Biodeterioration of natural stone with special reference to nitrifying bacteria. *Biodegradation* **1998**, *9*, 47–64.

60. Sakai, K.; Ikehata, Y.; Ikenaga, Y.; Wakayama, M.; Moriguchi, M. Nitrite oxidation by heterotrophic bacteria under various nutritional and aerobic conditions. *J. Ferment. Bioeng.* **1996**, *82*, 613–617.

61. Schippers, A.; Rohwerder, T.; Sand, W. Intermediary sulfur compounds in pyrite oxidation: Implications for bioleaching and biodepyritization of coal. *Appl. Microbiol. Biotechnol.* **1999**, *52*, 104–110.

62. Pronk, J.T.; Meulenberg, R.; Hazeu, W.; Bos, P.; Kuenen, J.G. Oxidation of reduced inorganic sulfur compounds by acidophilic thiobacilli. *FEMS Microbiol. Rev.* **1990**, *75*, 293–306.

63. McNamara, C.J.; Mitchell, R. Microbial deterioration of historic stone. *Front. Ecol. Environ.* **2005**, *3*, 445–451.

64. De la Torre, M.A.; Gómez-Alarcón, G.; Vizcaino, C.; Garcia, M.T. Biochemical mechanisms of stone alteration carried out by filamentous fungi living in monuments. *Biogeochemistry* **1993**, *19*, 129–147.

65. Fischer, K.; Bipp, H.P. Removal of heavy metals from soil components and soils by natural chelating agents: Part II. Soil extraction by sugar acids. *Water Air Soil Pollut.* **2002**, *138*, 271–288.

66. Uroz, S.; Calvaruso, C.; Turpault, M.P.; Fley-Klett, P. Mineral weathering by bacteria: Ecology, actors and mechanisms. *Trends Microbiol.* **2009**, *17*, 378–387.

67. Uroz, S.; Calvaruso, C.; Turpault, M.P.; Sarniguet, A.; de Boer, W.; Leveau, J.H.J.; Frey-Klett, P. Efficient mineral weathering is a distinctive functional trait of the bacterial genus *Collimonas*. *Soil Biol. Biochem.* **2009**, *41*, 2178–2186.

68. Bennett, P.C.; Melcer, M.E.; Siegel, D.I.; Hassett, J.P. The dissolution of quartz in dilute aqueous solutions of organic acids at 25 °C. *Geochim. Cosmochim. Acta* **1988**, *52*, 1521–1530.

69. Blake, R.E.; Walter, L.M. Effects of organic acids on the dissolution of orthoclase at 80 °C and pH 6. *Chem. Geol.* **1996**, *132*, 91–102.

70. Hausrath, E.M.; Neaman, A.; Brantley, S.L. Elemental release rates from dissolving basalt and granite with and without organic ligands. *Am. J. Sci.* **2009**, *309*, 633–660.

71. Welch, S.A.; Taunton, A.E.; Banfield, J.F. Effect of microorganisms and microbial metabolites on apatite dissolution. *Geomicrobiol. J.* **2002**, *19*, 343–367.

72. Zhang, H.; Bloom, P.R. Dissolution kinetics of hornblende in organic acid solutions. *Soil Sci. Soc. Am. J.* **1999**, *63*, 815–822.

73. Di Bonaventura, M.P.; Gallo, M.D.; Cacchio, P.; Ercole, C.; Lepidi, A. Microbial formation of oxalate films on monument surfaces: Bioprotection or biodeterioration? *Geomicrobiol. J.* **1999**, *16*, 55–64.

74. Scheerer, S.; Ortega-Morales, O.; Gaylarde, C. Microbial deterioration of stone monuments—An updated overview. *Adv. Appl. Microbiol.* **2009**, *66*, 97–139.

75. Sterflinger, K. Fungi as geologic agents. *Geomicrobiol. J.* **2000**, *17*, 97–124.

76. Alibhai, K.A.K.; Dudeney, A.W.L.; Leak, D.J.; Agatzini, S.; Tzeferis, P. Bioleaching and bioprecipitation of nickel and iron from laterites. *FEMS Microbiol. Rev.* **1993**, *22*, 87–93.

77. Simate, G.S.; Ndlovu, S.; Walubita, L.F. The fungal and chemolithotrophic leaching of nickel laterites—Challenges and opportunities. *Hydrometallurgy* **2010**, *103*, 150–157.

78. Neilands, J.B. Siderophores: Structure and function of microbial iron transport compounds. *J. Biol. Chem.* **1995**, *270*, 26723–26726.

79. Boukhalfa, H.; Reiley, S.D.; Michalczyk, R.; Iyer, S.; Neu, M.P. Iron(III) coordination properties of a pyoverdin siderophore produced by *Pseudomonas putida* ATCC 33015. *Inorg. Chem.* **2006**, *45*, 5607–5616.

80. Duckworth, O.W.; Bargar, J.R.; Sposito, G. Coupled biogeochemical cycling of iron and manganese as mediated by microbial siderophores. *Biometals* **2009**, *22*, 605–613.

81. Kraemer, S.M. Iron oxide dissolution and solubility in the presence of siderophores. *Aquat. Sci.* **2004**, *66*, 3–18.

82. Dehner, C.A.; Awaya, J.D.; Maurice, P.A.; du Bois, J.L. Roles of siderophores, oxalate and ascorbate in mobilization of iron from hematite by the aerobic bacterium *Pseudomonas mendocina*. *Appl. Environ. Microbiol.* **2010**, *76*, 2041–2048.

83. Saal, L.B.; Duckworth, O.W. Synergistic dissolution of manganese oxides as promoted by siderophores and small organic acids. *Soil Sci. Soc. Am. J.* **2010**, *74*, 2021–2031.

84. Loukola-Ruskeeniemi, K.; Heino, T.; Talvitie, J.; Vanne, J. Base-metal-rich metamorphosed black shales associated with Proterozoic ophiolites in the Kainuu schist belt, Finland: A genetic link with the Outokumpu rock assemblage. *Miner. Depos.* **1991**, *26*, 143–151.

85. Loukola-Ruskeeniemi, K. Geochemical evidence for hydrothermal origin of sulphur, base metals and gold in Proterozoic metamorphosed black shales, Kainuu and Outokumpu areas, Finland. *Miner. Depos.* **1991**, *26*, 152–164.

86. Tariq, M.; Aziz, A.; Shafiq, M.; Sajjid, M.; Iqbal, M.M.; Muhammad, S. Characterization of black shale of Chimiari Khyber Pakthunkhawa region of Pakistan for its potential as multiminerals. *Int. J. Sci. Res.* **2013**, *2*, 231–238.

87. Gouin, J.; Auge, T.; Bailly, L.; d'Hugues, P. Organic and mineral characteristics of Kupferschiefer ore from Lubin mine (Poland): Implications for bioleaching of the ore. In *Digging Deeper*; Andrew, C.J., Ed.; IAEG: Dublin, Ireland, 2007.

88. Kamradt, A.; Borg, G.; Schaefer, J.; Kruse, S.; Fiedler, M.; Romm, P.; Schippers, A.; Gorny, R.; du Bois, M.; Bieligk, C.; *et al.* An integrated process for innovative extraction of metals from kupferschiefer mine dumps. *Chem. Ing. Tech.* **2012**, *84*, 1694–1703.

89. Matlakowska, R.; Ruszkowski, D.; Sklodowska, A. Microbial transformation of fossil organic matter of Kupferschiefer black shale—Elements mobilization from metalloorganic compounds and metalloporphyrins by a community of indigenous microorganisms. *Physicochem. Probl. Miner. Process.* **2013**, *49*, 223–231.

90. Farbiszewska-Kiczma, J.; Farbiszewska, T.; Bak, M. Bioleaching of metals from Polish black shale in neutral medium. *Physicochem. Probl. Miner. Process.* **2004**, *38*, 273–280.

91. Farbiszewska-Kiczma, J.; Farbiszewska, T. Isolation of bacteria that degrade organometallic compounds from metallic wastes. *Physicochem. Probl. Miner. Process.* **2005**, *39*, 263–267.

92. Matlakowska, R.; Hallberg, K.B.; Sklodowska, A. Isolation and characterisation of microorganisms from copper-bearing black shale of Lubin mine, Poland. In *Biohydrometallurgy: From Single Cell to the Environment*; Schippers, A., Sand, W., Glombitza, F., Willscher, S., Eds.; TransTech Publications: Zurich, Switzerland, 2007; p. 580.

93. Matlakowska, R.; Sklodowska, A. The culturable bacteria isolated from organic-rich black shale potentially useful in biometallurgical procedures. *J. Appl. Microbiol.* **2009**, *107*, 858–866.

94. Matlakowska, R.; Narkiewicz, W.; Sklodowska, A. Biotransformation of organic-rich copper-bearing black shale by indigenous microorganisms isolated from Lubin mine (Poland). *Environ. Sci. Technol.* **2010**, *44*, 2433–2440.

95. Itävaara, M.; Nyyssönen, M.; Kapanen, A.; Nousiainen, A.; Ahonen, L.; Kukkonen, I. Characterization of bacterial diversity to a depth of 1500 m in the Outokumpu deep borehole, Fennoscandian Shield. *FEMS Microbiol. Ecol.* **2011**, *77*, 295–309.

96. Li, J.; Sun, W.; Wang, S.; Sun, Z.; Lin, S.; Peng, X. Bacteria diversity, distribution and insight into their role in S and Fe biogeochemical cycling during black shale weathering. *Environ. Microbiol.* **2014**, *16*, 3533–3547.

97. Petsch, S.T.; Edwards, K.J.; Eglinton, T.I. Microbial transformations of organic matter in black shales and implications for global biogeochemical cycles. *Palaeogeogr. Palaeoclimatol. Palaeoecol.* **2005**, *219*, 157–170.

98. Sadowski, Z.; Szubert, A.; Maliszewska, I.; Jazdzyk, E. A view on the organic matter and metalloporphyrins biodegradation as characteristic component of black shale ore. In *Biohydrometallurgy: From Single Cell to the Environment*; Schippers, A., Sand, W., Glombitza, F., Willscher, S., Eds.; TransTech Publications: Zurich, Switzerland, 2007; pp. 95–98.

99. Matlakowska, R.; Sklodowska, A. Uptake and degradation of copper and cobalt porphyrins by indigenous microorganisms of Kupferschiefer (Fore-Sudetic Monocline, Poland). *Hydrometallurgy* **2010**, *104*, 501–505.

100. Matlakowska, R.; Sklodowska, A. Biodegradation of organic matter and release of heavy metals from the copper bearing black shale of Fore Sudetic Monocline (Poland). In *Biohydrometallurgy: From Single Cell to the Environment*; Schippers, A., Sand, W., Glombitza, F., Willscher, S., Eds.; TransTech Publications: Zurich, Switzerland, 2007; pp. 238–239.

101. Matlakowska, R.; Sklodowska, A. Biodegradation of Kupferschiefer black shale organic matter (Fore-Sudetic Monocline, Poland) by indigenous microorganisms. *Chemosphere* **2011**, *83*, 1255–1261.

102. Ehrlich, H.L. Microbes and metals. *Appl. Microbiol. Biotechnol.* **1997**, *48*, 687–692.

103. Lyalikova, N.N.; Vedenina, I.Y.; Romanova, A.K. Assimilation of carbon dioxide by a culture of *Stibiobacter senarmontii*. *Microbiology* **1976**, *45*, 476–477.

104. Ilyaletdinov, A.N.; Abdrashitova, S.A. Autotrophic arsenic oxidation by *Pseudomonas arsenitoxidans* culture. *Mikrobiologiia* **1981**, *50*, 197–204.

105. Johnson, D.B.; Kanao, T.; Hedrich, S. Redox transformations of iron at extremely low pH: Fundamental and applied aspects. *Front. Microbiol.* **2012**, *3*, 96, doi:10.3389/fmicb.2012.00096.

106. Johnson, D.B.; Grail, B.M.; Hallberg, K.B. A new direction for biomining: Extraction of metals by reductive dissolution of oxidized ores. *Minerals* **2013**, *3*, 49–58.

107. Johnson, D.B. Selective solid media for isolating and enumerating acidophilic bacteria. *J. Microbiol. Methods* **1995**, *23*, 205–218.

108. Johnson, D.B.; Hallberg, K.B. Techniques for detecting and identifying acidophilic mineral-oxidising microorganisms. In *Biomining*; Rawlings, D.E., Johnson, D.B., Eds.; Springer-Verlag: Berlin, Germany, 2007; pp. 237–262.

109. Valenzuela, L.; Chi, A.; Beard, S.; Orell, A.; Guiliani, N.; Shabanowitz, J.; Hunt, D.F.; Jerez, C.A. Genomics, metagenomics and proteomics in biomining microorganisms. *Biotechnol. Adv.* **2006**, *24*, 197–211.

110. Shiers, D.W.; Ralph, D.E.; Watling, H.R. A comparative study of substrate utilisation by *Sulfobacillus* species in mixed ferrous ion and tetrathionate growth medium. *Hydrometallurgy* **2010**, *104*, 363–369.

111. Shiers, D.W.; Ralph, D.E.; Watling, H.R. The effects of nitrate on substrate utilisation by some iron(II)- and sulfur-oxidising Bacteria and Archaea. *Hydrometallurgy* **2014**, *150*, 259–268.

112. Suzuki, I.; Takeuchi, T.L.; Yuthasastrakosol, T.D.; Oh, J.K. Ferrous iron and sulfur oxidation and ferric iron reduction activities of *Thiobacillus ferrooxidans* are affected by growth on ferrous iron, sulfur or a sulfide ore. *Appl. Environ. Microbiol.* **1990**, *56*, 1620–1626.

113. Sugio, T.; Tsujita, Y.; Katagiri, T.; Inagaki, K.; Tano, T. Reduction of Mo^{6+} with elemental sulfur by *Thiobacillus ferrooxidans*. *J. Bacteriol.* **1988**, *170*, 5956–5959.

114. Sugio, T.; Tsujita, Y.; Inagaki, K.; Tano, T. Reduction of cupric ions with elemental sulfur by *Thiobacillus ferrooxidans*. *Appl. Environ. Microbiol.* **1990**, *56*, 693–696.

115. Bridge, T.A.M.; Johnson, D.B. Reduction of soluble iron and reductive dissolution of ferric-iron containing minerals by moderately thermophilic iron-oxidizing bacteria. *Appl. Environ. Microbiol.* **1998**, *64*, 2181–2186.

116. Hallberg, K.B.; Grail, B.M.; du Plessis, C.A.; Johnson, D.B. Reductive dissolution of ferric iron minerals: A new approach for Bioprocessing nickel laterites. *Miner. Eng.* **2011**, *24*, 620–624.

117. Johnson, D.B.; Stallwood, B.; Kimura, S.; Hallberg, K.B. Isolation and characterization of *Acidicaldus organivorus*, gen. nov., sp. nov.: A novel sulfur-oxidizing, ferric iron-reducing thermoacidophilic heterotrophic *Proteobacterium*. *Arch. Microbiol.* **2006**, *185*, 212–221.

118. Coupland, K.; Johnson, D.B. Evidence that the potential for dissimilatory ferric iron reduction is widespread among acidophilic heterotrophic bacteria. *FEMS Microbiol. Lett.* **2008**, *279*, 30–35.

119. Bridge, T.A.M.; Johnson, D.B. Reductive dissolution of ferric iron minerals by *Acidiphilium* SJH. *Geomicrobiol. J.* **2000**, *17*, 193–206.

120. Muyzer, G.; Stams, A.J.M. The ecology and biotechnology of sulphate-reducing bacteria. *Nat. Rev. Microbiol.* **2008**, *6*, 441–454.

121. PAQUES Metal and Mining. Available online: http://en.paques.nl/your-sector/other/metal-and-mining (accessed on 22 October 2014).

122. Watling, H. Adaptability of biomining organisms in hydrometallurgical processes. In *Biohydrometallurgical Processes: A Practical Approach*; Santos Sobral, L.G., Monteiro de Oliveira, D., Gomes de Souza, C.E., Eds.; CETEM/MCTI: Rio de Janeiro, Brazil, 2011; pp. 39–70.

123. Hackl, R.P.; Wright, F.R.; Bruynesteyn, A. Bacteria for Oxidizing Multimetallic Sulphide Ores. US Patent 5,089,412, 18 February 1992.

124. Van Aswegen, P.C.; Haines, A.K.; Marais, H.J. Design and operation of a commercial bacterial oxidation plant at Fairview. In Proceedings of the Randol Gold Conference, Perth, Australia; 28 October–1 November 1988; pp. 144–147.

125. Williams, T.L. Factors affecting bacterial population dynamics at the Youanmi bacterial oxidation plant. In *Biotechnology Comes of Age*; Australian Minerals Foundation: Glenside, SA, Australia, 1997.

126. Battaglia, F.; Morin, D.; Ollivier, P. Dissolution of cobaltiferous pyrite by *Thiobacillus ferrooxidans* and *Thiobacillus thiooxidans*: Factors influencing bacterial leaching efficiency. *J. Biotechnol.* **1994**, *32*, 11–16.

127. D'Hughes, P.; Cezac, P.; Cabral, T.; Battaglia, F.; Truong-Meyer, X.M.; Morin, D. Bioleaching of a cobaltiferous pyrite: A continuous laboratory-scale study at high solids concentration. *Miner. Eng.* **1997**, *10*, 507–527.

128. Heinzle, T.; Miller, D.; Nagel, V. Results of an integrated pilot plant operation using the BioNIC® process to produce nickel metal. In *Biomine' 99 and Water Management in Metallurgical Operations' 99*; AusIMM: Melbourne, Australia, 1999; pp. 16–25.

129. Du Plessis, C.A.; Batty, J.D.; Dew, D.W. Commercial applications of thermophile bioleaching. In *Biomining*; Rawlings, D.E., Johnson, D.B., Eds.; Springer-Verlag: Berlin, Germany, 2007; pp. 57–80.

130. Duncan, D.W.; Landesman, J.; Walden, C.C. Role of *Thiobacillus ferrooxidans* in the oxidation of sulfide minerals. *Can. J. Microbiol.* **1967**, *13*, 397–403.

131. Shiers, D.W.; Blight, K.R.; Ralph, D.E. Sodium sulphate and sodium chloride effects on batch culture of iron-oxidising bacteria. *Hydrometallurgy* **2005**, *80*, 75–82.

132. Suzuki, I.; Lee, D.; McKay, B.; Harahuc, L.; Oh, J.K. Effect of various anions, pH, and osmotic pressure on oxidation of elemental sulfur by *Thiobacillus thiooxidans*. *Appl. Environ. Microbiol.* **1999**, *65*, 5163–5168.

133. Zammit, C.M.; Mutch, L.A.; Watling, H.R.; Watkin, E.L.J. The characterization of salt tolerance in biomining microorganisms and the search for novel salt tolerant strains. In *Biohydrometallurgy: A Meeting Point between Microbial Ecology, Metal Recovery Processes and Environmental Remediation*; Donati, E.R., Viera, M.R., Tavani, E.L., Giaveno, M.A., Lavalle, T.L., Chiaccharini, P.A., Eds.; TransTech Publications: Zurich, Switzerland, 2009; pp. 283–286.

134. Gahan, C.S.; Sundkvist, J.E.; Dopson, M.; Sandström, Å. Effect of chloride on ferrous iron oxidation by a Leptospirillum ferriphilum-dominated chemostat culture. *Biotechnol. Bioeng.* **2010**, *106*, 422–431.

135. Williams, T.L. BioHeap™ bacterial leaching of the Sherlock Bay Nickel Mine primary nickel-sulphide ore in saline water. In Proceedings of the ALTA Nickel-Cobalt Conference, Perth, Australia, 15–17 May 2006.

136. Rautenbach, G.F.; Davis-Belmar, C.S.; Demergasso, C.S. A Method of Treating Sulphide Mineral. World Patent, WO2010/009481-A, 21 January 2010.

137. Patel, B.C.; Tipre, D.R.; Dave, S.R. Optimization of copper and zinc extraction from polymetallic bulk concentrate and ferric iron bioregeneration under metallic stress. *Hydrometallurgy* **2012**, *117–118*, 18–23.

138. Corbillon, M.S.; Olazabal, M.A.; Madariaga, J.M. Potentiometric study of aluminium fluoride complexation equilibria and definition of the thermodynamic model. *J. Solut. Chem.* **2008**, *37*, 567–579.

139. Radic, N.; Bralic, M. Aluminium fluoride complexation and its ecological importance in the aquatic environment. *Sci. Total Environ.* **1995**, *172*, 237–243.

140. Sicupira, L.; Veloso, T.; Reis, F.; Leao, V. Assessing metal recovery from low-grade copper ores containing fluoride. *Hydrometallurgy* **2011**, *109*, 202–210.

141. Soli, A.L.; Byrne, R.H. The hydrolysis and fluoride complexation behaviour of Fe(III) at 25 °C and 0.68 molal ionic strength. *J. Solut. Chem.* **1996**, *25*, 773–785.

142. Sundkvist, J.E.; Sandström, Å.; Gunneriusson, L.; Lindstrom, E.B. Fluorine toxicity in bioleaching systems. In Proceedings of the 16th International Biohydrometallurgy Symposium, Cape Town, South Africa, 25–29 September 2005; Harrison, S.T.L., Rawlings, D.E., Petersen, J., Eds.; IBS: Cape Town, South Africa, 2005; pp. 19–28.

143. Razzell, W.E.; Trussell, P.C. Isolation and properties of an iron-oxidizing *Thiobacillus*. *J. Bacteriol.* **1963**, *85*, 595–603.

144. Harahuc, L.; Lizama, H.; Suzuki, I. Effect of anions on selective solubilisation of zinc and copper in bacterial leaching of sulfide ores. *Biotechnol. Bioeng.* **2000**, *69*, 196–203.

145. Andrews, G.F.; Noah, K.S.; Glenn, A.Q.; Stevens, C.J. Combined physical/microbial beneficiation of coal using the flood/drain bioreactor. *Fuel Process. Technol.* **1994**, *40*, 283–296.

146. Andrews, G.F.; Noah, K.S. The flood/drain bioreactor for coal and mineral processing. In *Mineral Bioprocessing II*; Holmes, D.S., Smith, R.W., Eds.; TMS: Warrendale, PA, USA, 1995; pp. 219–230.

147. Andrews, G.F.; Stevens, C.J.; Glenn, A.; Noah, K.S. Microbial coal depyritization: Why and how? In *Biohydrometallurgical Technologies*; Torma, A.E., Wey, J.E., Lakshmanan, V.L., Eds.; TMS: Warrendale, PA, USA, 1993; pp. 381–391.

148. Andrews, G. The optimal design of bioleaching processes. *Miner. Process. Extr. Metall. Rev.* **1998**, *19*, 149–165.

149. Tyagi, R.D.; Couillard, D.; Tran, F.T. Comparative study of bacterial leaching of metals from sewage sludge in continuous stirred tank and airlift reactors. *Process Biochem.* **1991**, *26*, 47–54.

150. Chen, S.Y.; Lin, J.G. Bioleaching of heavy metals from contaminated sediment by indigenous sulfur-oxidizing bacteria in an air-lift bioreactor: Effects of sulfur concentration. *Water Res.* **2004**, *38*, 3205–3214.

151. Puhakka, J.; Tuovinen, O.H. Biological leaching of sulfide minerals with the use of shake flask, aerated column, airlift reactor, and percolation techniques. *Acta Biotechnol.* **1986**, *6*, 345–354.

152. Miller, P.C.; Huberts, R.; Livesey-Goldblatt, E. The semicontinuous bacterial agitated leaching of nickel sulphide material. In *Fundamental and Applied Biohydrometallurgy*; Lawrence, R.W., Branion, R.M.R., Ebner, H.G., Eds.; Elsevier: Amsterdam, The Netherlands, 1986; pp. 23–42.

153. Sethurajan, M.; Aruliah, R.; Karthikeyan, O.P.; Balasubramanian, R. Bioleaching of copper from black shale ore using mesophilic mixed populations in an airlift reactor. *Environ. Eng. Manag. J.* **2012**, *11*, 1839–1848.

154. Slavkina, O.V.; Fomchencko, N.V.; Biriukov, V.V.; Arkhipov, M.Y. Study on bacterial leaching of copper-zinc ore concentrate. 3. Experimental trial of two-stage recirculation technology of copper-zinc concentrate leaching. *Biotekhnologiya* **2005**, *3*, 48–54.

155. Li, D.; Li, D.W.; Zhang, S.J. The study of toxic elements removal and valuable metals recovery from mine tailings in gas-liquid-solid internal circulation bioreactor. *Res. J. Chem. Environ.* **2011**, *15*, 990–993.

156. Loi, G.; Mura, A.; Trois, P.; Rossi, G. Bioreactor performance *vs.* Solids concentration in coal biodepyritization. *Fuel Process. Technol.* **1994**, *40*, 251–260.

157. Loi, G.; Trois, P.; Rossi, G. Biorotor®: A new development for biohydrometallurgical processing. In *Biohydrometallurgical Processing*; Vargas, T., Jerez, C.A., Wiertz, J.V., Toledo, H., Eds.; University of Chile: Santiago, Chile, 1995; Volume I, pp. 263–271.

158. Rossi, G. The design of bioreactors. *Hydrometallurgy* **2001**, *59*, 217–231.

159. Loi, G.; Rossi, A.; Trois, P.; Rossi, G. Continuous revolving barrel bioreactor tailored to the bioleaching microorganisms. *Miner. Metall. Process.* **2006**, *23*, 196–202.

160. Groudev, S.; Groudeva, V.I. Microbial communities in four industrial copper dump leaching operations in Bulgaria. *FEMS Microbiol. Rev.* **1993**, *11*, 261–267.

161. Brierley, J.A.; Brierley, C.L. Present and future commercial applications of biohydrometallurgy. *Hydrometallurgy* **2001**, *59*, 233–239.

162. Duncan, D.W.; Bruynesteyn, A. Enhancing bacterial activity in a uranium mine. *Can. Min. Metall. Bull.* **1971**, *64*, 32–36.

163. McCready, R.G.L.; Gould, W.D. Bioleaching of uranium at Denison Mines. In *Biohydrometallurgy*; Sally, J., McCready, R.G.L., Wichlacz, P.L., Eds.; Canmet: Ottawa, ON, Canada, 1989; pp. 477–485.

164. Domic, E. A review of the development and current status of copper bioleaching operations in Chile: 25 years of successful commercial implementation. In *Biomining*; Rawlings, D.E., Johnson, D.B., Eds.; Springer-Verlag: Berlin, Germany, 2007; pp. 81–95.

165. Brierley, C.L.; Brierley, J.A. Progress in bioleaching: Part B: Applications of microbial processes by the minerals industries. *Appl. Microbiol. Biotechnol.* **2013**, *97*, 7543–7552.

166. Logan, T.; Seal, T.; Brierley, J.A. Whole-ore heap biooxidation of sulfidic gold-bearing ores. In *Biomining*; Rawlings, D.E., Johnson, D.B., Eds.; Springer-Verlag: Berlin, Germany, 2007; pp. 113–138.

167. Readett, D.; Fox, F. Commercialisation of Ni heap leaching at Murrin Murrin operations. In Proceedings of the XXV International Mineral Processing Congress, Brisbane, Australia, 6–10 September 2010; AusIMM: Melbourne, Australia, 2010; pp. 3611–3616.

168. Fernandes, H.M.; Lamego Simoes Filho, F.F.; Perez, V.; Ramalho Franklin, M.; Gomiero, L.A. Radioecological characterization of a uranium mining site located in a semi-arid region of Brazil. *J. Environ. Radioact.* **2006**, *88*, 140–157.

169. Lizama, H.M.; Harlamovs, J.R.; Belanger, S.; Brienne, S.H. The Teck Cominco Hydrozinc™ Process. In *Hydrometallurgy 2003*; Young, C., Alfantazi, A.M., Anderson, C.G., Dreisinger, D.B., Harris, B., James, A., Eds.; TMS: Warrendale, PA, USA, 2003; Volume 2, pp. 1503–1516.

170. Harvey, T.J.; Bath, M. The Geobiotics GEOCOAT® technology—Progress and challenges. In *Biomining*; Rawlings, D.E., Johnson, D.B., Eds.; Springer-Verlag: Berlin, Germany, 2007; pp. 97–112.

171. Gunn, M.; Tittes, P.; Harvey, P.; Carretero, E.; da Silva, P.M.; de Souza, J.P. Laboratory and demonstration-scale operation of the Caraiba heap leach using GEOCOAT®. In *Hydrocopper 2009*; Casas, J., Domic, E., Eds.; Gecamin: Santiago, Chile, 2009; pp. 352–364.

172. Williams, T.L.; Gunn, M.J.; Jaffer, A.; Harvey, P.I.; Tittes, P.R. The application of Geobiotics LLC GEOCOAT® technology to the bacterial oxidation of a refractory arsenopyrite gold concentrate. In *Hydrometallurgy 2008*; Young, C.A., Taylor, P.R., Anderson, C.G., Choi, Y., Eds.; SME: Littleton, CO, USA, 2008; pp. 474–483.

173. Sampson, M.I.; van der Merwe, J.W.; Harvey, T.J.; Bath, M.D. Testing the ability of a low-grade sphalerite concentrate to achieve autothermality during biooxidation heap leaching. *Miner. Eng.* **2005**, *18*, 427–437.

174. Soleimani, M.; Petersen, J.; Roostaazad, R.; Hosseini, S.; Mousavi, S.M.; Najafi, A.; Kazemi Vasiri, A. Leaching of a zinc ore and concentrate using the Geocoat™ technology. *Miner. Eng.* **2011**, *24*, 64–69.

175. Petersen, J.; Dixon, D.G. Competitive bioleaching of pyrite and chalcopyrite. *Hydrometallurgy* **2006**, *83*, 40–49.

176. Mwase, J.M.; Petersen, J.; Eksteen, J.J. A conceptual flow-sheet for heap leaching of platinum group metals (PGMs) from a low-grade ore concentrate. *Hydrometallurgy* **2012**, *111–112*, 129–135.

177. Cope, L.W. Vat leaching—An overlooked process. *Eng. Min. J.* **1999**, *200*, 17–24.

178. Mackie, D.; Trask, F. Continuous vat leaching—First copper pilot trials. In *ALTA Copper-X, Perth, Australia*; ALTA Metallurgical Services: Melbourne, Australia, 2009.

179. Schlitt, J.; Johnston, A. The Marcobre vat leach system: A new look at an old process. In Proceedings of Copper 2010, Hamburg, Germany, 6–10 June 2010; GDMB: Clausthal-Zellerfeld, Germany, 2010; Volume 5, pp. 2039–2057.

180. Van Aswegen, P.C.; van Niekerk, J.; Olivier, W. The BIOX™ process for the treatment of refractory gold concentrates. In *Biomining*; Rawlings, D.E., Johnson, D.B., Eds.; Springer-Verlag: Berlin, Germany, 2007; pp. 1–33.

181. Miller, P.C. The design and operating practice of bacterial oxidation plant using moderate thermophiles (The BacTech Process). In *Biomining: Theory, Microbes and Industrial Processes*; Rawlings, D.E., Ed.; Springer-Verlag: Berlin, Germany, 1997; pp. 81–102.

182. Rorke, G.V.; Basson, P.; Miller, D.M. Advancements in thermophile bioleaching technology. In *ALTA Nickel/Cobalt-7 Technical Proceedings*; ALTA: Melbourne, Australia, 2001.

183. Batty, J.D.; Rorke, G.V. Development and commercial demonstration of the BioCOP™ thermophile process. *Hydrometallurgy* **2006**, *83*, 83–89.

184. Wang, S. Copper leaching from chalcopyrite concentrates. *JOM* **2005**, *57*, 48–51.

185. Miller, D.M.; Dew, D.W.; Norton, A.E.; Johns, M.W.; Cole, P.M.; Benetis, G.; Dry, M. The BioNIC Process: Description of the process and presentation of pilot plant results. In *Nickel/Cobalt 97*; Cooper, W.C., Mihaylov, I., Eds.; CIM: Montreal, QC, Canada, 1997; Volume 1, pp. 97–110.

186. Gilbertson, B.P. Creating value through innovation: Biotechnology in mining. *Miner. Process. Extr. Metall.* **2000**, *109*, 61–67.

187. Morin, D.H.R.; d'Hugues, P. Bioleaching of a cobalt-containing pyrite in stirred tank reactors: A case study from laboratory scale to industrial application. In *Biomining*; Rawlings, D.E., Johnson, D.B., Eds.; Springer-Verlag: Berlin, Germany, 2007; pp. 35–55.

188. Romero, R.; Palencia, I.; Carranza, F. Silver catalysed IBES process: Application to a Spanish copper-zinc sulphide concentrate: Part 3. Selection of the operational parameters for a continuous pilot plant. *Hydrometallurgy* **1998**, *49*, 75–86.

189. Frias, C.; Carranza, F.; Sanchez, F.; Mazuelos, A.; Frades, M.; Romero, R.; Diaz, G.; Iglesias, N. New developments in indirect bioleaching of zinc and lead sulphide concentrates. In *Hydrometallurgy 2008*; Young, C.A., Taylor, P.R., Anderson, C.G., Choi, Y., Eds.; SME: Littleton, CO, USA, 2008; pp. 497–505.

190. Zafiratos, J.G.; Agatzini-Leonardou, S. Aerobic and anaerobic leaching of manganese. In *Biohydrometallurgy: A Sustainable Technology in Evolution*; Tsezos, M., Hatzikioseyian, A., Remoundaki, E., Eds.; National Technical University of Athens: Athens, Greece, 2004; pp. 41–54.

191. Ritcey, G.M. Solvent extraction in hydrometallurgy: Present and future. *Tsinghua Sci. Technol.* **2006**, *11*, 137–152.

192. Alexandros, S.D. Ion-exchange resins: A Retrospective from Industrial and Engineering Chemistry Research. *Ind. Eng. Chem. Res.* **2009**, *48*, 388–398.

193. Sole, K.C.; Cole, P.M.; Feather, A.M.; Kotze, M.H. Solvent extraction and ion exchange applications in Africa's resurging uranium industry: A review. *Solvent Extr. Ion Exch.* **2011**, *29*, 868–899.

194. Van Deventer, J. Selected ion exchange applications in the hydrometallurgical industry. *Solvent Extr. Ion Exch.* **2011**, *29*, 695–718.

195. Torma, A.E. Microbiological oxidation of synthetic cobalt, nickel and zinc sulfides by *Thiobacillus ferrooxidans*. *Rev. Can. Biol.* **1971**, *30*, 209–216.

196. Third, K.A.; Cord-Ruwisch, R.; Watling, H.R. The role of iron-oxidizing bacteria in stimulation or inhibition of chalcopyrite bioleaching. *Hydrometallurgy* **2000**, *57*, 225–233.

197. Fowler, T.A.; Crundwell, F.K. Leaching of zinc sulfide by *Thiobacillus ferrooxidans*: Bacterial oxidation of the sulfur product layer increases the rate of zinc sulfide dissolution at high concentrations of ferrous ions. *Appl. Environ. Microbiol.* **1999**, *65*, 5285–5292.

198. Zhang, G.; Fang, Z. The contribution of direct and indirect actions in bioleaching of pentlandite. *Hydrometallurgy* **2005**, *80*, 59–66.

199. King, A.J.; Budden, J.R. The bacterial oxidation of nickel and cobalt polymetallic concentrates. In Proceedings of the Nickel/Cobalt Pressure Leaching and Hydrometallurgy Forum, Perth, Australia, 13–14 May 1996; ALTA Metallurgical Services: Melbourne, Australia, 1996.

200. Gómez, C.; Blázquez, M.L.; Ballester, A. Bioleaching of a Spanish complex sulphide ore bulk concentrate. *Miner. Eng.* **1999**, *12*, 93–106.

201. Pradhan, D.; Kim, D.J.; Chaudhury, G.R. Dissolution kinetics of complex sulfides acidophilic microorganisms. *Jpn. Inst. Met. Mater. Trans.* **2010**, *51*, 413–419.

202. Kim, D.J.; Pradhan, D.; Roy Chaudhury, G.; Ahn, J.G.; Lee, S.W. Bioleaching of complex sulfides concentrate and correlation of leaching parameters using multivariate data analysis technique. *Jpn. Inst. Met. Mater. Trans.* **2009**, *50*, 2318–2322.

203. Wang, Y.; Su, L.; Zeng, W.; Qiu, G.; Wan, L.; Chen, X.; Zhou, H. Optimization of copper extraction for bioleaching of complex Cu-polymetallic concentrate by moderate thermophiles. *Trans. Nonferr. Met. Soc. China* **2014**, *24*, 1161–1170.

204. Uryga, A.; Sadowski, Z.; Grotowski, A. Bioleaching of cobalt from mineral products. *Physicochem. Probl. Miner. Process.* **2004**, *38*, 291–299.

205. Tasa, A.; Garcia, O.; Bigham, J.M.; Vuorinen, A.; Tuovinen, O.H. Acid and biological leaching of a black shale from Toolse, Estonia. In *Biohydrometallurgical Processing*; Vargas, T., Jerez, C.A., Wiertz, J.V., Toledo, H., Eds.; University of Chile: Santiago, Chile, 1995; Volume I, pp. 229–238.

206. Abdollahi, H.; Shaefaei, S.Z.; Noaparast, M.; Manafi, Z.; Aslan, N. Bio-dissolution of Cu, Mo and Re from molybdenite concentrate using a mix mesophilic microorganism in shake flask. *Trans. Nonferr. Met. Soc. China* **2013**, *23*, 219–230.

207. Gericke, M.; Muller, H.H.; van Staden, P.J.; Pinches, A. Development of a tank bioleaching process for the treatment of complex Cu-polymetallic concentrates. *Hydrometallurgy* **2008**, *94*, 23–28.

208. Mehta, K.D.; Pandey, B.D. Bioleaching of a copper sulphide concentrate by two different strains of acidophilic bacteria. *Int. J. Metall. Eng.* **2012**, *1*, 83–86.

209. Norris, P.R.; Burton, N.P.; Clark, D.A. Mineral sulfide concentrate leaching in high temperature bioreactors. *Miner. Eng.* **2013**, *48*, 10–19.

210. Fomchenko, N.V.; Biryukov, V.V. A two-stage technology for bacterial and chemical leaching of copper-zinc raw materials by Fe^{3+} ions with their subsequent regeneration by chemolithotrophic bacteria. *Appl. Biochem. Microbiol.* **2009**, *1*, 56–60.

211. Fomchenko, N.V.; Muravyev, M.I.; Kondrat'eva, T.F. Two-stage bacterial-chemical oxidation of refractory gold-bearing sulfidic concentrates. *Hydrometallurgy* **2010**, *101*, 28–34.

212. Patel, B.C.; Tipre, D.R.; Dave, S.R. Development of *Leptospirillum ferriphilum* dominated consortium for ferric iron regeneration and metal bioleaching under extreme stresses. *Bioresour. Technol.* **2012**, *118*, 483–489.

213. Spolaore, P.; Joulian, C.; Gouin, J.; Ibáñez, A.; Augé, T.; Morin, D.; d'Hugues, P. Bioleaching of an organic-rich polymetallic concentrate using stirred-tank technology. *Hydrometallurgy* **2009**, *99*, 137–143.

214. Conic, V.T.; Cvetkovski, V.B.; Stanojevich Simsic, Z.S.; Dragulovic, S.S.; Ljubomirovic, Z.S.; Cvetkovska, M.; Vukovic, M.N. Bioleaching of Zn-Pb-Ag sulphidic concentrate. In Proceedings of the 15th International Research/Expert Conference "Trends in the Development of Machinery and Associated Technology" (TMT 2011), Prague, Czech Republic, 12–18 September 2011.

215. Askari Zamani, M.A.; Hiroyoshi, N.; Tsunekawa, M.; Vaghar, R.; Oliazadeh, M. Bioleaching of Sarcheshmeh molybdenite concentrate for extraction of rhenium. *Hydrometallurgy* **2005**, *80*, 23–31.

216. Askari Zamani, M.A.; Vaghar, R.; Oliazadeh, M. Selective copper dissolution during bioleaching of molybdenite concentrate. *Int. J. Miner. Process.* **2006**, *81*, 105–112.

217. Langhans, D.L.; Baglin, E.G. Biological oxidation of a platinum-group metal flotation concentrate and converter matte. In *Biohydrometallurgical Technologies*; Torma, A.E., Wey, J.E., Lakshmanan, V.L., Eds.; TMS: Warrendale, PA, USA, 1993; pp. 315–325.

218. Romano, P.; Blázquez, M.L.; Alguacil, F.J.; Muñoz, J.A.; Ballester, A.; González, F. Comparative study on the selective chalcopyrite bioleaching of a molybdenite concentrate with mesophilic and thermophilic bacteria. *FEMS Microbiol. Lett.* **2001**, *196*, 71–75.

219. Dymov, I.; Ferron, C.J.; Philips, W. The development of a hybrid biological leaching—Pressure oxidation process for auriferous arsenopyrite/pyrite feedstocks. In *Biohydrometallurgy: A Sustainable Technology in Evolution*; Tsezos, M., Hatzikioseyian, A., Remoundaki, E., Eds.; National Technical University of Athens: Athens, Greece, 2004; pp. 377–386.

220. Tipre, D.R.; Vora, S.B.; Dave, S.R. Comparison of air-lift and stirred tank batch and semicontinuous bioleaching of polymetallic bulk concentrate. In *Biohydrometallurgy: A Sustainable Technology in Evolution*; Tsezos, M., Hatzikioseyian, A., Remoundaki, E., Eds.; National Technical University of Athens: Athens, Greece, 2004; pp. 211–218.

221. Tipre, D.R.; Dave, S.R. Bioleaching process for Cu-Pb-Zn bulk concentrate at high pulp density. *Hydrometallurgy* **2004**, *75*, 37–43.

222. Okibe, N.; Gericke, M.; Hallberg, K.B.; Johnson, D.B. Enumeration and characterization of acidophilic microorganisms isolated from a pilot plant stirred-tank bioleaching operation. *Appl. Environ. Microbiol.* **2003**, *69*, 1936–1943.

223. Pivovarova, T.A.; Melamud, V.S.; Savari, E.E.; Sedel'nikova, G.V.; Kondrat'eva, T.F. Species and strain composition of microbial associations oxidizing different types of gold-bearing concentrates. *Appl. Microbiochem. Microbiol.* **2012**, *46*, 497–504.

224. D'Hugues, P.; Joulian, C.; Spolaore, P.; Michel, C.; Garrido, F.; Morin, D. Continuous bioleaching of a pyrite concentrate in stirred reactors: Population dynamics and exopolysaccharides production *vs.* Bioleaching performances. *Hydrometallurgy* **2008**, *94*, 34–41.

225. Hao, C.; Wang, L.; Dong, H.; Zhang, H. Succession of acidophilic bacterial community during bio-oxidation of refractory gold-containing sulfides. *Geomicrobiol. J.* **2010**, *27*, 683–691.

226. Wang, J.; Zhao, H.; Zhuang, T.; Qin, W.; Zhu, S.; Qiu, G. Bioleaching of Pb-Zn-Sn chalcopyrite concentrate in tank bioreactor and microbial community succession analysis. *Trans. Nonferr. Met. Soc. China* **2013**, *23*, 3758–3762.

227. Attia, Y.A.; El-Zeky, M. Effects of galvanic interactions on sulfides on extraction of precious metals from refractory complex sulfides by bioleaching. *Int. J. Miner. Process.* **1990**, *30*, 99–111.

228. D'Hugues, P.; Norris, P.R.; Hallberg, K.B.; Sanchez, F.; Langwaldt, J.; Grotowski, A.; Chmielewski, T.; Groudev, S.; Bioshale Consortium. Bioshale FP6 European project: Exploiting black shale ores using biotechnologies? *Miner. Eng.* **2008**, *21*, 111–120.

229. Langwaldt, J. Bioleaching of multimetal black shale by thermophilic micro-organisms. In *Biohydrometallurgy: From Single Cell to the Environment*; Schippers, A., Sand, W., Glombitza, F., Willscher, S., Eds.; TransTech Publications: Zurich, Switzerland, 2007; p. 167.

230. Spolaore, P.; Joulian, C.; Gouin, J.; Sanchez, F.; Auge, T.; Morin, D.; d'Hugues, P. Continuous bioleaching of a polymetallic black shale concentrate using the stirred tank technology. In Proceedings of the XXIV International Mineral Processing Congress, Beijing, China, 24–28 September 2008; Wang, D., Sun, C., Wang, F., Zhan, L., Han, L., Eds.; Science Press: Beijing, China, 2008; pp. 2576–2584.

231. Spolaore, P.; Joulian, C.; Ménard, Y.; d'Hugues, P. Non-traditional operating conditions for a copper concentrate continuous bioleaching. In *BioHydromet 10*; MEI: Falmouth, UK, 2010; pp. 156–174.

232. Beane, R.; Ramey, D. *In-situ* leaching at the San Manuel porphyry copper deposit, Arizona, USA. In *Copper 95—Cobre 95*; Cooper, W.C., Dreisinger, D.B., Dutrizac, J.E., Hein, H., Ugarte, G., Eds.; MetSoc CIM: Montreal, QC, Canada, 1995; Volume III, pp. 363–375.

233. Rossi, G. Potassium recovery through leucite bioleaching: Possibilities and limitations. In *Metallurgical Applications of Bacterial Leaching and Related Microbiological Phenomena*; Murr, L.E., Torma, A.E., Brierley, J.A., Eds.; Academic Press: New York, NY, USA, 1978; pp. 297–319.

234. Gallant, A.; Wadden, D.G. The in-place leaching or uranium at Denison Mines. *Can. Metall. Q.* **1984**, *24*, 127–134.

235. Groudeva, V.I.; Groudev, S.N. Combined bacterial and chemical leaching of a polymetallic sulfide ore. In *Mineral Bioprocessing*; Smith, R.W., Misra, M., Eds.; TMS: Warrendale, PA, USA, 1991; pp. 153–161.

236. Korehi, H.; Schippers, A. Bioleaching of a marine hydrothermal sulfide ore with mesophiles, moderate thermophiles and thermophiles. In *Integrating Scientific and Industrial Knowledge on Biohydrometallurgy*; Guiliani, N., Demergasso, C., Quatrini, R., Remonsellez, F., Davis-Belmar, C., Levican, G., Parada, P., Barahona, C., Zale, R., Eds.; TransTech Publications: Zurich, Switzerland, 2013; pp. 229–232.

237. Hunter, C. BioHeap™ leaching of a primary nickel-copper sulphide ore. In *Nickel-Cobalt-8 Technical Proceedings, Perth, Australia*; ALTA Metallurgical Services: Melbourne, Australia, 2002.

238. Riekkola-Vanhanen, M. Talvivaara Sotkamo mine—Bioleaching of a polymetallic nickel ore in subarctic climate. *Nova Biotechnol.* **2010**, *10*, 7–14.

239. Wen, J.K.; Ruan, R.; Guo, X.J. Heap leaching—An option of treating nickel sulfide ore and laterite. In Proceedings of the Nickel/Cobalt Conference, Perth, Australia, 15–17 May 2006; ALTA Metallurgical Services: Melbourne, Australia, 2006.

240. Norton, A.E.; Coetzee, J.J.; Barnett, C.C. BioNIC®: An economically competitive process for the biological extraction of nickel. In Proceedings of the Nickel/Cobalt Pressure Leaching and Hydrometallurgy Forum, Perth, Australia, 25–27 May 1998; Metallurgical Services: Melbourne, Australia, 1998.

241. Watling, H.R.; Elliot, A.D.; Maley, M.; van Bronswijk, W.; Hunter, C. Leaching of a low-grade, copper-nickel sulfide ore. 1. Key parameters impacting on Cu recovery during column bioleaching. *Hydrometallurgy* **2009**, *97*, 204–212.

242. Maley, M.; van Bronswijk, W.; Watling, H.R. Leaching of a low-grade, copper-nickel sulfide ore. 2. Impact of aeration and pH on Cu recovery during abiotic leaching. *Hydrometallurgy* **2006**, *98*, 66–72.

243. Maley, M.; van Bronswijk, W.; Watling, H.R. Leaching of a low-grade, copper-nickel sulfide ore. 3. Interactions of Cu with selected sulfide minerals. *Hydrometallurgy* **2009**, *98*, 73–80.

244. Elliot, A.D.; Watling, H.R. Chalcopyrite formation through the metathesis of pyrrhotite with aqueous copper. *Geochim.Cosmochim. Acta* **2011**, *75*, 2103–2118.

245. Cameron, R.A.; Lastra, R.; Mortazavi, S.; Bedard, P.L.; Morin, L.; Gould, W.D.; Kennedy, K.J. Bioleaching of a low-grade ultramafic nickel sulphide ore in stirred-tank reactors at elevated pH. *Hydrometallurgy* **2009**, *97*, 213–220.

246. Cameron, R.A.; Lastra, R.; Mortazavi, S.; Gould, W.D.; Thibault, Y.; Bédard, P.L.; Morin, L.; Kennedy, K.J. Elevated-pH bioleaching of a low-grade ultramafic nickel sulphide ore in stirred-tank reactors at 5 to 45 °C. *Hydrometallurgy* **2009**, *99*, 77–83.

247. Cameron, R.A.; Yeung, C.W.; Greer, C.W.; Gould, W.D.; Mortazavi, S.; Bédard, PL.; Morin, L.; Lortie, L.; Dinardo, O.; Kennedy, K.J.; *et al*. The bacterial structure during bioleaching of a low-grade nickel sulphide ore in stirred tank reactors at different combinations of temperature and pH. *Hydrometallurgy* **2010**, *104*, 207–215.

248. Cameron, R.A.; Lastra, R.; Gould, W.D.; Mortazavi, S.; Thibault, Y.; Bédard, P.L.; Morin, L.; Koren, D.W.; Kennedy, K.J. Bioleaching of six nickel sulphide ores with differing mineralogies in stirred tank reactors at 30 °C. *Miner. Eng.* **2013**, *49*, 172–183.

249. Qin, W.; Zhen, S.; Yan, Z.; Campbell, M.; Wang, J.; Liu, K.; Zhang, Y. Heap bioleaching of a low-grade nickel-bearing sulfide ore containing high levels of magnesium as olivine, chlorite and antigorite. *Hydrometallurgy* **2009**, *98*, 58–65.

250. Miller, P.C. Large scale bacterial leaching of a copper-zinc ore *in situ*. In *Fundamental and Applied Biohydrometallurgy*; Lawrence, R.W., Branion, R.M.R., Ebner, H.G., Eds.; Elsevier: Amsterdam, The Netherlands, 1986; pp. 215–239.

251. Oros, V.; Peterfi, M.; Bivolaru, M.; Kovacs, S.; Straut, I.; Jelea, M.; Hudrea, I. Production of copper and zinc by microbial *in situ* stope leaching at Ilba mine, (Romania). In Proceedings of the 9th International Symposium on Biohydrometallurgy, Troia, Portugal, 9–13 September 1991; Duarte, J.C., Lawrence, R.W., Eds.; Forbitec: Queluz, Portugal, 1991.

252. Sand, W.; Hallmann, R.; Rohde, K.; Sobotke, B.; Wentzien, S. Controlled microbiological *in-situ* stope leaching of a sulphidic ore. *Appl. Microbiol. Biotechnol.* **1993**, *40*, 421–426.

253. Krafft, C.; Hallberg, R.O. Bacterial leaching of two Swedish zinc sulfide ores. *FEMS Microbiol. Rev.* **1993**, *11*, 121–128.

254. Brauckmann, B.; Poppe, W.; Beyer, W.; Lerche, R.; Steppke, H.D. Investigations of increased biological *in-situ* leaching of the "Old Deposit" of the Preussag Rammelsberg ore mine. In *Biohydrometallurgy*; Norris, P.R., Kelly, D.P., Eds.; Science and Technology Letters: Kew, UK, 1988; pp. 521–523.

255. Burton, C.; Cowman, S.; Heffernan, J.; Thorne, B. *In-situ* bioleaching of sulphide ores at Avoca, Ireland: Part I. Development, characterization and operation of a medium-scale (6000 t) experimental leach site. In *Recent Progress in Biohydrometallurgy*; Rossi, G., Torma, A.E., Eds.; Associazone Mineraria Sarda: Iglesias, Italy, 1983; pp. 243–264.

256. Zhen, S.; Qin, W.; Yan, Z.; Zhang, Y.; Wang, J.; Ren, L. Bioleaching of low-grade nickel sulfide mineral in column reactor. *Trans. Nonferr. Met. Soc. China* **2008**, *18*, 1480–1484.

257. Giaveno, A.; Lavalle, L.; Chiacchiarini, P.; Donati, E. Bioleaching of zinc from low-grade complex sulfide ores in an airlift by isolated *Leptospirillum ferrooxidans*. *Hydrometallurgy* **2007**, *89*, 117–126.

258. Chen, J.W.; Gao, C.J.; Zhang, Q.X.; Xiao, L.S.; Zhang, G.Q. Leaching of nickel-molybdenum sulfide ore in membrane biological reactor. *Trans. Nonferr. Met. Soc. China* **2011**, *21*, 1395–1401.

259. Lizama, H.M.; Oh, J.K.; Takeuchi, T.L.; Suzuki, I. Bacterial leaching of copper and zinc from a sulfide ore by a mixed culture of *Thiobacillus ferrooxidans* and *Thiobacillus thiooxidans* in laboratory scale and pilot plant scale columns. In *Biohydrometallurgy*; Salley, J., McCready, R.G.L., Wichlacz, P.L., Eds.; Canmet: Ottawa, ON, Canada, 1989; pp. 519–531.

260. Groudev, S.N. Complex utilization of polymetallic sulphide ores by means of combined bacterial and chemical leaching. In *Harnessing Biotechnology for the 21st Century*; Ladsich, M.R., Bose, A., Eds.; American Chemical Society: Washington, DC, USA, 1992; pp. 454–457.

261. Sandström, Å.; Petersson, S. Bioleaching of a complex sulphide ore with moderate thermophilic and extreme thermophilic microorganisms. *Hydrometallurgy* **1997**, *46*, 181–190.

262. Liao, M.X.; Deng, T.L. Zinc and lead from complex sulfides by sequential bioleaching and acidic brine leach. *Miner. Eng.* **2004**, *17*, 17–22.

263. Radio Hill and Sholl Heap Leaching Project. Available online: www.foxresources.com.au/radio-hill-sholl-heap.asp (accessed on 22 October 2014).

264. Lippmaa, E.; Maremae, E.; Pihlak, A.T. Resources, production and processing of Baltoscandian multimetal black shales. *Oil Shale* **2011**, *28*, 68–77.

265. Ketris, M.P.; Yudovich, Y.E. Estimations of Clarkes for carbonaceous biolithes: World averages for trace element contents in black shales and coals. *Int. J. Coal Geol.* **2009**, *78*, 135–148.

266. Watling, H.R.; Collinson, D.M.; Shiers, D.W.; Bryan, C.G.; Watkin, E.L.J. Effects of temperature and solids loading on microbial community structure during batch culture on a polymetallic ore. *Miner. Eng.* **2013**, *48*, 68–76.

267. Watling, H.R.; Collinson, D.M.; Fjastad, S.; Kaksonen, A.H.; Li, J.; Morris, C.; Perrot, F.A.; Rea, S.M.; Shiers, D.W. Column bioleaching of a polymetallic ore: Effects of pH and temperature on metal extraction and microbial community structure. *Miner. Eng.* **2014**, *58*, 90–99.

268. Choi, M.S.; Cho, K.S.; Kim, D.S.; Ryu, H.W. Bioleaching of uranium from low-grade schists by *Acidithiobacillus ferrooxidans*. *World J. Microbiol. Biotechnol.* **2005**, *21*, 377–380.

269. Anjum, F.; Bhatti, H.N.; Ambreen, A. Bioleaching of black shale by *Acidithiobacillus ferrooxidans*. *Asian J. Chem.* **2009**, *21*, 5251–5266.

270. Pal, S.; Pradhan, D.; Das, T.; Sukla, L.B.; Roy Chaudhury, G. Bioleaching of low-grade uranium ore using *Acidithiobacillus ferrooxidans*. *Indian J. Microbiol.* **2010**, *50*, 70–75.

271. Riekkola Vanhanen, M.; Heimala, S. Study of the bioleaching of a nickel-containing black-schist ore. In *Biohydrometallurgy and the Environment toward the Mining of the 21st Century*; Amils, R., Ballester, A., Eds.; Elsevier: Amsterdam, The Netherlands, 1999; pp. 533–542.

272. Technical Report on the Polymetallic Black Shale SBH Property, Birch Mountains, Athabasca Region, Alberta, Canada. Available online: www.dnimetals.com/pdf/TechRpt_SBH-pty-AB_Dumont-2008.pdf (accessed on 22 October 2014).

273. Anjum, F.; Shahid, M.; Akcil, A. Biohydrometallurgy techniques of low grade ores: A review on black shale. *Hydrometallurgy* **2012**, *117–118*, 1–12.

274. Aziz, A.; Sajjad, M.; Mohammad, B. Elemental characterization of black shales of Khyber Pakthunkhawa (KPK) region of Pakistan using AAS. *Chin. J. Geochem.* **2013**, *32*, 248–251.

275. Developing High Margin Uranium Projects. Available online: www.auraenergy.com.au/assets/ Aura_Roadshow_Presentation_January_2014.pdf (accessed on 22 October 2014).

276. Jowitt, S.M.; Keays, R.R. Shale-hosted Ni–(Cu–PGE) mineralisation: A global overview. *Trans. Inst. Min. Metall. Appl. Earth Sci.* **2011**, *120*, 187–197.

277. Major Base Metal Districts Favourable for Future Bioleaching Technologies. Final Report to the Bioshale Project WP6. Available online: http://infoterre.brgm.fr/rapports/RP-55610-FR.pdf (accessed 22 October 2014).

278. Pašava, J.; Hladiková, J.; Dobeš, P. Origin of Proterozoic metal-rich black shales from the Bohemian Massif, Czech Republic. *Econ. Geol.* **1996**, *91*, 63–79.

279. Pašava, J.; Zaccharini, F.; Aiglsperger, T.; Vymazalová, A. Platinum group elements (PGE) and their principal carriers in metal-rich black shales: An overview with new data from Mo-Ni-PGE black shales (Zunyi region, Guizhou Province, south China). *J. Geosci.* **2013**, *58*, 209–216.

280. Norris, P.R.; Brown, C.F.; Caldwell, P.E. Ore column leaching with thermophiles: II, polymetallic sulfide ore. *Hydrometallurgy* **2012**, *127–128*, 70–76.

281. Norris, P.R.; Gould, O.; Ogden, T. Anaerobic microbial growth to enhance iron removal from a polymetallic sulfide ore from 30 to 75 °C. In Proceedings of Biohydrometallurgy 2014, Falmouth, UK, 9–11 June 2014.

282. Kalinowski, B.E.; Oskarsson, A.; Albinsson, Y.; Arlinger, J.; Ödegaard-Jensen, A.; Andlid, T.; Pederson, K. Microbial leaching of uranium and other trace elements from shale mine tailings at Ranstad. *Geoderma* **2004**, *122*, 177–194.

283. Mishra, A.; Pradhan, N.; Kar, R.N.; Sukla, L.B.; Mishra, B.K. Microbial recovery of uranium using native fungal strains. *Hydrometallurgy* **2009**, *95*, 175–177.

284. Abd El Wahab, G.M.; Amin, M.M.; Aita, S.K. Bioleaching of uranium-bearing material from Abu Thor area, West Central Sinai, Egypt, for recovering uranium. *Arab J. Nucl. Sci. Appl.* **2012**, *45*, 169–178.

285. Anjum, F.; Bhatti, H.N.; Asgher, M.; Shahid, M. Leaching of metal ions from black shale by organic acids produced by *Aspergillus niger*. *Appl. Clay Sci.* **2010**, *47*, 356–361.

286. Bhatti, H.N.; Sarwar, S.; Ilyas, S. Effect of organic acids produced by *Penicillium notatum* on the extraction of metals ions from brown shale. *J. Chem. Soc. Pak.* **2012**, *34*, 1040–1047.

287. Nouren, S.; Bhatti, H.N.; Ilyas, S. Bioleaching of copper, aluminium, magnesium and manganese from brown shale by *Ganoderma lucidum*. *Afr. J. Biotechnol.* **2011**, *10*, 10664–10673.

288. Anjum, F.; Bhatti, H.N.; Ghauri, M.A.; Bhatti, I.A.; Asgher, M.; Asi, M.R. Bioleaching of copper, cobalt and zinc from black shale by *Penicillium notatum*. *Afr. J. Biotechnol.* **2009**, *8*, 5038–5045.

289. Anjum, F.; Bhatti, H.N.; Ghauri, M.A. Enhanced bioleaching of metals from black shale using ultrasonics. *Hydrometallurgy* **2010**, *100*, 122–128.

290. Sjoberg, V.; Grandin, A.; Karlsson, L.; Karlsson, S. Bioleaching of shale—Impact of carbon source. In *The New Uranium Mining Boom*; Merkel, B., Schipek, M., Eds.; Springer-Verlag: Berlin, Germany, 2011; pp. 449–454.

291. Hsu, K.J.; Sun, S.; Li, J.L.; Chen, H.H.; Pen, H.P.; Sengor, A.M.C. Mesozoic overthrust tectonics in south China. *Geology* **1988**, *16*, 418–421.

292. Morin, D.H.R.; Pinches, T.; Huisman, J.; Frias, C.; Norberg, A.; Forssberg, E. Progress after three years of BioMinE-Research and Technological Development project for a global assessment of biohydrometallurgical processes applied to European non-ferrous metal resources. *Hydrometallurgy* **2008**, *94*, 58–68.

293. Loukola-Ruskeeniemi, K. Geochemistry and genesis of the black-shale hosted Ni-Cu-Zn deposit at Talvivaara, Finland. *Econ. Geol.* **1996**, *91*, 80–110.

294. Puhakka, J.A.; Kaksonen, A.H.; Riekkola-Vanhanen, M. Heap leaching of black schist. In *Biomining*; Rawlings, D.E., Johnson, D.B., Eds.; Springer: Berlin, Germany, 2007; pp. 139–151.

295. Salo-Zieman, V.L.A.; Kinnunen, P.H.M.; Puhakka, J.A. Bioleaching of acid-consuming low-grade nickel ore with elemental sulfur addition and subsequent acid generation. *J. Chem. Technol. Biotechnol.* **2006**, *81*, 34–40.

296. Halinen, A.K.; Rahunen, N.; Kaksonen, A.H.; Puhakka, J.A. Heap bioleaching of a complex sulfide ore. Part I: Effect of pH on metal extraction and microbial composition in pH controlled columns. *Hydrometallurgy* **2009**, *98*, 92–100.

297. Halinen, A.K.; Rahunen, N.; Kaksonen, A.H.; Puhakka, J.A. Heap bioleaching of a complex sulfide ore. Part II: Effect of temperature on base metal extraction and bacterial compositions. *Hydrometallurgy* **2009**, *98*, 101–107.

298. Bhatti, T.M.; Bigham, J.M.; Riekkola-Vanhanen, M.; Tuovinen, O.H. Altered mineralogy associated with stirred tank bioreactor leaching of a black schist ore. *Hydrometallurgy* **2010**, *100*, 181–184.

299. Bhatti, T.M.; Bigham, J.M.; Vuorinen, A.; Tuovinen, O.H. Chemical and bacterial leaching of metals from black schist sulfide minerals in shake flasks. *Int. J. Miner. Process.* **2012**, *110–111*, 25–29.

300. Bhatti, T.M.; Vuorinen, A.; Tuovinen, O.H. Dissolution of non-sulfide phases during the chemical and bacterial leaching of a sulfidic black schist. *Hydrometallurgy* **2012**, *117–118*, 32–35.

301. Wakeman, K.; Auvinen, H.; Johnson, D.B. Microbiological and geochemical dynamics in simulated-heap leaching of a polymetallic sulfide ore. *Biotech. Bioeng.* **2008**, *101*, 739–750.

302. Halinen, A.K.; Beecroft, N.J.; Määttä, K.; Nurmi, P.; Laukkanen, K.; Kaksonen, A.H.; Riekkola-Vanhanen, M.; Puhakka, J.A. Microbial community dynamics during a demonstration-scale bioheap leaching operation. *Hydrometallurgy* **2013**, *125–126*, 34–41.

303. Updated Resource Estimate and Preliminary Economic Assessment Estimate Viken Project NPV at US $1 Billion. Available online: www.CZQminerals.com/news-releases/page/2/ (accessed on 22 October 2014).

304. Investment Presentation: High Impact Exploration in a World-Class Producing Copper-Nickel Region. Available online: www.finnaustmining.com/ (accessed on 22 October 2014).

305. Alberta Black Shale Metals Projects. Available online: www.dnimetals.com/properties/black_shales.htm (accessed on 22 October 2014).

306. Vaughan, D.J.; Sweeney, M.; Friedrich, G.; Diedel, R.; Haranczyk, C. The Kupferschiefer: An overview with an appraisal of the different types of mineralization. *Econ. Geol.* **1989**, *84*, 1003–1027.

307. Sadowski, Z.; Szubert, A. Comparison of kinetics of black shale bioleaching process using stationary and agitated systems. *Physicochem. Probl. Miner. Process.* **2007**, *41*, 387–395.

308. Grobelski, T.; Farbiszewska-Kiczma, J.; Farbiszewska, T. Bioleaching of Polish black shale. *Physicochem. Probl. Miner. Process.* **2007**, *41*, 259–264.

309. Groudev, S.N.; Spasova, I.I.; Nicolova, M.V.; Georgiev, P.S. Bacterial leaching of black shale copper ore. *Adv. Mater. Res.* **2007**, *50*, 187–190.

310. Farbiszewska, T.; Farbiszewska-Kiczma, J.; Bak, M. Biological extraction of metals from a Polish black shale. *Physicochem. Probl. Miner. Process.* **2003**, *37*, 51–56.

311. BioMOre: An Alternative Mining Concept—Raw Materials Commitment. Available online: https://ec.europa.eu/eip/raw-materials/en/content/biomore-alternative-mining-concept-raw-materials-commitment (accessed on 22 October 2014).

312. Taylor, A. Laterites—Still a Frontier of Nickel Process Development. Paper Presented at TMS2013, San Antonio, Texas, USA, 3–7 March 2013. Available online: http://www.altamet.com.au/wp-content/uploads/2013/04/laterites-still-a-frontier-of-nickel-process-development1.pdf (accessed on 22 October 2014).

313. Zhang, W.; Cheng, C.Y. Manganese metallurgy review. Part 1: Leaching of ores/secondary materials and recovery of electrolytic/chemical manganese dioxide. *Hydrometallurgy* **2007**, *89*, 137–159.

314. Das, A.P.; Sukla, L.B.; Pradhan, N.; Nayak, S. Manganese biomining: A review. *Bioresour. Technol.* **2011**, *102*, 7381–7387.

315. Bacon, W.G.; Dalvi, A.D.; Rochon, B.A.; Selby, M. Nickel outlook—2000–2010. *Cim Bull.* **2002**, *95*, 47–52.

316. Tzeferis, P.G.; Agatzini-Leonardou, S. Leaching of nickel and iron from Greek non-sulphide nickeliferous ores by organic acids. *Hydrometallurgy* **1994**, *36*, 345–360.

317. Tzeferis, P.G.; Agatzini, S.; Nerantzis, E.T. Mineral leaching of non-sulphide nickel ores using heterotrophic microorganisms. *Lett. Appl. Microbiol.* **1994**, *18*, 209–213.

318. Agatzini, S.; Tzeferis, P. Bioleaching of nickel-cobalt oxide ores. *AusIMM Proc.* **1997**, *1*, 9–15.

319. Sukla, L.B.; Panchanadikar, V.V.; Kar, R.N. Microbial leaching of lateritic nickel ore. *World J. Microbiol. Biotechnol.* **1993**, *9*, 255–257.

320. Sukla, L.B.; Panchanadikar, V.V. Bioleaching of lateritic nickel ore using an indigenous microflora. In *Biohydrometallurgical Technologies*; Torma, A.E., Wey, J.E., Lakshmanan, V.L., Eds.; TMS: Warrendale, PA, USA, 1993; Volume 1, pp. 373–380.

321. Sukla, L.B.; Panchanadikar, V. Bioleaching of lateritic nickel ore using a heterotrophic micro-organism. *Hydrometallurgy* **1993**, *32*, 373–379.

322. Sukla, L.B.; Swamy, K.M.; Narayana, K.L.; Kar, R.N.; Panchanadikar, V.V. Bioleaching of Sukinda laterite using ultrasonics. *Hydrometallurgy* **1995**, *37*, 387–391.

323. Sukla, L.B.; Kar, R.N.; Panchanadikar, V.V.; Choudhury, S.; Mishra, R.K. Bioleaching of lateritic nickel ore using *Penicillium* sp. *Trans. Indian Inst. Met.* **1995**, *48*, 103–106.

324. Valix, M.; Tang, J.Y.; Cheung, W.H. The effects of mineralogy on the biological leaching of nickel laterite ores. *Miner. Eng.* **2001**, *14*, 1629–1635.

325. Valix, M.; Tang, J.Y.; Malik, R. Heavy metal tolerance of fungi. *Miner. Eng.* **2001**, *14*, 499–505.

326. Valix, M.; Loon, L.O. Adaptive tolerance behaviour of fungi in heavy metals. *Miner. Eng.* **2003**, *16*, 193–198.

327. Valix, M.; Thangavalu, V.; Ryan, D.; Tang, J. Using halotolerant *Aspergillus foetidus* in bioleaching nickel laterite ore. *Int. J. Environ. Waste Manag.* **2009**, *3*, 253–264.

328. Tang, J.A.; Valix, M. Leaching of low grade limonite and nontronite ores by fungi metabolic acids. *Miner. Eng.* **2006**, *19*, 1274–1279.

329. Tang, J.; Valix, M. Leaching kinetics of limonite and nontronite ores. *Int. J. Environ. Waste Manag.* **2009**, *3*, 244–252.

330. Thangavalu, V.; Tang, J.; Ryan, D.; Valix, M. Effect of saline stress on fungi metabolism and biological leaching of weathered saprolite ores. *Miner. Eng.* **2006**, *19*, 1266–1273.

331. Simate, G.S.; Ndlovu, S. Characterisation of factors in the bacterial leaching of nickel laterites using statistical design of experiments. In *Biohydrometallurgy: From Single Cell to the Environment*; Schippers, A., Sand, W., Glombitza, F., Willscher, S., Eds.; TransTech Publications: Zurich, Switzerland, 2007; pp. 66–69.

332. Simate, G.S.; Ndlovu, S.; Gericke, M. Bacterial leaching of nickel laterites using chemolithotrophic microorganisms: Process optimisation using response surface methodology and central composite rotatable design. *Hydrometallurgy* **2009**, *98*, 241–246.

333. Watling, H.R.; Elliot, A.D.; Fletcher, H.M.; Robinson, D.J.; Sully, D.M. Ore mineralogy of nickel laterites: Controls on processing characteristics under simulated heap-leach conditions. *Aust. J. Earth Sci.* **2011**, *58*, 725–744.

334. Imai, K. On the mechanism of bacterial leaching. In *Metallurgical Applications of Bacterial Leaching and Related Phenomena*; Murr, L.E., Torma, A.E., Brierley, J.A., Eds.; Academic Press: New York, NY, USA, 1978; pp. 275–295.

335. Nakazawa, H.; Sato, H. Bacterial leaching of cobalt-rich ferromanganese crusts. *Int. J. Miner. Process.* **1995**, *43*, 255–265.

336. Belyi, A.V.; Pustochilov, P.P.; Gurevich, Y.L.; Kadochnikova, G.G.; Ladygina, V.P. Bacterial leaching of manganese ores. *Appl. Biochem. Microbiol.* **2006**, *42*, 289–292.

337. Kai, T.; Taniguchi, S.; Ikeda, S.I.; Takahashi, T. Effect of *Thiobacillus ferrooxidans* during redox leaching of manganese nodules and nickel sulfide. *Resour. Environ. Biotechnol.* **1996**, *1*, 99–1009.

338. Konishi, Y.; Asai, S. Leaching of marine manganese nodules by acidophilic bacteria growing on elemental sulfur. *Metall. Mater. Trans. B* **1997**, *28*, 25–32.

339. Ndlovu, S.; Simate, G.S.; Gericke, M. The microbial assisted leaching of nickel laterites using a mixed culture of chemolithotrophic microorganisms. In *Biohydrometallurgy: A Meeting Point between Microbial Ecology, Metal Recovery Processes and Environmental Remediation*; Donati, E.R., Viera, M.R., Tavani, E.L., Giaveno, M.A., Lavalle, T.L., Chiaccharini, P.A., Eds.; TransTech Publications: Zurich, Switzerland, 2009; pp. 493–496.

340. Kumari, A.; Natarajan, K.A. Bioleaching of ocean manganese nodules in the presence of reducing agents. *Eur. J. Miner. Process. Environ. Prot.* **2001**, *1*, 10–24.

341. Jain, N.; Sharma, D.K. Biohydrometallurgy for nonsulfidic minerals—A review. *Geomicrobiol. J.* **2004**, *21*, 135–144.

342. Dave, S.R.; Natarajan, K.A.; Bhat, J.V. Leaching of copper and zinc from oxidised ores by fungi. *Hydrometallurgy* **1981**, *7*, 235–242.

343. Groudev, S.N.; Groudeva, V.I. Biological leaching of aluminium from clays. In *Biotechnology for the Mining, Metal-Refining, and Fossil Fuel Processing Industries*; Ehrlich, H.L., Holmes, D.S., Eds.; John Wiley and Sons: New York, NY, USA, 1986; pp. 91–96.

344. Mohapatra, S.; Bohidar, S.; Pradhan, N.; Kar, R.N.; Sukla, L.B. Microbial extraction of nickel from Sukinda chromite overburden by *Acidithiobacillus ferrooxidans* and *Aspergillus* strains. *Hydrometallurgy* **2007**, *85*, 1–8.

345. Das, C.; Mehta, K.D.; Pandey, B.D. Biochemical leaching of metals from Indian Ocean nodules by *Aspergillus niger*. *Miner. Process. Technol.* **2005**, *8*, 460–467.

346. McKenzie, D.I.; Denys, L.; Buchanan, A. The solubilisation of nickel, cobalt and iron from laterites by means of organic chelating agents at low pH. *Int. J. Miner. Process.* **1987**, *21*, 275–292.

347. Das, A.P.; Swain, S.; Panda, S.; Pradhan, N.; Sukla, L.B. Reductive acid leaching of low-grade manganese ores. *Geomaterials* **2012**, *2*, 70–72.

348. Burgstaller, W.; Schinner, F. Leaching of metals with fungi. *J. Biotechnol.* **1993**, *27*, 91–116.

349. Le, L.; Tang, J.; Ryan, D.; Valiz, M. Bioleaching nickel laterite ores using multi-metal tolerant *Aspergillus foetidus* organism. *Miner. Eng.* **2006**, *19*, 1259–1265.

350. Veglio, F.; Beolchini, F.; Gasbarro, A.; Toro, L.; Ubaldini, S.; Abbruzzese, C. Batch and semi-continuous tests in the bioleaching of manganiferous minerals by heterotrophic mixed microorganisms. *Int. J. Miner. Process.* **1997**, *50*, 255–273.

351. Mukherjee, A.; Raichur, A.M.; Modak, J.M.; Natarajan, K.A. A novel bio-leaching process to recover valuable metals from Indian Ocean nodules using a marine isolate. In *Biohydrometallurgy: A Sustainable Technology in Evolution*; Tsezos, M., Hatzikioseyian, A., Remoundaki, E., Eds.; National Technical University of Athens: Athens, Greece, 2004; pp. 25–33.

352. Lovley, D.R. Organic matter mineralization with the reduction of ferric iron: A review. *Geomicrobiol. J.* **1987**, *5*, 375–399.

353. Lovley, D.R. Microbial reduction of iron, manganese and other metals. *Adv. Agron.* **1995**, *54*, 175–231.

354. Lovley, D.R.; Coates, J.D. Novel forms of anaerobic respiration of environmental relevance. *Curr. Opin. Microbiol.* **2000**, *3*, 252–256.

355. Eisele, T.C.; Gabby, K.L. Review of reductive leaching of iron by anaerobic bacteria. *Miner. Process. Extr. Metall. Rev.* **2014**, *35*, 75–105.

356. Lee, E.Y.; Noh, S.R.; Cho, K.S. Leaching of Mn, Co and Ni from manganese nodules using an anaerobic bioleaching method. *J. Biosci. Bioeng.* **2001**, *92*, 354–359.

357. Agate, A.D. Recent advances in microbial mining. *World J. Microbiol. Biotechnol.* **1996**, *12*, 487–495.

358. Ehrlich, H.L. Ocean manganese nodules: Biogenesis and bioleaching possibilities. *Miner. Metall. Process.* **2000**, *17*, 121–128.

359. Urrutia, M.M.; Roden, E.E.; Fredrickson, J.K.; Zachara, J.M. Microbial and surface chemistry controls on reduction of synthetic Fe(III) oxide minerals by the dissimilatory iron-reducing bacterium *Shewanella alga*. *Geomicrobiol. J.* **1998**, *15*, 269–291.

360. Cummings, D.E.; Caccavo, F.; Fendorf, S.; Rosenzweig, R.F. Arsenic mobilization by the dissimilatory Fe(III) reducing bacterium *Shewanella alga* BrY. *Environ. Sci. Technol.* **1999**, *33*, 723–729.

361. Liu, C.; Gorby, Y.A.; Zachara, J.M.; Fredrickson, J.K.; Brown, C.F. Reduction kinetics of Fe(III), Co(III), U(VI) and Tc(VII) in cultures of dissimilatory metal-reducing bacteria. *Biotechnol. Bioeng.* **2002**, *80*, 637–649.

362. Konishi, Y.; Saitoh, N.; Ogi, T. A new biohydrometallurgical method for processing of deep-sea mineral resources. In Proceedings of the ASME 2009 28th International Conference on Ocean, Offshore and Arctic Engineering (OMAE2009), Honolulu, HI, USA, 31 May–5 June 2009; ASME: New York, NY, USA, 2009; Volume 4B, pp. 1319–1324.

363. Brand, N.W.; Butt, C.R.M.; Elias, M. Nickel laterites: Classification and features. *AGSO J. Aust. Geol. Geophys.* **1998**, *17*, 81–88.

364. Freyssinet, P.; Butt, C.R.M.; Morris, R.C.; Piantone, P. Ore-forming processes related to laterite weathering. In *Economic Geology 100th Anniversary Volume*; Hedenquist, J.W., Thomson, J.F.H., Goldfarb, R.J., Richards, J.P., Eds.; Economic Geology Publishing Company: New Haven, CT, USA, 2005; pp. 681–722.

365. Du Plessis, C.A.; Slabbert, W.; Hallberg, K.B.; Johnson, D.B. Ferredox: A biohydrometallurgical processing concept for limonitic nickel laterites. *Hydrometallurgy* **2011**, *109*, 221–229.

366. Dold, B. Sustainability in metal mining: From exploration, over processing to mine waste management. *Rev. Environ. Sci. Biotechnol.* **2008**, *7*, 275–285.

367. Mudd, G.M. Sustainability and Mine Waste Management—A Snapshot of Mining Waste Issues. Available online: http://users.monash.edu.au/~gmudd/files/2007-WasteMment-Sustainability-v-MineWastes.pdf (accessed on 22 October 2014).

368. Shaw, R.A.; Petravratzi, E.; Bloodworth, A.J. Resource recovery from mine waste. In *Waste as a Resource*; Hester, R.M., Harrison, R.M., Eds.; Royal Society of Chemistry: Cambridge, UK, 2013; pp. 44–65.

369. Olson, G.J.; Brierley, C.L.; Briggs, A.P.; Calmet, E. Biooxidation of thiocyanate-containing refractory gold tailings from Minacalpa, Peru. *Hydrometallurgy* **2006**, *81*, 159–166.

370. Thomas, J.; Subramanian, S.; Riyaz Ulla, M.S.; Louis, K.T.; Gundewar, C.S. Studies on the biodissolution of cobaltic pyrite from copper tailings. In Proceedings of the XXIII International Mineral Processing Congress, Istanbul, Turkey, 3–8 September 2006; Önal, G., Acarkan, N., Çelik, M.S., Arslan, F., Ateşok, G., Güney, A., Sirkeci, A.A., Yüce, A.E., Perek, K.T., Eds.; Promed Advertising Agency: Istanbul, Turkey, 2006; pp. 1329–1333.

371. Kitobo, W.; Ilunga, A.; Frenay, J.; Gaydardzhiev, S.; Basti, D. Bacterial leaching of complex sulphides from mine tailings altered by acid drainage. In *Hydrocopper 2009*; Casas, J., Domic, E., Eds.; Gecamin: Santiago, Chile, 2009; pp. 365–373.

372. Muñoz, A.; Bevilaqua, D.; Garcia, O. Leaching of Ni and Cu from mine wastes (tailings and slags) using acid solutions of A. ferrooxidans. In *Biohydrometallurgy: A Meeting Point between Microbial Ecology, Metal Recovery Processes and Environmental Remediation*; Donati, E.R., Viera, M.R., Tavani, E.L., Giaveno, M.A., Lavalle, T.L., Chiaccharini, P.A., Eds. TransTech Publications: Zurich, Switzerland, 2009; pp. 425–428.

373. Zhu, J. Bacterial leaching research of Cu-Ni flotation tailing. In *Biohydrometallurgy: Biotech Key to Unlock Mineral Resources Values*; Qiu, G., Jiang, T., Qin, W., Liu, X., Yang, Y., Wang, H., Eds.; Central South University Press: Changsha, China, 2011; pp. 821–823.

374. Kondrat'eva, T.F.; Pivovarova, T.A.; Bulaev, A.G.; Melamud, V.S.; Muravyov, M.I.; Usoltsev, A.V.; Vasil'ev, E.A. Percolation bioleaching of copper and zinc and gold recovery from flotation tailings of the sulfide complex ores of the Ural region, Russia. *Hydrometallurgy* **2012**, *111–112*, 82–86.

375. Bulaev, A.G.; Muravyov, M.I.; Pivovarova, T.A.; Fomochenko, N.V. Bioprocessing of mining and metallurgical wastes containing non-ferrous and precious metals. In *Integrating Scientific and Industrial Knowledge on Biohydrometallurgy*; Guiliani, N., Demergasso, C., Quatrini, R., Remonsellez, F., Davis-Belmar, C., Levican, G., Parada, P., Barahona, C., Zale, R., Eds.; TransTech Publications: Zurich, Switzerland, 2013; pp. 301–304.

376. Galvez-Cloutier, R.; Mulligan, C.N.; Ouattra, A. Biolixiviation of Cu, Ni, Pb and Zn using organic acids by *Aspergillus niger* and *Penicillium simplicissinum*. In *Biohydrometallurgy: A Sustainable Technology in Evolution*, Tsezos, M., Hatzikioseyian, A., Remoundaki, E., Eds.; National Technical University of Athens: Athens, Greece, 2004; pp. 175–184.

377. Bryan, C.G.; Hallberg, K.B.; Johnson, D.B. Mobilisation of metals in mineral tailings at the abandoned Sao Domingos copper mine (Portugal) by indigenous acidophilic bacteria. *Hydrometallurgy* **2006**, *83*, 184–194.

378. Willscher, S.; Pohle, C.; Sitte, J.; Werner, P. Solubilization of heavy metals from a fluvial AMD generating tailings sediment by heterotrophic microorganisms. Part I: Influence of pH and solid content. *J. Geochem. Explor.* **2007**, *92*, 177–185.

379. Tan, G.L.; Shu, W.S.; Hallberg, K.B.; Li, F.; Lan, C.Y.; Zhou, W.H.; Huang, L.N. Culturable and molecular phylogenetic diversity of microorganisms in an open-dumped, extremely acidic Pb/Zn mine tailings. *Extremophiles* **2008**, *12*, 657–664.

380. Tan, G.L.; Shu, W.S.; Hallberg, K.B.; Li, F.; Lan, C.Y.; Huang, L.N. Cultivation-dependent and cultivation-independent characterization of the microbial community in acid mine drainage associated with acidic Pb/Zn tailings at Lechang, Guandong, China. *FEMS Microbiol. Ecol.* **2007**, *59*, 118–126.

381. Kock, D.; Schippers, A. Quantitative microbial community analysis of three different sulfidic mine tailing dumps generating acid mine drainage. *Appl. Environ. Microbiol.* **2008**, *74*, 5211–5219.

382. Costa, M.C.; Carvalho, N.; Iglesias, N.; Palencia, I. Bacterial leaching studies of a Portuguese flotation tailing. In *Biohydrometallurgy: A Sustainable Technology in Evolution*; Tsezos, M., Hatzikioseyian, A., Remoundaki, E., Eds.; National Technical University of Athens: Athens, Greece, 2004; pp. 75–84.

383. Rivera-Santillan, R.E.; Becerril-Reyes, V. Comparison of the bioleaching effect of mesophilic (35 °C) and thermophilic (45 °C) bacteria on the Tizapa tailings. In *Biohydrometallurgy: From the Single Cell to the Environment*; Schippers, A., Sand, W., Glombitza, F., Willscher, S., Eds.; TransTech Publications: Zurich, Switzerland, 2007; pp. 34–37.

384. Nagy, A.A.; Gock, E.D.; Melcher, F.; Atmaca, T.; Hahn, L.; Schippers, A. Biooxidation and cyanidation for gold and silver recovery from acid mine drainage generating tailings (Ticapampa, Peru) In *Biohydrometallurgy: From Single Cell to the Environment*; Schippers, A., Sand, W., Glombitza, F., Willscher, S., Eds.; TransTech Publications: Zurich, Switzerland, 2007; pp. 91–94.

385. Schippers, A.; Nagy, A.A.; Kock, D.; Melcher, F.; Gock, E.D. The use of FISH and real time PCR to monitor the biooxidation and cyanidation for gold and silver recovery from mine tailings concentrate (Ticapampa, Peru). *Hydrometallurgy* **2008**, *94*, 77–81.

386. Tong, L.; Yang, H.; Liu, C. Bioleaching of Cu-Zn tailings. In *Biohydrometallurgy: Biotech Key to Unlock Mineral Resources Values*; Qiu, G., Jiang, T., Qin, W., Liu, X., Yang, Y., Wang, H., Eds.; Central South University Press: Changsha, China, 2011; pp. 793–797.

387. Liu, Y.G.; Zhou, M.; Zeng, G.M.; Li, X.; Xu, W.H.; Fan, T. Effect of solids concentration on removal of heavy metals from mine tailings via bioleaching. *J. Hazard. Mater.* **2007**, *141*, 202–208.

388. Liu, Y.G.; Zhou, M.; Zeng, G.M.; Wang, X.; Li, X.; Fan, T.; Xu, W.H. Bioleaching of heavy metals from mine tailings by indigenous sulfur-oxidizing bacteria: Effects of substrate concentration. *Bioresour. Technol.* **2008**, *99*, 4124–4129.

389. Nguyen, V.K.; Lee, J.U. Catalytic effect of activated charcoal on microbial extraction of arsenic and heavy metals from mine tailings. *Geosci. J.* **2014**, *18*, 355–363.

390. Mugire, F. Changing Lives: Chinese Investment in Uganda's Copper Mining Gives Hope to Thousands. Available online: http://english.people.com.cn/business/8568960.html (accessed on 22 October 2014).

391. Tollinsky, N. Bioleaching Plant Proposed for Cobalt. Available online: www.Sudburyminingsolutions.com/bioleaching-plant-proposed-for-Cobalt.html (accessed on 22 October 2014).

392. Miller, P. The use of bioleaching for cobalt/arsenic tailings remediation in Ontario Canada. In *ALTA Ni-Co 2009, Perth, Australia*; ALTA Metallurgical Services: Melbourne, Australia, 2009.

393. McEwan, K.N.; Ralph, D.E.; Savage, C.J. *In-Situ* Bio-Oxidation of Low-Grade Refractory Sulphide Minerals. World Patent 2002061155 A1, 30 January 2014.

394. McEwan, K.; Ralph, D. The i-BOTM process and related treatments for mine waste remediation. In *Brownfield Sites*; Brebbia, C.A., Almorza, D., Klapperich, H., Eds.; WIT Press: Southampton, UK, 2002; pp. 525–532.

395. Seh-Bardan, B.J.; Othman, R.; Ab Wahib, S.; Husin, A.; Sadegh-Zadeh, F. Column bioleaching of arsenic and heavy metals from gold mine tailings by *Aspergillus fumigatus*. *Clean Soil Air Water* **2012**, *40*, 607–614.

396. Hernández, I.; Galizia, F.; Coto, O.; Donati, E. Improvement in metal recovery from laterite tailings by bioleaching. In *Biohydrometallurgy: A Meeting Point between Microbial Ecology, Metal Recovery Processes and Environmental Remediation*; Donati, E.R., Viera, M.R., Tavani, E.L., Giaveno, M.A., Lavalle, T.L., Chiaccharini, P.A., Eds.; TransTech Publications: Zurich, Switzerland, 2009; pp. 489–492.

397. Coto, O.; Galizia, F.; Hernandez, I.; Marrero, J.; Donati, E. Cobalt and nickel recoveries from laterite tailings by organic and inorganic bio-acids. *Hydrometallurgy* **2008**, *94*, 18–22.

398. Coto, O.; Schippers, A. Influence of sulfur source on nickel and cobalt bioleaching from laterite tailings using *Acidithiobacillus thiooxidans*. In *Biohydrometallurgy: Biotech Key to Unlock Mineral Resources Values*; Qiu, G., Jiang, T., Qin, W., Liu, X., Yang, Y., Wang, H., Eds.; Central South University Press: Changsha, China, 2011; pp. 579–583.

399. Cabrera, G.; Gomez, J.M.; Hernandez, I.; Coto, O.; Cantero, D. Different strategies for recovering metals from CARON process residue. *J. Hazard. Mater.* **2011**, *189*, 836–842.

400. Hernández Diaz, I.; Galizia, F.; Coto Perez, O. Reduction of heavy-metal content in overburden material by bacterial action. In *Biohydrometallurgy: A Meeting Point between Microbial Ecology, Metal Recovery Processes and Environmental Remediation*; Donati, E.R., Viera, M.R., Tavani, E.L., Giaveno, M.A., Lavalle, T.L., Chiaccharini, P.A., Eds.; TransTech Publications: Zurich, Switzerland, 2009; pp. 653–656.

401. Mulligan, C.N.; Galvez-Cloutier, R. Bioremediation of metal contamination. *Environ. Monit. Assess.* **2003**, *84*, 45–60.

402. Mulligan, C.N.; Kamali, M.; Gibbs, B.F. Bioleaching of copper and other metals from low-grade oxidized mining ores by *Aspergillus niger*. *J. Chem. Technol. Biotechnol.* **2003**, *78*, 497–503.

403. Behera, S.K.; Panda, P.P.; Singh, S.; Pradhan, N.; Sukla, L.B.; Mishra, B.K. Study on reaction mechanism of bioleaching of nickel and cobalt from lateritic chromite overburdens. *Int. Biodeterior. Biodegrad.* **2011**, *65*, 1035–1042.

404. Behera, S.K.; Sukla, L.B. Microbial extraction of nickel from chromite overburdens in the presence of surfactant. *Trans. Nonferr. Met. Soc. China* **2012**, *22*, 2840–2845.

405. Behera, S.K.; Panda, S.K.; Pradhan, N.; Sukla, L.B.; Mishra, B.K. Extraction of nickel by microbial reduction of laterite chromite overburden of Sukinda, India. *Bioresour. Technol.* **2012**, *125*, 17–22.

406. Johnson, D.B. Recent developments in microbiological approaches for securing mine wastes and for recovering metals from mine waters. *Minerals* **2014**, *4*, 279–292.

4

Quantum-Mechanical Methods for Quantifying Incorporation of Contaminants in Proximal Minerals

Lindsay C. Shuller-Nickles [1,*], **Will M. Bender** [2], **Sarah M. Walker** [2] **and Udo Becker** [2]

[1] Department of Environmental Engineering and Earth Science, Clemson University, 342 Computer Court, Anderson, SC 29625-6510, USA

[2] Department of Earth and Environmental Sciences, University of Michigan, 1100 North University Avenue, Ann Arbor, MI 48109-1005, USA; E-Mails: benderwm@umich.edu (W.M.B.); smwalk@umich.edu (S.M.W.); ubecker@umich.edu (U.B.)

* Author to whom correspondence should be addressed; E-Mail: lshulle@clemson.edu;

Abstract: Incorporation reactions play an important role in dictating immobilization and release pathways for chemical species in low-temperature geologic environments. Quantum-mechanical investigations of incorporation seek to characterize the stability and geometry of incorporated structures, as well as the thermodynamics and kinetics of the reactions themselves. For a thermodynamic treatment of incorporation reactions, a source of the incorporated ion and a sink for the released ion is necessary. These sources/sinks in a real geochemical system can be solids, but more commonly, they are charged aqueous species. In this contribution, we review the current methods for *ab initio* calculations of incorporation reactions, many of which do not consider incorporation from aqueous species. We detail a recently-developed approach for the calculation of incorporation reactions and expand on the part that is modeling the interaction of periodic solids with aqueous source and sink phases and present new research using this approach. To model these interactions, a systematic series of calculations must be done to transform periodic solid source and sink phases to aqueous-phase clusters. Examples of this process are provided for three case studies: (1) neptunyl incorporation into studtite and boltwoodite: for the layered boltwoodite, the incorporation energies are smaller (more favorable) for reactions using environmentally relevant source and sink phases (*i.e.*, ΔE_{rxn}(oxides) > ΔE_{rxn}(silicates) > ΔE_{rxn}(aqueous)). Estimates of the solid-solution behavior of Np^{5+}/P^{5+}- and U^{6+}/Si^{4+}-boltwoodite and Np^{5+}/Ca^{2+}- and U^{6+}/K^{+}-boltwoodite solid solutions are used to predict the limit of Np-incorporation into boltwoodite (172 and 768 ppm at 300 °C, respectively); (2) uranyl and neptunyl incorporation into carbonates and sulfates: for both carbonates and sulfates, it was found that actinyl incorporation into a defect site is more favorable than incorporation into defect-free periodic structures. In addition, actinyl incorporation into carbonates with aragonite structure is more favorable than into carbonates with calcite structure; and (3) uranium incorporation into magnetite: within the configurations tested that preserve charge neutrality ($U^{6+} \rightarrow 2Fe^{3+}_{oct/tet}$ or $U^{4+} \rightarrow Fe^{2+}_{oct}$), uranium incorporation into magnetite is most favorable when U^{6+} replaces octahedral Fe^{3+} with charge balancing accomplished by an octahedral Fe^{3+} iron vacancy. At the end of this article, the limitations of this method and important sources of error inherent in these calculations (e.g., hydration energies) are discussed. Overall, this method and examples may serve as a guide for future studies of incorporation in a variety of contexts.

Keywords: incorporation; quantum-mechanical calculations; radionuclide

---

1. Introduction

1.1. Incorporation Reactions Important in Low Temperature Mineralogy

The incorporation of atoms into a host phase is an important phenomenon in mineralogy and materials science. Understanding the fundamental thermodynamics and kinetics of the incorporation mechanism, including charge-balanced substitutions, final incorporated geometries, and the thermodynamic stability of substituted phases, is of importance for many geochemical and mineralogical studies including biomineralization, acid mine drainage, and long-term predictions of geologic alteration of nuclear waste.

Incorporation of contaminants into minerals is a primary mechanism for immobilization of radionuclides at a contaminated site (e.g., Hanford Site in Richland, Washington) or potential immobilization after failure of a geologic repository (e.g., Waste Isolation Pilot Plant in Carlsbad, New Mexico). The study of incorporation of radionuclides will enable more refined risk assessment models for long-term evaluation of geologic repositories. Risk assessment of a geologic repository requires an understanding of the breakdown of natural and engineered barriers leading to the transport of radionuclides. Significant mineral sink phases occur as a result of the oxidation of the fuel itself, oxidation of engineered spent fuel canisters, occur in the surrounding geology, or are added to a geologic repository as a backfill. Table 1 lists minerals available as a sink for radionuclides along with the source of the mineral. The examples used in the current review are related to immobilization of radioactive waste, particularly actinides.

Table 1. Examples of incorporation reactions studied for better understanding of the transport of radionuclides in the environment.

Mineral (formula)	Source of Mineral	Incorporated Elements	Reference
Studtite $[UO_2O_2(H_2O)_2](H_2O)_2$	Oxidation of fuel or uranium deposit	Np	[1]
Boltwoodite $KUO_2SiO_4(H_2O)_{1.5}$	Oxidation of fuel or uranium deposit	Np	[2]
Garnet	Natural occurrence, potential waste form	U, Pu	[3–6]
$(Ca,Ba,Pb,Sr)SO_4$ [a]	Natural occurrence	U, Np	[7]
Gypsum $(CaSO_4 \cdot 2H_2O)$	Natural occurrence	U, Np	[7]
$(Ca,Ba,Pb,Sr)CO_3$ (arag [b])	Natural occurrence	U, Np	[7]
Calcite $(CaCO_3)$	Natural occurrence	U, Np	[7]
Fe-oxides (magnetite Fe_3O_4, hematite Fe_2O_3, goethite FeOOH)	Natural occurrence, canister oxidation	U, Tc	[8–10], this paper
Muscovite	Natural occurrence, backfill material	Cs	[11]
Rutile TiO_2	Natural occurrence, ceramic waste forms	Tc, Ru	[12]

Notes: [a] This item comprises the sulfate minerals anhydrite (Ca), barite (Ba), anglesite (Pb), celestine (Sr);

[b] Carbonates with aragonite structure: aragonite (Ca), witherite (Ba), cerussite (Pb), strontianite (Sr).

1.2. Quantum-Mechanical Incorporation Calculations

Quantum-mechanical calculations can be used to identify atomistic scale phenomena that impact incorporation, including coupled substitutions and charge transfer. The use of quantum-mechanical calculations to study low-temperature geochemical and mineralogical phenomena has been ever increasing since the implementation of density functionals, such as the local density approximation (LDA) and the generalized gradient approximation (GGA) [13], proving the Hohenberg-Sham theorem that the total energy of a system can be determined according to the electron density. Taking advantage of Bloch's theorem and the periodicity of mineral structures, density functional theory (DFT) calculations using planewave basis sets and pseudopotential approximations have become common place in computational mineralogy [14]. However, dealing with incorporation at a quantum-mechanical level is challenging due to the combination of aqueous and solid systems. That is, the incorporated species typically comes from an aqueous solution, rather than a solid phase, while the final doped mineral is a solid. Thus, the subject of incorporation (*i.e.*, the charged, aqueous source species) would be best treated using a quantum-chemical cluster model, whereas the solid host phase is best treated using periodic boundary conditions. A similar challenge is posed by the reaction products which are the solid periodic host with the incorporated molecule or ion and the, typically charged, aqueous sink phase of the species that was released from the host mineral into the surrounding water. The challenge is, then, how to accurately model an incorporation reaction such that a uniform computational theory is used throughout the incorporation reaction. Details regarding computational challenges for calculating accurate incorporation energies are in Sections 2.1 and 4.

Incorporation energies are the reaction energy describing the incorporation reaction, where the final incorporation may range from a slight impurity to a complete solid solution. For the latter, the incorporation does not result in any structural changes *sans* modification of the unit cell parameters according to Vegard's law (e.g., [15]). In this case, the Gibbs free energy of mixing can be considered the incorporation energy. More traditionally, incorporation energies refer to the incorporation of an impurity into a bulk mineral phase. For example, Se is associated with pyrite and arsenopyrite [16]; thus, the energy associated with the "defect" Se in the Fe-sulfide is the incorporation energy.

From quantum-mechanical calculations, the kinetics of the incorporation (*i.e.*, the precise pathway of the incorporated ion) is not considered. For example, this type of calculation does not consider whether a cation sorbs on the mineral surface and diffuses into the mineral or sorbs on the surface and co-precipitates with the mineral, with the latter being a more likely scenario. Rather, the computational methodology presented here, allows one to evaluate the overall thermodynamic stability of the final incorporated phase. In contrast, the experimental incorporation limit may be dependent on the dominating process of the ones described above, because, e.g., co-precipitation maybe far from equilibrium with difficulty reaching equilibrium subsequently by diffusing out excess incorporation cations.

The incorporation energy is influenced by the mechanism, e.g., the incorporation site. On the experimental side, one might assume that the incorporated ion or molecule would find its most favorable position; however, computationally, different incorporation mechanisms/sites must be assessed separately due to the static nature of the quantum-mechanical incorporation calculations. For example, the incorporation of Ti into zircon, an important reaction for accurate thermochronological dating, is dependent on the location of the incorporated Ti. That is, for pressures below 3.5 GPa, Ti substitution at the Si site is more thermodynamically favored over substitutions at the Zr site and *vice versa* at higher pressures [17]. On the atomic-scale, incorporation of Ti into zircon is significantly altered by pressure conditions and corrections are required for the Ti-in-zircon geothermometer.

Substitutions of cations with different oxidation states than the host require a charge-balancing mechanism. A common mineralogical example is the plagioclase feldspars ($M^{x+}(Al_x,Si_{4-x})O_8$, $M = Na^+$, Ca^{2+}), where the substitution of Al^{3+} for Si^{4+} associated with the Ca^{2+} for Na^+ in the anorthite-albite solid solution [18,19]. The excess charge at the A site (Ca^{2+} for Na^+) is compensated in the tetrahedral framework (Al^{3+} for Si^{4+}). One can imagine, however, for incorporations that do not form a complete solid solution (e.g., Y^{3+} incorporation into zirconia, ZrO_2 [20]), the charge-balance mechanism is more complex. In addition, coupled-substitutions at other lattice positions are not always possible, depending on the mineral structure.

In the absence of secondary cation sites (e.g., SiO_4 tetrahedra) for coupled substitutions, interstitial impurities or oxygen vacancies are added to balance the total charge. The addition of an H^+ atom is commonly used as a theoretical charge-balancing mechanism owing to its small size and availability. For a quantum-mechanically calculated incorporation energy, the position of the additional H^+ atom in the substituted structure is an important parameter to consider. For example, for NpO_2^+ substitution of UO_2^{2+} in studtite ($[UO_2O_2(H_2O)_2](H_2O)_2$), three positions for the additional H^+ atom were considered, including bonded to the axial oxygen, peroxide oxygen, and interlayer water. The additional H^+ atom was found to favor the site on the axial oxygen by 0.8 to 1.3 eV with respect to the other possible sites. This H^+ substitution mechanism is also employed in the example below detailing NpO_2^+ incorporation into sulfate and carbonate minerals (see Section 3.1).

In the introduction of this study, we highlight previous approaches from the literature for calculating incorporation energies using quantum-mechanical methods (Section 2.1). In addition, a detailed description of our new method for calculating incorporation energies is presented (Section 2.2) along with examples of its implementation (Section 3). Subsections denoted by an * describe new research presented in this study. Finally, computational challenges that are addressed in this new method and those that remain are presented (Sections 4 and 5). Our intent is to help advance methodologies implemented in low-temperature computational mineralogy such that we can more accurately and predictively model environmental mineralogical phenomena, namely incorporation.

2. Methods for Calculating Incorporation Energy

2.1. Previous Approaches

Traditionally, optimized geometries and charge densities, rather than energies, have been evaluated for incorporated systems (e.g., [9,21,22]). Quantum-mechanically optimized geometries are compared with experimental measurements, such as extended X-ray absorption fine structure (EXAFS) measurements, to identify coordination environments and distances to nearest neighbors. In this way, impurity incorporation, as opposed to adsorption, and the impurity geometry within a host phase can be confirmed (e.g., [9]). In addition, changes in charge or electron density about an impurity can be identified along with associated changes in the electronic band structure. However, quantum-mechanical incorporation energy calculations are less common due to some inherent computational challenges.

Demichelis *et al.* [23] studied the incorporation of water into calcium carbonate to form hydrous carbonate minerals. Using a DFT-based approach, the authors characterized the thermodynamics of water incorporation reactions starting from low water content carbonates (calcite, aragonite) and adding water to create hydrated phases (monohydrocalcite, ikaite). A generalized carbonate hydration reaction is: $CaCO_3 + xH_2O \rightarrow CaCO_3 \cdot xH_2O$.

While the structures of both hydrated and anhydrous carbonates have been accurately modeled by DFT methods, the thermodynamics of the hydration reactions have not been reproducible with standard DFT techniques. A range of density and hybrid functionals (e.g., Perdew-Becke-Ernzerhof (PBE), Adamo-hybrid PBE (PBE0), Becke three-parameter Lee-Yang-Parr (B3LYP) [13,24–26]), as well as dispersion corrections for van der Waals interactions (e.g., B3LYP-D2, PBE-D2 [27,28]), were used to determine the most accurate computational approach. In addition, this study evaluated the treatment of water in two phases: gas and solid (ice). The computed values of enthalpy, entropy, and Gibbs free energy for the reactions were determined based on quantum-mechanical geometry and vibrational frequency calculations. By combining their calculated internal energies with experimental values for heat of fusion and heat capacity, the enthalpies of carbonate hydration reactions were significantly improved. There was considerable variability when different functionals were implemented, but dispersion corrections provided the closest match to experimental data.

Tsuchiya and Wang [29] used *ab initio* determination of thermodynamic properties to understand ferric iron incorporation into $MgSiO_3$ perovskite. In this study, the reaction energy of the incorporation of Fe^{3+} into Mg-perovskite was not explicitly determined. However, the, calculation of phonon frequencies, using a quasi-harmonic approximation, allowed for derivation of properties such as Helmholtz free energy, Gibbs free energy, bulk modulus, and heat capacity of the phase with the most favorable incorporated geometry. For these calculations, the authors employed an LSDA + U approach (the "S" in LSDA stands for the treatment of unpaired spins, whereas the "U" denotes the application of a Hubbard U parameter as an approximation for the localization of valence orbitals, [30]) combined with a direct LDA method [31] on an 80-atom supercell of Fe-incorporated $MgSiO_3$.

Another study by Kuo *et al.* [12] investigated solid phase reactions of technetium and ruthenium incorporation into rutile (TiO_2). This study examined the possible dopant configurations within the rutile host and the extent to which incorporation may occur as a function of temperature. Using two DFT-based codes, local Density functional calculations on Molecules ($DMol^3$) [32,33] and Vienna *Ab initio* Simulation Package (VASP) [34–37], the authors assessed a variety of clustering schemes for Tc and Ru within a large rutile supercell. The energetics of binary dopant clustering in adjacent and near neighbor configurations were compared to single defect incorporation. Overall, Tc single defects were shown to have lower energy than Ru ones, indicating that Tc would more readily enter the rutile structure. The extent of dopant incorporation and preferred cluster configuration was modeled over a range of temperatures and the computed results correlate well with experimental observations [38–40].

Technetium was also the subject of a study by Skomurski *et al.* [8] who examined the potential incorporation of Tc^{4+} and TcO_4^- into hematite. Incorporation of these two Tc species was modeled by coupled substitution of Tc^{4+} and Fe^{2+} for two ferric iron atoms and by TcO_4^- occupying a lattice vacancy. Energies were determined by using unrestricted Hartree-Fock calculations in Crystal06 [41]. Incorporation energies in this study consider only calculated electronic energies and do not include any thermochemical energy contributions. Ionic species in the reaction are calculated as charged clusters in the gas phase with no hydration component. Incorporated Tc^{4+} and Fe^{2+} geometries were determined over a range of Tc^{4+}-Fe^{2+} distances. The computed energies are lower at shorter distances, which is expected given a lower Coulomb repulsion between Tc^{4+} and Fe^{2+} than between two ferric irons. Overall, Tc^{4+} incorporation was shown to be energetically feasible, with ΔE_{incorp} on the order of −0.43 eV per Tc^{4+} ion for up to four Tc within the $4 \times 1 \times 1$ hematite supercell. In contrast, TcO_4^- incorporation into hematite is unlikely as it was shown to be energetically unfavorable and requiring a defect in the hematite lattice. These types of incorporation reaction are good candidates for using a combination of hydrated cluster and solid phases because the sources of Tc and sink of Fe is the surrounding aqueous environment. This computational treatment may provide more accurate information about the likelihood of this particular Tc immobilization pathway.

Contaminant incorporation has also been explored in relation to clay minerals. In the case of clays, incorporation typically occurs via cation exchange, where interlayer cations are replaced when in contact with aqueous solutions. This type of reaction has been modeled in several studies including one by Rosso *et al.* [11]. They simulated the replacement of interlayer K^+ with Cs^+ in muscovite, $KAl_2(AlSi_3)O_{10}(OH)_2$. Muscovite has a 2:1 tetrahedral-octahedral (T-O-T) layering ratio. In the ideal structure, the tetrahedral sites host Si^{4+}, while the octahedral sites host Al^{3+}. However, these sites can become disordered and lead to a net charge on the layer surface, commonly called layer charge (LC). This negative surface charge is then balanced by interlayer cations, like K^+ in the case of muscovite.

To examine the ability of muscovite to accept Cs into its structure, the authors set up the following reaction to explore using computational methods: $X(K) + Cs^0_{(g)} \rightarrow X(Cs) + K^0_{(g)}$, where X represents the two-formula unit ($Z = 2$) primitive unit-cell of muscovite with one of the two potential K interlayer sites being considered for exchange. Geometry optimizations and total energy calculations for these species were performed using the planewave-based DFT code Cambridge Serial Total Energy Package (CASTEP) operating with a GGA/PBE setup. Ultrasoft pseudopotentials were used to lessen the computational expense. Calculation of the gas-phase K and Cs species was conducted by placing the lone atoms in a periodic box. The box size was varied to ensure that energy obtained was unaffected by the periodicity. The energetics of the reactions were considered using gas-phase K and Cs, but the authors also applied experimentally-derived hydration energies and ionization potential from the literature. This change to $K^+_{(aq)}$ and $Cs^+_{(aq)}$ lowered the energy of exchange by ~3 kJ/mol for all reactions.

Exchanging Cs for K was explored for four different scenarios designed to separately evaluate the influence of LC, cation radii/interlayer geometry, and Al/Si substitution on the exchange energetics. Overall, they found that increasing LC promotes Cs exchange for K. The authors also discovered that K and Cs have opposing effects on the interlayer distance when the LC is high (K causes the distance to shrink, while Cs allows the layers to spread apart). The most favorable reaction, with an exchange energy of −5.3 kJ/mol, was achieved using aqueous cations and muscovite layers with no Al/Si substitution creating a net LC of −1.

These examples indicate that a comprehensive method is required to treat incorporation from an aqueous phase that allows for either a complete computational treatment of sub-equations or for usage of, e.g., experimental hydration energies, where available.

2.2. Method/Research for Calculating Incorporation Reactions Using Aqueous Source and Sink Phases

Quantum-mechanical calculations of reaction energies require calculations of each reactant and product using the same computational theories and parameters. Thus, as seen in previous studies, all reactants and products are calculated using either periodic boundary conditions or cluster models, but not a combination of the two. However, the reactant and product describing the source and sink phases for the substitution couple are typically charged aqueous species. Thus, calculations that are more focused on the actual geochemical process of incorporation at the mineral-water interface should combine periodic and cluster calculations such that the solid host phase and aqueous source and sink phases are each modeled with the best accuracy.

We have developed a method for evaluating incorporation of aqueous species using a combination of periodic and cluster calculations. One governing rule for this method is that all calculations within a single reaction are performed with the same computational method. Scheme 1 outlines the computational steps taken in our first study using charged aqueous source and sink phases for incorporation reactions. Reaction (1) describes the traditional reaction, where the sources for Np^{5+} and H^+ are Np_2O_5 and water (i.e., an H_2O molecule calculated in a $10 \times 10 \times 10$ Å^3 unit cell) and the sink for U^{6+} is UO_3. Reaction (2) describes the reaction using neutral source and sink molecules calculated using periodic boundary conditions in CASTEP [42]. While the overall reaction energy is negative, indicating favorable incorporation, the source and sink molecules are environmentally uncommon, even in carbonate-free water. Therefore, an extra step is taken to consider ionization and hydration of the neptunyl and uranyl (Reaction (3)). The ionization and hydration step is performed using Conductor-like Screening Model (COSMO) calculations as implemented in DMol^3 [43,32,33]. The energy contributions from the aqueous species in Reaction (3) are a combination of the total energy for the gaseous molecule and the hydration energy based on the COSMO calculation. Thus, the final combined reaction shows the cancelation of the periodic and cluster calculation for the gas phase molecule, as well as the hydration of that molecule. An example where ionization and hydration are treated separately and the influence of both of these can be evaluated is shown in Scheme 2 and, in more detail, in Appendix Table A1.

Scheme 1. Equations describing the method for calculating incorporation energies using charged aqueous species, where Reactions (1) and (2) are performed on neutral species using CASTEP and Reaction (3) is performed on neutral and charged aqueous molecules in DMol^3. In the reactions below, $K_2(UO_2)_2(SiO_4)_2(H_2O)_{3(s)}$ stands for boltwoodite, whereas $K_2(UO_2)(NpOOH)(SiO_4)_2(H_2O)_{3(s)}$ stands for the neptunyl-incorporated boltwoodite. The combined reaction is Reaction (2) minus Reaction (3) [1].

(1) Solid oxide source and sink phases with periodic boundary conditions

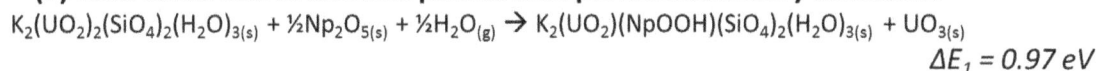

$$K_2(UO_2)_2(SiO_4)_2(H_2O)_{3(s)} + \tfrac{1}{2}Np_2O_{5(s)} + \tfrac{1}{2}H_2O_{(g)} \rightarrow K_2(UO_2)(NpOOH)(SiO_4)_2(H_2O)_{3(s)} + UO_{3(s)}$$

$$\Delta E_1 = 0.97\,eV$$

(2) Gaseous source and sink phases with periodic boundary conditions

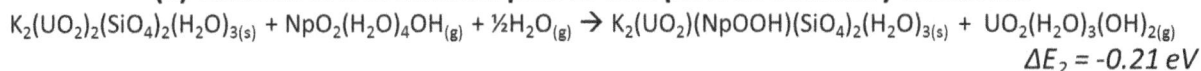

$$K_2(UO_2)_2(SiO_4)_2(H_2O)_{3(s)} + NpO_2(H_2O)_4OH_{(g)} + \tfrac{1}{2}H_2O_{(g)} \rightarrow K_2(UO_2)(NpOOH)(SiO_4)_2(H_2O)_{3(s)} + UO_2(H_2O)_3(OH)_{2(g)}$$

$$\Delta E_2 = -0.21\,eV$$

(3) Ionization and hydration

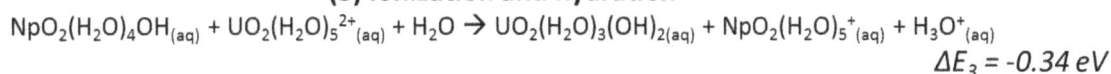

$$NpO_2(H_2O)_4OH_{(aq)} + UO_2(H_2O)_5{}^{2+}{}_{(aq)} + H_2O \rightarrow UO_2(H_2O)_3(OH)_{2(aq)} + NpO_2(H_2O)_5{}^+{}_{(aq)} + H_3O^+{}_{(aq)}$$

$$\Delta E_3 = -0.34\,eV$$

Combined reactions

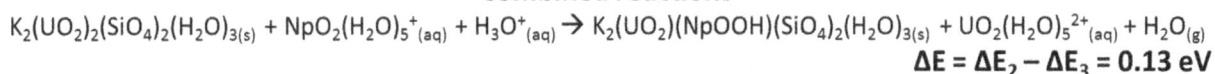

$$K_2(UO_2)_2(SiO_4)_2(H_2O)_{3(s)} + NpO_2(H_2O)_5{}^+{}_{(aq)} + H_3O^+{}_{(aq)} \rightarrow K_2(UO_2)(NpOOH)(SiO_4)_2(H_2O)_{3(s)} + UO_2(H_2O)_5{}^{2+}{}_{(aq)} + H_2O_{(g)}$$

$$\Delta E = \Delta E_2 - \Delta E_3 = 0.13\,eV$$

In our more recent investigations ([7], this study), this methodology has been further developed such that five primary reactions are calculated individually and combined for the total reaction energy (Scheme 2 modified from [7]). The series of reactions in Scheme 2 starts with using the mineral host (PbSO$_4$) and UO$_3$ as a solid host for UO$_2$$^{2+}$ and PbO as a solid sink for Pb^{2+} (Reaction (1)). This setup of the equation with only solid phases on both sides of the equation describes the relative stability of these solid phases and can be used for further solid solution calculations. The example given in Reaction (1) of Scheme 2 would be equivalent to going from the Pb end member to a quarter fraction of the uranyl end member. Theoretically, the incorporation energy using such a periodic approach should be independent of the actual program used (e.g., CASTEP, DMol^3, Crystal) as long as suitable computational parameters (such as, basis set, density functional, k-point density) are used. Our calculations indicate, however, that small differences are observed depending on the program used, on the order of a few tenth of an eV. In order to make the transition from solid to aqueous sources and sinks, solid phase and cluster calculations must be added together to get to an overall reaction of the periodic solid phase with aqueous contaminants. The first step (Reaction (1)) of this process describes incorporation using the traditional periodic solid oxide source and sink phase approach. All calculations for this reaction are performed using periodic boundary conditions. In addition, any H$_2$O, added as a source for hydrogen, is calculated as gaseous H_2O with periodic boundary conditions. Literature values were used for the condensation energy of water.

Scheme 2. Reactions describing the method for calculating incorporation energies using charged aqueous species. Reactions (1) and (2) are performed on neutral species using periodic boundary conditions in CASTEP and $DMol^3$. Reaction (3) is performed on neutral molecules with and without periodic boundary conditions in $DMol^3$. Reactions (4) and 5) are performed using cluster calculations in $DMol^3$. Hydration energy is implicitly included using a bulk dielectric fluid as implemented with COSMO in $DMol^3$, or better, by a combination of explicit water molecules in the 1st and 2nd hydration sphere and a dielectric fluid around. The combined reaction is the sum of Reactions (1)–(5) (modified from [7]). A more detailed set of reactions for neptunyl incorporation is provided in Appendix Table A1.

(1) Solid oxide source and sink phases with periodic boundary conditions

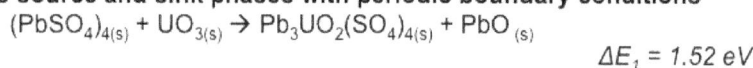

$$(PbSO_4)_{4(s)} + UO_{3(s)} \rightarrow Pb_3UO_2(SO_4)_{4(s)} + PbO_{(s)}$$

$$\Delta E_1 = 1.52\ eV$$

(2) Solid to gaseous molecule with periodic boundary conditions

$$UO_2(OH)_{2(pbc\ molc)} + PbO_{(s)} \rightarrow UO_{3(s)} + Pb(OH)_{2(pbc\ molc)}$$

$$\Delta E_2 = -0.97\ eV$$

(3) Transition to neutral molecular (gas phase) species

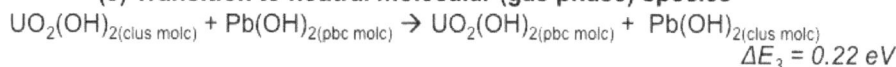

$$UO_2(OH)_{2(clus\ molc)} + Pb(OH)_{2(pbc\ molc)} \rightarrow UO_2(OH)_{2(pbc\ molc)} + Pb(OH)_{2(clus\ molc)}$$

$$\Delta E_3 = 0.22\ eV$$

(4) Dissociation and ionization

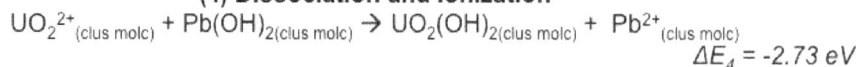

$$UO_2{}^{2+}{}_{(clus\ molc)} + Pb(OH)_{2(clus\ molc)} \rightarrow UO_2(OH)_{2(clus\ molc)} + Pb^{2+}{}_{(clus\ molc)}$$

$$\Delta E_4 = -2.73\ eV$$

(5) Hydration

$$UO_2{}^{2+}{}_{(aq)} + Pb^{2+}{}_{(clus\ molc)} \rightarrow UO_2{}^{2+}{}_{(clus\ molc)} + Pb^{2+}{}_{(aq)}$$

$$\Delta E_5 = 2.37\ eV$$

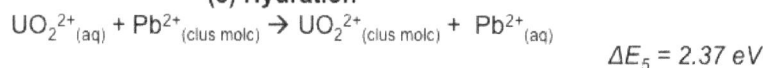

Combined reactions

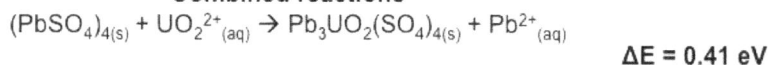

$$(PbSO_4)_{4(s)} + UO_2{}^{2+}{}_{(aq)} \rightarrow Pb_3UO_2(SO_4)_{4(s)} + Pb^{2+}{}_{(aq)}$$

$$\mathbf{\Delta E = 0.41\ eV}$$

Reaction (2) describes the conversion of the solid oxide source and sink phases to neutral gaseous species in a periodic arrangement. That is, the neutral molecules are calculated in a large unit cell ($\geq 10 \times 10 \times 10\ Å^3$ to minimize molecule-molecule interactions especially when they carry a dipole moment) using periodic boundary conditions. Reaction (3) describes the transition from a periodic to a neutral molecular species, where all calculations are performed using a program that can use consistent computational parameters (e.g., basis sets, density functionals, spin treatment, spin-orbit coupling if applied) for cluster and periodic approaches. The work presented here is done using $DMol^3$, but other programs are available that could be used in a similar fashion (e.g., Crystal09 [41,44]). Ideally, and this can be used as a test for computational consistency, the energy of transition from periodic to non-periodic treatment (ΔE_3 in Scheme 2) should be much less than the overall reaction energy (ΔE in Scheme 2). The magnitude of ΔE_3 is due to any dipole-dipole interaction between molecules in periodic unit cells. For most reactions in this study on actinyl incorporation, these neutral molecules are metal-hydroxide complexes (e.g., $NpO_2(OH)_2$). Reaction (4) describes the dissociation and ionization of the gaseous molecular species using cluster calculations. This reaction was previously considered as the ionization reaction, but in fact, an energetic component of dissociating the ligands from the cations is inherent to the calculation. Thus, it has been redefined as the dissociation and ionization energy. Reaction (5) describes the hydration of the ionized species. In this example, hydration was considered implicitly using the COSMO bulk dielectric fluid model as implemented in DMol^3. For even higher accuracy [45], the first and second hydration sphere are considered explicitly. Reactions (4) and (5) could be combined (and are in Scheme 1); however, their separation allows one to observe the impact of the hydration on the overall reaction. The sum of reaction energies (1) through (5) is the energy for the incorporation reaction based on the charged aqueous source and sink molecules.

3. Examples

3.1. Np-Incorporation into Uranyl Minerals

Quantum-mechanical calculations have been used to gain atomistic understanding of neptunyl incorporation into uranyl minerals, namely studtite ($UO_2O_2(H_2O)_2(H_2O)_2$) and boltwoodite ($KUO_2SiO_4(H_2O)_{1.5}$) [1,2]. Both NpO_2^+ and NpO_2^{2+} substitution for UO_2^{2+} in studtite were evaluated using the traditional solid oxide source and sink phases (*i.e.*, Np_2O_5 and UO_3), resulting in incorporation energies equal to 1.12 and 0.42 eV, respectively. Note that a high spin state was used for Np^{5+} (Np^{5+} has two unpaired 5f electrons) in all calculations and the spin ordering was evaluated in previous studies for Np_2O_5 [2]. For NpO_2^+ incorporation, an additional H^+ atom was added to the axial neptunyl oxygen for charge balance, and H_2O, calculated with periodic boundary conditions, was used as the source. A comparison of source and sink phases (e.g., oxide, silicate) was performed in the study of Np-incorporation into boltwoodite, which led to the development of the methodology presented in this paper for calculating incorporation reactions using charged aqueous molecules as source and sink molecules. Scheme 1 shows a subset of those results, where the incorporation energy for reactions using the charged aqueous molecules was 0.13 eV compared to 0.97 eV for the reaction based on solid oxide source and sink phases.

Boltwoodite is a uranyl sheet silicate composed of layers of uranyl hexagonal bipyramids connected by silica tetrahedra (Figure 1). Monovalent cations (K^+ and/or Na^+) and water are contained between the uranyl silicate sheets (in the interlayer). Due to the complexity of the boltwoodite structure, multiple charge-balancing incorporation mechanisms were evaluated, including (1) addition of a H^+ atom; (2) coupled substitution of an interlayer cation (divalent for monovalent); and (3) coupled substitution within the uranyl silicate sheet or intra-layer (PO_4^+ for SiO_4). The incorporation limit was estimated for the interlayer and intra-layer coupled substitutions by calculating the enthalpy of mixing for the boltwoodite/Np-modified boltwoodite solid solution series. That is, simple solid solution calculations were able to be performed because the theoretical Np-boltwoodite could be constructed in which all of the U sites were replaced by Np. Thus, the Gibbs free energy of mixing for the solid solution was approximated based on a Margules fit for the enthalpy of mixing and the $-T\Delta S$ part of ΔG was approximated by using the configurational entropy (Equation (2). The configurational entropy consists of two terms, the basic configurational entropy for a binary solid (*i.e.*, $\Delta S_{mix} = -R[x\ln(x) + (1-x)\ln(1-x)])$) and a term describing the probability of a configurational relationship between the two substituted cations. That is, for the single boltwoodite unit cell, there are two possible locations for Np and two locations for P substitution, in the case of intra-layer substitution. This entropy term can be fit based on the difference in the calculated enthalpy for the different configurations.

$$\Delta G_{mix} = Ax(1-x) + RT(x\ln(x) + (1-x)\ln(1-x) + x(1-x)((1-x)\ln Q_1 + \ln Q_2)) \tag{1}$$

$$\ln Q_1 = (-1+Z) + \ln(1-z); \ln Q_2 = -Z\ln Z; Z = \frac{e^{H_1-H_2}/RT}{1+e^{H_1-H_2}/RT} \tag{2}$$

$H_1 - H_2$ is the difference in total energies of the host mineral with the substitution in sites 1 and 2, respectively.

Figure 1. Atomic model showing the uranyl silicate sheets of boltwoodite $K(UO_2)SiO_3OH(H_2O)_{1.5}$, where the interlayer (between the sheets) is filled with K and H_2O. Colors for the atoms are as follows: U^{6+} blue, Si^{4+} yellow, K^+ purple, O^{2-} red, H^+ white.

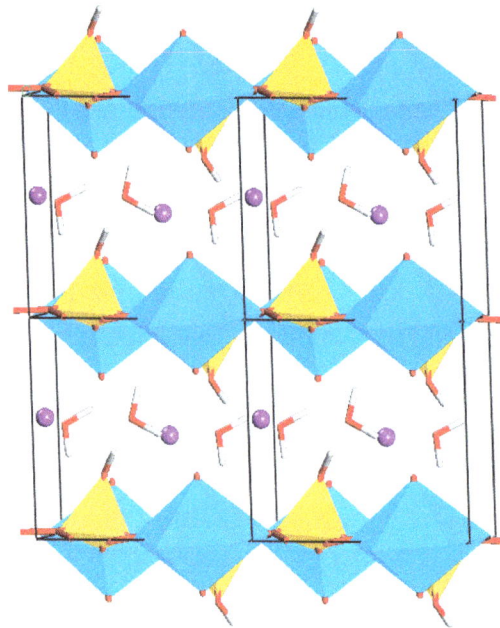

Figure 2 shows the approximated Gibbs free energy of mixing curves for the intra-layer and interlayer coupled substitution mechanisms. Based on our assumptions, the Gibbs free energy of mixing is always negative near the end-member compositions due to the infinite entropy contribution at that point (see inset of Figure 2).

The incorporation limit is calculated as the minimum of the Gibbs free energy of mixing. Thus, the resulting expression for the limit of incorporation is shown in Equation (3). Based on this approximation for the thermodynamics of mixing in the U-Np-boltwoodite solid solution, the approximate limit of Np incorporation into boltwoodite is 3.12×10^{-4} molar fraction Np at 300 °C (172 ppm Np/(boltwoodite)) for the interlayer coupled substitution and 1.39×10^{-3} molar fraction Np at 300 °C (768 ppm Np/(boltwoodite)) for the intra-layer coupled substitution. The resulting incorporation limit was in relative agreement with experimental results, which showed hundreds of ppm of Np incorporated into uranyl sheet silicates with charged interlayer cations [46].

$$x_{inc} = (1-x)e^{\frac{-A(1-2x)}{RT} - \ln Q_1(1-4x+3x^2) - \ln Q_2(2x-3x^2) - 2} \tag{3}$$

Figure 2. Gibbs free energy of mixing (ΔG_{mix}) *vs.* composition at 300 °C for the Np-U boltwoodite solid solution series based on the intra-layer (solid) and interlayer (dashed) coupled substitution mechanisms. The inset highlights the negative ΔG_{mix} near the end-member composition (modified from [1]).

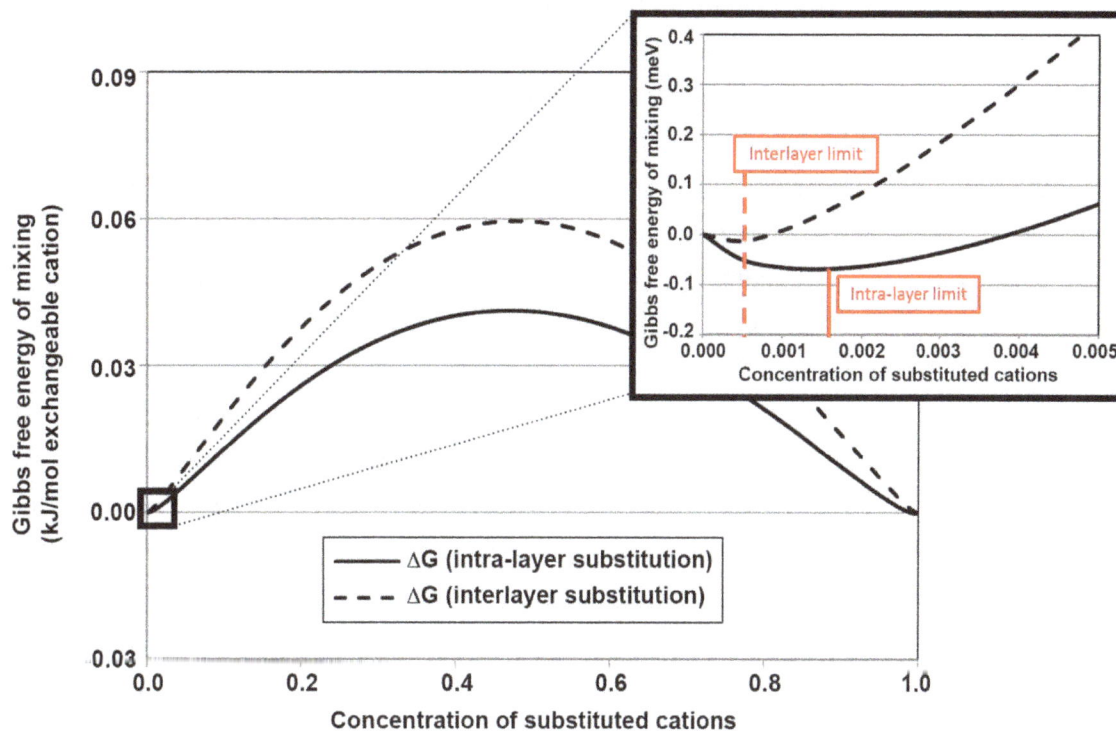

3.2. U-Incorporation into Sulfate and Carbonate Minerals *

One study implementing this new method was conducted by Walker and Becker [7] and investigates uranyl (UO_2^{2+}) and neptunyl (NpO_2^+) incorporation into carbonate and sulfate mineral phases. Five sulfates (anglesite, anhydrite, barite, celestine, gypsum) along with five carbonates (aragonite, calcite, cerussite, strontianite, witherite) were considered as host phases for the radionuclide-bearing ions. The reactions are set up such that uranyl and neptunyl replace cations in the crystal lattice. For neptunyl incorporation, the charge imbalance created by substitution of a monovalent species for a divalent cation is compensated by a nearby H^+.

Calculations to determine the energetics of incorporation reactions using solid sources of the actinide ions and solid sinks for the replaced cations were carried out using two quantum-mechanical codes: CASTEP and DMol^3. Both codes were used to calculate geometries and energies of periodic species (solid and molecular), while DMol^3 alone was used for cluster calculations. Computational parameters were held consistent throughout the calculations and included the use of the generalized gradient approximation (GGA) with PBE functional. For CASTEP calculations, a planewave cutoff energy of 500 eV was selected (for DMol^3 calculation, a double numerical basis set with polarization *d*-functions (DND) was applied [32]). In addition, ultrasoft pseudopotentials were used to approximate core electrons and lessen computational expense [47]. For neptunyl-bearing phases, a spin-polarized approach was used to account for the two unpaired 5f-like spins on Np^{5+}.

For the aqueous cluster calculations in DMol^3, a dielectric continuum model, COSMO, was used in combination with increasing numbers of explicit water molecules added to simulate hydration of charged aqueous species. It was found that when used as the sole hydration mechanism, COSMO underestimates hydration energies for the ions relevant to this study (e.g., Ca^{2+}, Ba^{2+}, Sr^{2+}, Pb^{2+}, etc.) by approximately 2–4 eV. This difference from experimental results is significant and can have a major influence on the overall ΔE of incorporation. The accuracy of computed hydration energies is improved when some degree of explicit hydration is applied in addition to COSMO. This approach gave hydration values within ~0.3 eV of experimental findings for UO_2^{2+} and NpO_2^{+} Additionally, literature values were used for the condensation energy of water and the hydration energy of smaller ions, like H^+.

Overall, calculated incorporation reaction energies show that gypsum and aragonite are the two most favorable hosts for uranyl and neptunyl. The incorporation energies for these reactions are as follows: $\Delta E_{gyp} = 0.19$ eV and $\Delta E_{arag} = 0.27$ eV for uranyl, $\Delta E_{gyp} = 0.17$ eV and $\Delta E_{arag} = -0.37$ eV for neptunyl. The least favorable structures for uranyl and neptunyl incorporation from these calculations are calcite, barite, and witherite. Optimized geometries for the incorporated structures were studied in detail and some trends were observed. In sulfate mineral hosts, sulfate groups rotate to accommodate uranyl and neptunyl species. Sulfate-group O atoms are moved out of alignment with the axial O on the actinyl ion. The coordination environment of the actinide species can also be observed and is composed by either four or five sulfate oxygen atoms. For carbonates, similar relationships are observed with one notable example being the uranyl-incorporated aragonite structure. In this structure, carbonate groups arrange themselves such that they achieve six-fold oxygen coordination in the equatorial plane. The stability of this structure gives aragonite the lowest (most favorable) calculated uranyl incorporation energy of 0.27 eV.

Incorporation reactions were also studied for selected systems with two types of lattice defects: vacancies and impurities. Vacancies in the carbonate and sulfate hosts were created by removing an adjacent cation-anion pair. Uranyl or neptunyl was then substituted for a cation near the vacancy. The presence of these vacancies increases the energy of the pure carbonate and sulfate phase and thus, makes incorporation energies more favorable, even negative for two of the tested uranyl incorporation reactions (celestine, cerussite). In addition to lattice vacancies, impurity ion defects were tested for a series of neptunyl incorporation reactions. For this type of reaction, a cation-anion pair in the sulfate or carbonate host was replaced by NH_4Cl. Ammonium chloride was chosen as it is a common reagent for precipitation experiments [48] because ammonium sulfates/carbonates and chlorides of divalent cations are more soluble than the respective host minerals chosen in this study. The incorporated ammonium chloride can be substituted with an actinyl and H^+ ion in solution. These reactions showed an increase in incorporation energy for some minerals (anglesite), while others became slightly more favorable (anhydrite, celestine).

Orbital projections and partial density of state (PDOS) spectra can be used to analyze the electronic configuration of incorporated mineral structures. For example, Figure 3 illustrates the electronic interaction between neptunium and the carbonate O atoms in neptunyl-incorporated aragonite. The neptunyl coordination environment is shown in Figure 3a. The neptunium is coordinated by 5 O atoms in the equatorial plane, and the hydrogen bond that forms between the neptunyl O and the charge-balancing H^+ is 1.94 Å. In Figure 3b, the highest occupied molecular orbital (HOMO) shows the eight-lobed shape of the Np 5f orbitals, not overlapping the O 2p orbital dumbbell. The HOMO is approximately

0.12 eV below the Fermi level (E_F) and is a non-bonding orbital. The orbital shown in Figure 3c, however, is a σ-bonding orbital, which is approximately 1.08 eV below the Fermi level (E_F = 2.24 eV) and shows the partial covalent character of the bond formed between the neptunium and the carbonate O atoms. Both the HOMO and the bonding orbital are located on the PDOS projection in Figure 3d (their contribution to the DOS is indicated by arrows). The contribution from the Np 5f orbitals fills in the top of the valence band just below the Fermi level. As compared with aragonite without neptunyl incorporation, these states with partial Np 5f character fill a portion of the aragonite bandgap of about 4 eV and reduce it to about 1.5 eV.

Figure 3. The coordination environment and electronic structure of neptunyl in aragonite. (**a**) Coordination environment of neptunyl with nearest-neighbor distances. One H^+ per unit cell is added to charge compensate the replacement of Ca^{2+} by NpO_2^+ The second H^+ ion comes from the adjacent unit cell (red = O, gray = C, white = H, blue = Np); (**b**) Wavefunction showing the non-bonding character of the HOMO (no overlap of the eight-lobed Np 5f and O 2p orbitals; (**c**) σ-bonding orbital (E = 1.15 eV or 1.08 eV below EF) between Np 5f and O 2p orbitals from two O atoms of the same carbonate molecule (the other three σ-bonding orbitals, not shown, have comparable electron binding energies, with one bond each from the three other CO_3^{2-}; and (**d**) Partial density of states (PDOS) projection for the energy range from −15.0 to 10 eV.

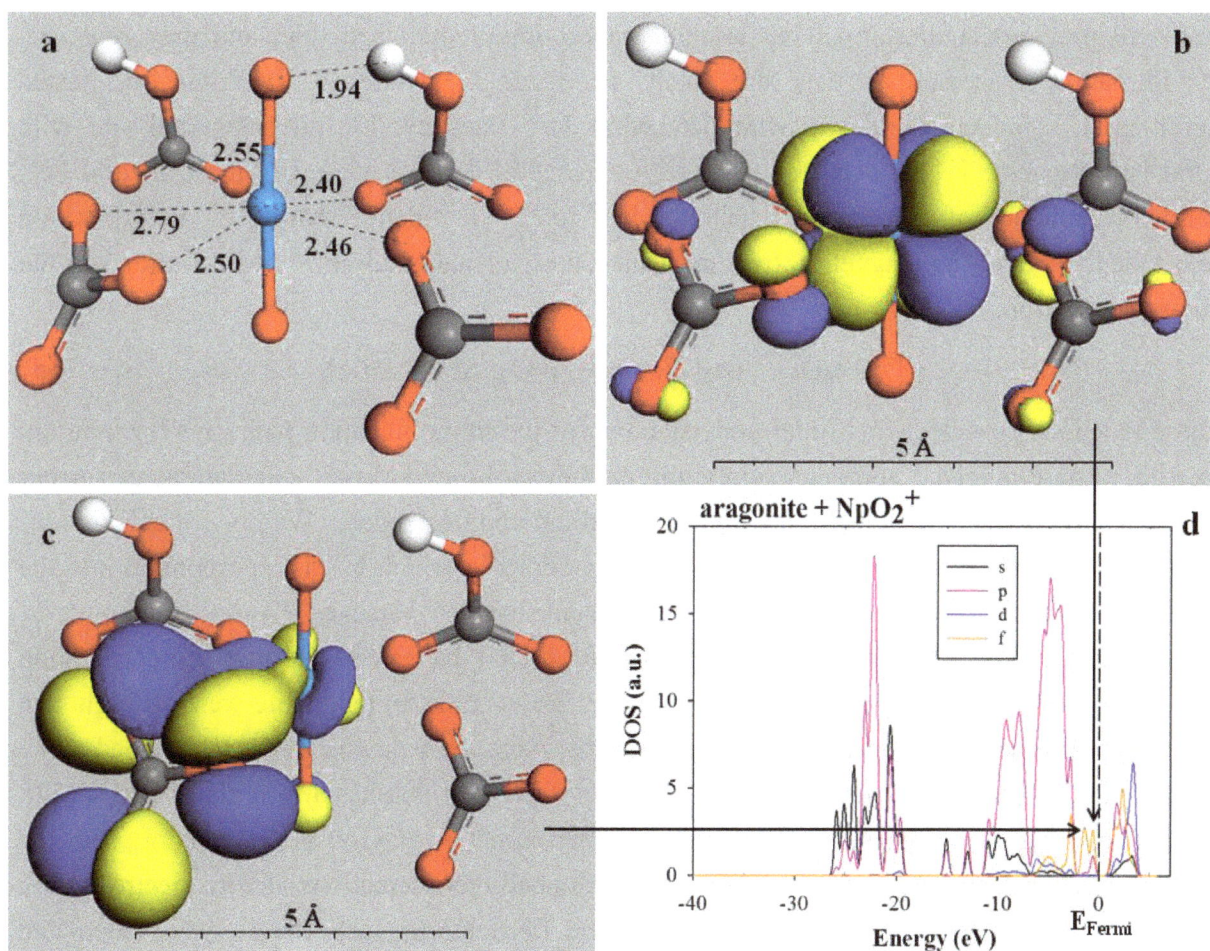

3.3. U-Incorporation into Fe-Oxide Minerals *

Aqueous uranium, commonly found as the UO_2^{2+} uranyl ion, can form hydroxide and water complexes which have been shown to sorb strongly to Fe^{3+}-oxide surfaces, including magnetite [49,50]. Magnetite (Fe_3O_4) is an interesting mineral to consider as it is a common corrosion product of steel, formed during anoxic water-steel interaction via the following reaction:

$$3Fe_{(s)} + 4H_2O \rightarrow Fe_3O_4 + 4H_{2(g)} \tag{4}$$

As a result, it is conceivable that magnetite may be one of the first materials to potentially immobilize uranium that has been previously released from a containment vessel.

This study initially considered incorporation of both U^{6+} and U^{4+} via a variety of solid phase reactions (*i.e.*, the source of U and the sink of Fe are both solid phases) using a quantum-mechanical periodic planewave code named CASTEP. A high-spin state was applied for U^{4+}, Fe^{2+}, and Fe^{3+}. The computational parameters used were GGA/PBE as a density functional, a planewave cutoff energy of 600 eV, 0.06 Å k-point spacing, ultrasoft pseudopotentials, and a convergence tolerance of the self-consistent field (SCF) procedure of 10^{-5} eV. Since both U^{6+} and U^{4+} have a higher positive charge than the replaced Fe^{2+} or Fe^{3+}, charge-balancing of the incorporated oxide phase was achieved by a structural vacancy of either Fe^{2+} or Fe^{3+} near the incorporation site (Figure 4). By comparing the energetics of vacancies in the tetrahedral and octahedral site, vacancies were found to be more energetically favorable in the octahedral lattice position. The energies of these reactions are comparable to previous studies involving uranium incorporation into hematite, and may represent vacancy formation processes as they are likely to occur in nature [51]. Within the tested configurations: U^{6+} in octahedral site with octahedral Fe^{3+} vacancy, U^{6+} in octahedral site with tetrahedral Fe^{3+} vacancy, U^{4+} in octahedral site with octahedral Fe^{2+} vacancy, U^{4+} in octahedral Fe^{3+} site along with reduction of neighboring octahedral Fe^{3+} to Fe^{2+} to balance charge; U^{6+} incorporation into the octahedral site of the magnetite structure balanced by a octahedral ferric iron vacancy was the most favorable reaction.

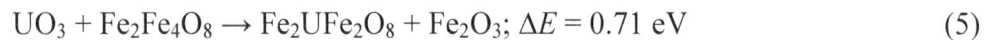

$$UO_3 + Fe_2Fe_4O_8 \rightarrow Fe_2UFe_2O_8 + Fe_2O_3; \; \Delta E = 0.71 \text{ eV} \tag{5}$$

while these reactions provide a first-order understanding of the immobilization pathways for uranium in magnetite, a more involved approach combining periodic solid and cluster calculations can better model the true environmental interaction of aqueous uranium and iron oxides. We present calculations using this new method for one of these incorporation reactions in which U^{6+} is incorporated into the ferric iron octahedral site of magnetite, balanced by an octahedral Fe^{3+} vacancy.

For these calculations, we have used DMol^3, an *ab initio* DFT-based code, to model incorporation into a 14-atom primitive cell of inverse-spinel magnetite. Iron cations in magnetite were all high spin and set in a ferrimagnetic arrangement. Tetrahedral sites were spin down, while all octahedral Fe atoms were spin up, giving a net moment of 8 for the primitive cell that has a formula of Fe_6O_8 [52]. Calculations were run using the generalized gradient approximation (GGA) in conjunction with the PBE functional. DFT semi-core pseudopotentials approximated the interactions of core electrons with valence shells. For oxygen, iron, and uranium, 6, 16, and 14 electrons, respectively, were calculated explicitly. An energy convergence tolerance of 2.7×10^{-4} eV was selected for both SCF and geometry optimization cycles. For our incorporation reactions, DMol^3 total energies were used to represent overall ΔE_{incorp}. One of the most critical terms in the overall thermodynamics for incorporation processes from the aqueous to the solid phase is the hydration energy of the aqueous species. While there are different ways to calculate these (combinations of explicit water molecules and homogeneous dielectric fluids surrounding them, static *vs.* molecular-dynamics approaches, force-field *vs.* quantum-mechanical methods), it is highly advisable to compare these calculated values with experimental results. However, an additional complication is that some experimental results are reported as ΔG values of hydration/solvation, others as ΔH. In the example described in Scheme 3, for some hydration energies, literature values from experiment have been used. For example, experimental values from the literature were used to describe the hydration energies of UO_2^{2+}, H_2O, and H_3O^+ [53,54]. The values for the uranyl ion and H_3O^+ are $\Delta G_{solvation}$ values, while the water value is the enthalpy of vaporization.

Scheme 3. Reactions describing U incorporation into magnetite using charged aqueous species. Reactions (1) and (2) are performed on neutral species using periodic boundary conditions in DMol^3. Reaction (3) is performed on neutral molecules with and without periodic boundary conditions in DMol^3. Reactions (4) and (5) are performed using cluster calculations in DMol^3. Hydration energy is implicitly included using a bulk dielectric fluid as implemented with COSMO in DMol^3, or better, by a combination of explicit water molecules in the 1st and 2nd hydration sphere and a dielectric fluid around. The combined reaction is the sum of Reactions (1)–(5). Periodic molecules and non-periodic clusters are labeled (pbc m) and (clus) respectively. A more detailed breakdown of the individual reactions is provided in Appendix Table A2.

(1) Solid oxide source and sink phases with periodic boundary conditions

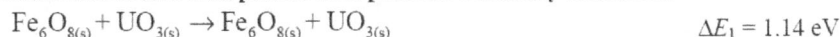

$$Fe_6O_{8(s)} + UO_{3(s)} \rightarrow Fe_6O_{8(s)} + UO_{3(s)} \qquad \Delta E_1 = 1.14 \text{ eV}$$

(2) Solid to gaseous molecule with periodic boundary conditions

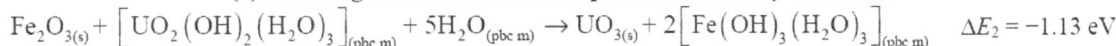

$$Fe_2O_{3(s)} + \left[UO_2(OH)_2(H_2O)_3\right]_{(pbc\,m)} + 5H_2O_{(pbc\,m)} \rightarrow UO_{3(s)} + 2\left[Fe(OH)_3(H_2O)_3\right]_{(pbc\,m)} \qquad \Delta E_2 = -1.13 \text{ eV}$$

(3) Transition to neutral molecular (gas phase) species

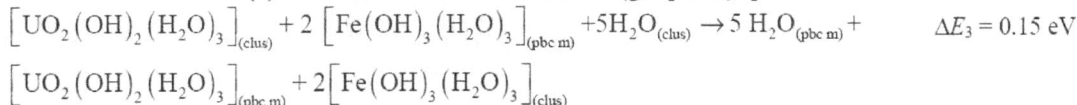

$$\left[UO_2(OH)_2(H_2O)_3\right]_{(clus)} + 2\left[Fe(OH)_3(H_2O)_3\right]_{(pbc\,m)} + 5H_2O_{(clus)} \rightarrow 5\,H_2O_{(pbc\,m)} + \qquad \Delta E_3 = 0.15 \text{ eV}$$
$$\left[UO_2(OH)_2(H_2O)_3\right]_{(pbc\,m)} + 2\left[Fe(OH)_3(H_2O)_3\right]_{(clus)}$$

(4) Dissociation and ionization

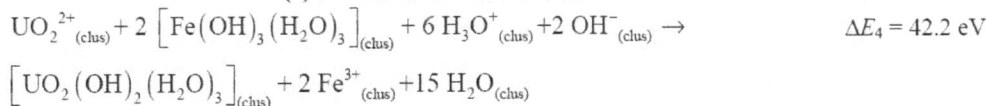

$$UO_2^{2+}{}_{(clus)} + 2\left[Fe(OH)_3(H_2O)_3\right]_{(clus)} + 6\,H_3O^+{}_{(clus)} + 2\,OH^-{}_{(clus)} \rightarrow \qquad \Delta E_4 = 42.2 \text{ eV}$$
$$\left[UO_2(OH)_2(H_2O)_3\right]_{(clus)} + 2\,Fe^{3+}{}_{(clus)} + 15\,H_2O_{(clus)}$$

(5) Hydration

$$[UO_2(H_2O)_5]^{2+}{}_{(aq)} + 2Fe^{3+}{}_{(clus)} + 7H_2O_{(clus)} \rightarrow UO_2^{2+}{}_{(clus)} + 2[Fe(H_2O)_6]^{3+}{}_{(aq)} \qquad \Delta E_5 = -78.7 \text{ eV}$$

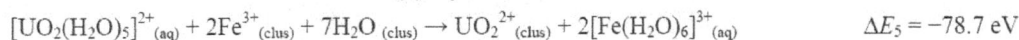

(6) Balancing of water and protons

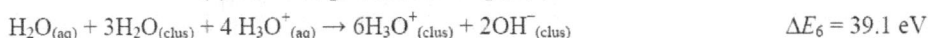

$$H_2O_{(aq)} + 3H_2O_{(clus)} + 4\,H_3O^+{}_{(aq)} \rightarrow 6H_3O^+{}_{(clus)} + 2OH^-{}_{(clus)} \qquad \Delta E_6 = 39.1 \text{ eV}$$

Combined reaction

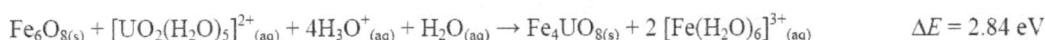

$$Fe_6O_{8(s)} + [UO_2(H_2O)_5]^{2+}{}_{(aq)} + 4H_3O^+{}_{(aq)} + H_2O_{(aq)} \rightarrow Fe_4UO_{8(s)} + 2\,[Fe(H_2O)_6]^{3+}{}_{(aq)} \qquad \Delta E = 2.84 \text{ eV}$$

The overall reaction of periodic magnetite with aqueous uranyl is achieved by moving systematically from solid source and sink phases, to neutral clusters and then to hydrated, charged clusters. In this reaction, in the presence of hydrated uranyl, water, and hydronium ion, magnetite is able to incorporate uranium into its structure and release hydrated ferric iron (ΔE_{incorp} = 2.84 eV). The multiple steps required to get to this overall solid-aqueous reaction are shown in Scheme 3. It is important to note that this energy cannot be considered the ΔG of the overall reaction, which would be necessary in order to convert the reaction energy into an equilibrium constant, because not all aspects of the entropy change are included. The entropy of the water molecules surrounding an ion in solution decreases during hydration and part of this process is considered in the treatment of hydration energies, while the treatment of changes in the vibrational entropies of all species involved is extremely computationally expensive. Work by Walker and Becker [7], which has been summarized in the previous section, has shown that the $-T\Delta S$ contribution to the ΔG change of incorporation reaction is on the order of 0.1 to 0.3 eV (typically the entropy of the solids goes up during incorporation because the perfect order of the host mineral phase is disturbed by the incorporation process). We can assume for our solid phases containing lattice vacancies that the variations may be slightly larger, but overall, the energy contribution may be small but significant if the conversion to equilibrium constants is made.

4. Challenges in Validating Calculated Results against Experiments

One particular challenge is the comparison between calculation and experiment. Ideally, one would like to compare the thermodynamics of incorporation, a closely related property which is the thermodynamic limit of incorporation at equilibrium, the structure and its distortion of the lattice around the incorporation site, and, e.g., potential changes to the electronic structure ("doping") as a result of incorporation. However, for most systems, hardly any thermodynamic data are available from experiment for incorporation reactions, in particular at the ppm (and lower) scale. For some systems, data are available for incorporation limits, either from experiment or from natural samples (see, e.g., Table 1 in [55] for actinide incorporation into zircon, or U incorporation into natural garnets [4]). Incorporation limits from experiment, e.g., from co-precipitation experiments, often suffer from the fact that equilibrium has not been reached during the time of incorporation. Another potential difficulty in co-precipitation experiments is that it is not always trivial to distinguish structural incorporation of single ions from incorporation of a different, e.g., nanoparticulate phase. While nature typically gives "samples" more time to reach equilibrium, even natural materials may not be fully equilibrated, especially if no "lubricating" medium, such as water, is available to help with the transport of ions that have been incorporated in excess. On the lower end of incorporation concentration, *i.e.*, below the thermodynamic limit, it is possible that nature did not provide enough of the species to be incorporated to reach the highest possible incorporation concentration. Thus, in many cases, computational results need to be compared with structural data from, e.g., EXAFS experiments. While a good comparison between properties such as coordination number and coordination distances are a necessary requirement to judge if the calculations are on the right track, they are by no means sufficient to evaluate changes in the thermodynamics upon incorporation.

For incorporation of foreign species, interesting and measurable changes of electronic properties may occur. For example, incorporation may lower the width of the bandgap, and it may also change the character of the semiconducting host, *i.e.*, if it is a *p*- or *n*-type semiconductor [3–6]. If the computations compare well with the experiment in terms of the electronic parameters that can be verified by experiments, such as conductivity, Hall effect, or inverse ultraviolet photoelectron spectroscopy, then, the calculations can offer significantly more insight into electronic property changes resulting from incorporation, such as changes in the band structure, the density of states, redistribution of the electron and spin density, the shape of individual bonding, non-bonding, and anti-bonding orbitals between the incorporated species and the host phase. Incorporation at higher concentrations may also generate specific vibrational modes of the incorporated species in the host that are different from host modes or of the species in solution. If these are calculated, they can be compared with Raman or IR spectroscopy results.

Since thermodynamic data, e.g., as obtainable from calorimetric measurements may be hard or impossible to determine, in order to compare experimental and theoretical results as comprehensibly as possible and to test the validity of the approach described in this study, experiments are most promising on systems that allow for incorporation at the percent scale, reach equilibrium within a suitable amount of time, and structure, electronic, and vibrational properties should be obtained. The combination of these would allow for the highest degree of validation.

Incorporation of uranium into iron (hydr)oxides has been investigated experimentally (e.g., [10,49,56]) and the results of these studies can help us validate or improve our computational approach. Unfortunately, these studies have yet to tackle the determination of incorporation energetics, a key focus of our quantum-mechanical calculations. At this point, these experimental studies can confirm certain minerals as possible hosts for contaminant species and also provide comparison for the incorporated geometries. For example, Nico *et al.* [10] studied the incorporation of U^{6+} into goethite and magnetite. Their batch chemistry experiments showed evidence for the precipitation of solid uranyl phases (hydroxides and carbonates), but also for incorporation (and sorption) of U into the structures of iron (hydr)oxides.

The incorporation of oxidized uranium into iron (hydr)oxide phases was confirmed by analyzing the mineral substrates after a thorough extraction process to remove adsorbed U. The incorporated geometries were characterized using X-ray Absorption Near-Edge Spectroscopy (XANES) and EXAFS analyses. The fitting of XANES and EXAFS is complex; for example, it is not trivial to distinguish the exact percentages of, e.g., aqueous *vs.* adsorbed *vs.* incorporated U in goethite or magnetite [10]. The derived bond distances and coordination numbers in the three simultaneously-occurring phases come with a certain amount of error, such that any comparison to computationally derived geometries must be mindful of potential discrepancies in bond distances. The U-incorporated magnetite structures calculated in this paper with U being in an octahedral Fe site have U-Fe distances with four Fe surrounding a U site (for charge compensation, two Fe in the coordination sphere were removed) are 3.08 Å compared to the "ideal" structure in [10] listed as 2.97 Å which would mean a difference of about 0.1 Å. For U-O distances in the first coordination sphere (Figure 4), the calculated values (2.13 Å for four equatorial O, 2.26 Å for the two axial O) compare well with "ideal" values in [10] of 2.09 Å (in the experimental paper, no distinction was made between equatorial and axial oxygen).

Figure 4. Atomic model showing the primitive cell for U^{6+} incorporated magnetite (Fe_4UO_8). Uranium has replaced a Fe^{3+} in an octahedral site and a second Fe^{3+} octahedral site is now vacant to balance charge. Bond distances (in Angstroms) for the first U-O sphere are indicated. Colors for the atoms are as follows: U^{6+} blue, Fe^{2+} purple, Fe^{3+} green, O^{2-} red.

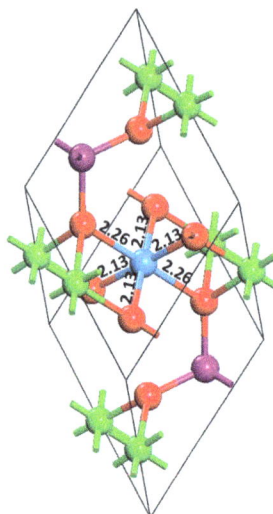

5. Theoretical Challenges

While the method presented in this paper is consistent in terms of combining aqueous charged species from solution and neutral host and incorporated phases with periodic boundary conditions, it may be worth trying to analyze the potentially greatest sources of error. In quantum-mechanical calculations, such as those presented for incorporation of species into mineral phases, it is always advantageous to apply the highest possible accuracy in terms of basis sets (high energy cutoff for planewave approaches or high-quality basis set for atomistic basis sets), k-point density, suitable DFT or hybrid functional, and other corrections (e.g., relativistic and spin-orbit coupling effects, localized f orbital treatment that can be treated using the Hubbard U formalism [3–6] or computationally-expensive hybrid DFT functionals such as the Heyd-Scuseria-Ernzerhof (HSE) functional [57]). All of these computational settings control incorporation energy changes as long as incorporation is treated using solid sink and source phases, the incorporated and released ion have the same charge and form similar complexes (e.g., actinyl ions are linear dioxides while Fe ions that may be released are just single ions, either in solids or in the aqueous phase, such that the latter criterion for replacement is not fulfilled), and the incorporation mechanism for two replacements to be compared is the same. However, typically, not all of these criteria are fulfilled such that other aspects of the overall suite of equations to be considered may control the error of the overall incorporation energy.

If the source and sink phases to be considered are aqueous species, the most significant source of error tends to be the treatment of hydration energy. First, there are, in principle, a number of computational treatments of hydration energy. One way is to "bond" a certain number of water molecules to the ion of interest in a static way according to:

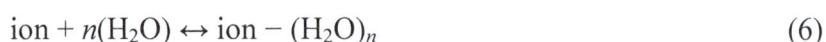

$$\text{ion} + n(H_2O) \leftrightarrow \text{ion} - (H_2O)_n \tag{6}$$

Equation (6) can be evaluated using a quantum-mechanical (which is advised due to the peculiar behavior of actinide valence orbitals) or a classical force-field approach. The reaction energy in

Equation (6) tends to become more and more negative with an increasing number of hydrating water molecules used (n) due to the static character of the calculation (at a finite temperature, water molecules further away from the central cation would have an increasingly less ordered orientation). One way to overcome the static character is using a (quantum-mechanical or classical) molecular-dynamics calculation [58–60]. Since there are hydrogen bonds between the original water molecules (Note: This is liquid water and not water from the gas phase!), the $n(H_2O)$ part on the left can be corrected by adding in the heat of condensation of water (on the order of -0.4 eV per water molecule) or by simulating the water as a droplet or water in a periodic box. Similar to the approach shown in Equation (6), the ion can also be "dipped" in a dielectric fluid using, e.g., a COSMO, polarizable continuum model (PCM) or conductor-like PCM (C-PCM) [61–64] approach. Ideally, one would combine the first or first two hydration spheres using explicit water molecules with a dielectric-fluid model in a static, or even better, dynamic approach. In either case, it may be necessary, and highly advisable, to benchmarks the atomistic calculations against experimental values, where available, or other computational values from the literature because discrepancies of a few eV (which is a lot especially once energies are converted to equilibrium constants) are not uncommon. When comparing work from different sources, special attention has to be applied to the nature of such hydration energy values because they may be ΔG, ΔH, or values of some other aspect of the hydration/solvation energy. This often depends on the nature of the calculation (e.g., static $vs.$ dynamic) or the way experimental hydration energy values were obtained. The question of the thermodynamic nature of the hydration energy, in particular the entropy changes involved, leads to a more general consideration about the nature of the overall thermodynamics of incorporation with all related aspects of changes in the entropy. Ideally, one would like to obtain ΔG values in order to judge, how much incorporation is thermodynamically possible. In addition, in order to obtain ΔG values for the hydration energy, other sources of entropy changes need to be evaluated. Even if solid source and sink phases are used, there may be a change in the vibrational entropies. Since pure mineral phases may have less vibrational modes than phases with incorporated species, the $-T\Delta S$ term tends to decrease by a few tens of eV, making incorporation somewhat less energetically uphill and, thus, somewhat more likely. Another (sometimes major) source of entropy changes may be a change from the number of reactants to the number of products; if the number of species goes up, so does the entropy, and $vice\ versa$.

In this paper, we have presented an approach to combine charged aqueous species as source and sink phases for incorporation reactions into solid mineral phases. This approach is widely applicable to a plethora of petrological, mineralogical, materials, environmental, or nuclear science applications. While the usual thoroughness of choosing computational parameters is indicated, special care has to be applied to the treatment of hydration energies, their distinction in terms of ΔG and ΔH values, and the treatment of other sources of entropy changes (vibrational, change in the number of species on both sides of the reaction). With these issues in mind, it should also be possible to create a database that allows for a widespread usage of this modular way of combining equations. This is because a lot of the equations (e.g., hydration energies) can be "re-used" with future users being able to focus on the energetics of the host, the host with the incorporated ion, and the (easier to calculate) solid source and sink phases. With such a publically available dataset, this approach may be a powerful tool to evaluate the thermodynamics of incorporation reactions, and aid in the benchmarking of experiments that tend to be subject to kinetic limitations.

Acknowledgments

Lindsay C. Shuller-Nickles acknowledges the U.S. Nuclear Regulatory Commission, Jr. Faculty Award #NRC-38-10-921, to Clemson University for support. The work on actinide incorporation into magnetite (Will M. Bender and Udo Becker) was supported the U.S. Department of Energy Biological and Environmental Research (DOE-BER) grant SC-0004883, the one on carbonates and sulfates (Sarah M. Walker and Udo Becker), by the U.S. Department of Energy Basic Energy Sciences (DOE-BES) grant DE-FG02-06ER15783.

Author Contributions

Lindsay C. Shuller-Nickles and Udo Becker developed the backbone of the method for calculating incorporation energies for neptunyl incorporation into uranyl minerals. Sarah M. Walker and Udo Becker further expanded the method for uranyl and neptunyl incorporation into carbonate and sulfate systems. Will M. Bender applied the method with a focus on uranium incorporation into magnetite. All authors contributed substantially to the writing of this manuscript.

Appendix

Table A1. Neptunyl (NpO_2^+) Incorporation into Anglesite ($PbSO_4$).

Solid Oxide Source and Sink Phase Reaction	ΔE (eV)
$(PbSO_4)_{4(s)} + \frac{1}{2}Np_2O_{5(s)} + \frac{1}{2}H_2O_{(pbc\ molc.)} \rightarrow Pb_3NpO_2(HSO_4)(SO_4)_{3(s)} + PbO_{(s)}$	2.99
Solid and Sink Conversion to Gas-Phase Periodic Molecules	
$NpO_2(OH)_{(pbc\ m)} \rightarrow \frac{1}{2}Np_2O_{5(s)} + \frac{1}{2}H_2O_{(pbc\ m)}$	−3.22
$PbO_{(s)} + H_2O_{(pbc\ m)} \rightarrow Pb(OH)_{2(pbc\ m)}$	0.62
Transition from Periodic Molecules to Gas-Phase Clusters	
$NpO_2(OH)_{(clus)} \rightarrow NpO_2(OH)_{(pbc\ m)}$	−0.04
$Pb(OH)_{2(pbc\ m)} + H_2O_{(clus)} \rightarrow Pb(OH)_{2(clus)} + H_2O_{(pbc\ m)}$	0.02
Dissociation and Ionization of Clusters	
$NpO_2^+{}_{(clus)} + OH^-{}_{(clus)} \rightarrow NpO_2(OH)_{(clus)}$	−10.40
$Pb(OH)_{2(clus)} \rightarrow Pb^{2+}{}_{(clus)} + 2OH^-{}_{(clus)}$	26.54
$H_3O^+{}_{(clus)} + OH^-{}_{(clus)} \rightarrow 2H_2O_{(clus)}$	−10.71
Hydration	
$NpO_2^+{}_{(aq)} \rightarrow NpO_2^+{}_{(clus)}$	7.50
$Pb^{2+}{}_{(clus)} \rightarrow Pb^{2+}{}_{(aq)}$ (calculation)	−14.77
$Pb^{2+}{}_{(clus)} \rightarrow Pb^{2+}{}_{(aq)}$ (experiment [a])	−16.27
$H_3O^+{}_{(aq)} \rightarrow H_3O^+{}_{(clus)}$	4.7 [b]
$H_2O_{(clus)} \rightarrow H_2O_{(aq)}$	−0.41 [c]
Overall Reaction	
$(PbSO_4)_{4(s)} + NpO_2^+{}_{(aq)} + H_3O^+{}_{(aq)} \rightarrow Pb_3NpO_2(HSO_4)(SO_4)_{3(s)} + Pb^{2+}{}_{(aq)} + H_2O_{(aq)}$	1.32

Notes: pbc m = molecule/cluster with periodic boundary conditions, "molecule in a box" clus = cluster/molecule in vacuum (gas phase); [a] q = hydrated (non-periodic) cluster, aqueous iona experimental value from [65] which has been used for the sum in the overall equation rather than the calculated value above; [b] from [53], ΔG value; [c] enthalpy of condensation for water.

Table A2. U^{6+} incorporation into octahedral Fe^{3+} site of magnetite (Fe_3O_4), charge balanced by octahedral Fe^{3+} vacancy.

Solid Oxide Source and Sink Phase Reaction	ΔE (eV)
$Fe_6O_{8(s)} + UO_{3(s)} \rightarrow Fe_4UO_{8(s)} + Fe_2O_{3(s)}$	1.14
Source/Sink Conversion to Gas-Phase Periodic Molecules	
$[UO_2(OH)_2(H_2O)_3]_{(pbc\,m)} \rightarrow UO_{3(s)} + 4H_2O_{(pbc\,m)}$	0.71
$Fe_2O_{3(s)} + 9H_2O_{(pbc\,m)} \rightarrow 2[Fe(OH)_3(H_2O)_3]_{(pbc\,m)}$	-1.84
Transition from Periodic Molecules to Gas-Phase Clusters	
$5H_2O_{(clus)} \rightarrow 5H_2O_{(pbc\,m)}$	-0.02
$[UO_2(OH)_2(H_2O)_3]_{(clus)} \rightarrow [UO_2(OH)_2(H_2O)_3]_{(pbc\,m)}$	0.18
$2[Fe(OH)_3(H_2O)_3]_{(pbc\,m)} \rightarrow 2[Fe(OH)_3(H_2O)_3]_{(clus)}$	-0.01
Dissociation and Ionization of Clusters	
$UO_2{}^{2+}{}_{(clus)} + 2OH^-{}_{(clus)} + 3H_2O_{(clus)} \rightarrow [UO_2(OH)_2(H_2O)_3]_{(clus)}$	-31.7
$2[Fe(OH)_3(H_2O)_3]_{(clus)} + 6H_3O^+{}_{(clus)} \rightarrow 2Fe^{3+}{}_{(clus)} + 18H_2O_{(clus)}$	73.9
Hydration and Balancing of Protons	
$[UO_2(H_2O)_5]^{2+}{}_{(aq)} \rightarrow UO_2{}^{2+}{}_{(clus)} + 5H_2O_{(clus)}$	17.2 [a]
$2Fe^{3+}{}_{(clus)} + 12H_2O_{(clus)} \rightarrow 2[Fe(H_2O)_6]^{3+}{}_{(aq)}$	-95.8 [b]
$4H_3O^+{}_{(aq)} \rightarrow 4H_3O^+{}_{(clus)}$	19.1 [c]
$4H_2O_{(clus)} \rightarrow 2H_3O^+{}_{(clus)} + 2OH^-{}_{(clus)}$	19.6 [d]
$H_2O_{(aq)} \rightarrow H_2O_{(clus)}$	0.41 [e]
Overall Reaction	
$Fe_6O_{8(s)} + [UO_2(H_2O)_5]^{2+}{}_{(aq)} + 4H_3O^+{}_{(aq)} + H_2O_{(aq)} \rightarrow Fe_4UO_{8(s)} + 2[Fe(H_2O)_6]^{3+}{}_{(aq)}$	2.84

Notes: pbc m = molecule/cluster with periodic boundary conditions, "molecule in a box" clus = cluster/molecule in vacuum (gas phase); [a] q = hydrated (non-periodic) cluster, aqueous ion [a] experimental value from [66]; [b] from calculation (47.9 eV for one Fe^{3+}), compared with experimental result of [67], 46.2 eV, and computational result of [58], 47.7 eV; [c] from [53], ΔG value; [d] from DMol^3 calculations; [e] enthalpy of condensation for water.

Conflicts of Interest

The authors declare no conflict of interest.

References

1. Shuller, L.C.; Ewing, R.C.; Becker, U. Quantum-mechanical evaluation of Np-incorporation into studtite. *Am. Mineral.* **2010**, *95*, 1151–1160.

2. Shuller, L.C.; Ewing, R.C.; Becker, U. Np-incorporation into uranyl phases: A quantum-mechanical evaluation. *J. Nucl. Mater.* **2013**, *434*, 440–450.

3. Rak, Z.; Ewing, R.C.; Becker, U. Electronic structure and thermodynamic stability of uranium-doped yttrium iron garnet. *J. Phys. Condens. Matter* **2013**, *25*, doi:10.1088/0953-8984/25/49/495502.

4. Rak, Z.; Ewing, R.C.; Becker, U. Ferric garnet matrices for immobilization of actinides. *J. Nucl. Mater.* **2013**, *436*, 1–7.

5. Rak, Z.; Ewing, R.C.; Becker, U. Role of iron in the incorporation of uranium in ferric garnet matrices. *Phys. Rev. B* **2011**, *84*, doi:10.1103/PhysRevB.84.155128.

6. Rak, Z.; Ewing, R.C.; Becker, U. First-principles investigation of Ca_3(Ti, Zr, Hf, Sn)$_2Fe_2SiO_{12}$ garnet structure for incorporation of actinides. *Phys. Rev. B* **2011**, *83*, doi:10.1103/PhysRevB.83.155123.

7. Walker, S.M.; Becker, U. First-principles insights into uranyl(VI) and neptunyl(V) incorporation in carbonate and sulfate minerals. In prepration.

8. Skomurski, F.N.; Rosso, K.M.; Krupka, K.M.; McGrail, B.P. Technetium incorporation into hematite (α-Fe_2O_3). *Environ. Sci. Technol.* **2010**, *44*, 5855–5861.

9. Kerisit, S.; Felmy, A.R.; Ilton, E.S. Atomistic simulations of uranium incorporation into iron (hydr)oxides. *Environ. Sci. Technol.* **2011**, *45*, 2770–2776.

10. Nico, P.S.; Stewart, B.D.; Fendorf, S. Incorporation of oxidized uranium into Fe (hydr)oxides during Fe(II) catalyzed remineralization. *Environ. Sci. Technol.* **2009**, *43*, 7391–7396.

11. Rosso, K.M.; Rustad, J.R.; Bylaska, E.J. The Cs/K exchange in muscovite interlayers: An *ab initio* treatment. *Clays Clay Miner.* **2001**, *49*, 500–513.

12. Kuo, E.Y.; Qin, M.J.; Thorogood, G.J.; Whittle, K.R.; Lumpkin, G.R.; Middleburgh, S.C. Technetium and ruthenium incorporation into rutile TiO_2. *J. Nucl. Mater.* **2013**, *441*, 380–389.

13. Perdew, J.P.; Burke, K.; Ernzerhof, M. Generalized gradient approximation made simple. *Phys. Rev. Lett.* **1996**, *77*, 3865–3868.

14. Winkler, B. An introduction to "Computational Crystallography". *Z. Krist.* **1999**, *214*, 506–527.

15. Shuller, L.C.; Ewing, R.C.; Becker, U. Thermodynamic properties of $Th_xU_{1-x}O_2$ ($0 < x < 1$) based on quantum-mechanical calculations and Monte-Carlo simulations. *J. Nucl. Mater.* **2011**, *412*, 13–21.

16. Diener, A.; Neumann, T. Synthesis and incorporation of selenide in pyrite and mackinawite. *Radiochim. Acta* **2011**, *99*, 791–798.

17. Ferriss, E.D.A.; Essene, E.J.; Becker, U. Computational study of the effect of pressure on the Ti-in-zircon geothermometer. *Eur. J. Mineral.* **2008**, *20*, 745–755.

18. Carpenter, M.A.; McConnell, J.D.C.; Navrotsky, A. Enthalpies of ordering in the plagioclase feldspar solid-solution. *Geochim. Cosmochim. Acta* **1985**, *49*, 947–966.

19. Vinograd, V.L.; Putnis, A.; Kroll, H. Structural discontinuities in plagioclase and constraints on mixing properties of the low series: A computational study. *Mineral. Mag.* **2001**, *65*, 1–31.

20. Li, P.; Chen, I.W.; Pennerhahn, J.E. Effect of dopants on zirconia stabilization—An X-ray-absorption study: I, Trivalent dopants. *J. Am. Ceram. Soc.* **1994**, *77*, 118–128.

21. Kelly, S.D.; Newville, M.G.; Cheng, L.; Kemner, K.M.; Sutton, S.R.; Fenter, P.; Sturchio, N.C.; Spotl, C. Uranyl incorporation in natural calcite. *Environ. Sci. Technol.* **2003**, *37*, 1284–1287.

22. Pingitore, N.E.; Lytle, F.W.; Davies, B.M.; Eastman, M.P.; Eller, P.G.; Larson, E.M. Mode of incorporation of Sr^{2+} in calcite—Determination by X-ray absorption spectroscopy. *Geochim. Cosmochim. Acta* **1992**, *56*, 1531–1538.

23. Demichelis, R.; Raiteri, P.; Gale, J.D.; Dovesi, R. Examining the accuracy of density functional theory for predicting the thermodynamics of water incorporation into minerals: The hydrates of calcium carbonate. *J. Phys. Chem. C* **2013**, *117*, 17814–17823.

24. Adamo, C.; Barone, V. Toward reliable density functional methods without adjustable parameters: The PBE0 model. *J. Chem. Phys.* **1999**, *110*, 6158–6170.

25. Becke, A.D. Density-functional thermochemistry. III. The role of exact exchange. *J. Chem. Phys.* **1993**, *98*, 5648–5652.

26. Lee, C.T.; Yang, W.T.; Parr, R.G. Development of the Colle-Salvetti correlation-energy formula into a functional of the electron-density. *Phys. Rev. B* **1988**, *37*, 785–789.

27. Grimme, S. Accurate description of van der Waals complexes by density functional theory including empirical corrections. *J. Comput. Chem.* **2004**, *25*, 1463–1473.

28. Grimme, S. Semiempirical GGA-type density functional constructed with a long-range dispersion correction. *J. Comput. Chem.* **2006**, *27*, 1787–1799.

29. Tsuchiya, T.; Wang, X.L. *Ab initio* investigation on the high-temperature thermodynamic properties of Fe^{3+}-bearing $MgSiO_3$ perovskite. *J. Geophys. Res. Solid Earth* **2013**, *118*, 83–91.

30. Cococcioni, M.; de Gironcoli, S. Linear response approach to the calculation of the effective interaction parameters in the LDA+U method. *Phys. Rev. B* **2005**, *71*, doi:10.1103/PhysRevB.71.035105.

31. Alfe, D. PHON: A program to calculate phonons using the small displacement method. *Comput. Phys. Commun.* **2009**, *180*, 2622–2633.

32. Delley, B. An all-electron numerical method for solving the local density functional for polyatomic molecules. *J. Chem. Phys.* **1990**, *92*, 508–517.

33. Delley, B. From molecules to solids with the $DMol^3$ approach. *J. Chem. Phys.* **2000**, *113*, 7756–7764.

34. Kresse, G.; Furthmuller, J. Efficient iterative schemes for ab initio total-energy calculations using a plane-wave basis set. *Phys. Rev. B* **1996**, *54*, 11169–11186.

35. Kresse, G.; Furthmuller, J. Efficiency of *ab-initio* total energy calculations for metals and semiconductors using a plane-wave basis set. *Comput. Mater. Sci.* **1996**, *6*, 15–50.

36. Kresse, G.; Hafner, J. Ab inito molecular-dynamics for liquid-metals. *Phys. Rev. B* **1993**, *47*, 558–561.

37. Kresse, G.; Hafner, J. *Ab initio* molecular-dynamics simulation of the liquid-metal amorphous-semiconductor transition in germanium. *Phys. Rev. B* **1994**, *49*, 14251–14269.

38. Carter, M.L.; Stewart, M.W.A.; Vance, E.R.; Begg, B.D.; Moricca, S.; Tripp, J. HIPed Tailored Cermaic Waste Forms for the Immobilization of Cs, Sr and Tc. In *Global 2007: Advanced Nuclear Fuel Cycles and Systems*; American Nuclear Society: Boise, ID, USA, 2007; pp. 1022–1028.

39. Rodriguez, E.E.; Poineau, F.; Llobet, A.; Sattelberger, A.P.; Bhattacharjee, J.; Waghmare, U.V.; Hartmann, T.; Cheetham, A.K. Structural studies of TcO_2 by neutron powder diffraction and first-principles calculations. *J. Am. Chem. Soc.* **2007**, *129*, 10244–10248.

40. Bolzan, A.A.; Fong, C.; Kennedy, B.J.; Howard, C.J. Structural studies of rutile-type metal dioxides. *Acta Crystallogr. Sect. B* **1997**, *53*, 373–380.

41. Dovesi, R.; Saunders, V.R.; Roetti, C.; Orlando, R.; Zicovich-Wilson, C.M.; Pascale, F.; Civalleri, B.; Doll, K.; Harrison, N.M.; Bush, I.J.; *et al. CRYSTAL09 User's Manual*; University of Torino: Torino, Italy, 2009.

42. Segall, M.D.; Lindman, P.J.K.; Probert, M.J.; Pickard, C.J.; Hasnip, P.J.; Clark, S.J.; Payne, M.C. First-principles simulation: Ideas, illustrations and the CASTEP code. *J. Phys.* **2002**, *14*, 2717–2744.

43. Klamt, A.; Schuurmann, G. COSMO: A new approach to dielectric screening in solvents with explicit expressions for the screening energy and its gradient. *J. Chem. Soc. Perkin Trans.* **1993**, *2*, 799–805.

44. Dovesi, R.; Orlando, R.; Civalleri, B.; Roetti, C.; Saunders, V.R.; Zicovich-Wilson, C.M. CRYSTAL: A computational tool for the ab initio study of the electronic properties of crystals. *Z. Krist.* **2005**, *220*, 571–573.

45. Gutowski, K.E.; Dixon, D.A. Predicting the energy of the water exchange reaction and free energy of solvation for the uranyl ion in aqueous solution. *J. Phys. Chem. A* **2006**, *110*, 8840–8856.

46. Burns, P.C.; Klingensmith, A.L. Uranium mineralogy and neptunium mobility. *Elements* **2006**, *2*, 351–356.

47. Vanderbilt, D. Soft self-consistent pseudopotentials in a generalized eigenvalue formalism. *Phys. Rev. B* **1990**, *41*, 7892–7895.

48. Reeder, R.J.; Elzinga, E.J.; Tait, C.D.; Rector, K.D.; Donohoe, R.J.; Morris, D.E. Site-specific incorporation of uranyl carbonate species at the calcite surface. *Geochim. Cosmochim. Acta* **2004**, *68*, 4799–4808.

49. Dodge, C.J.; Francis, A.J.; Gillow, J.B.; Halada, G.P.; Eng, C.; Clayton, C.R. Association of uranium with iron oxides typically formed on corroding steel surfaces. *Environ. Sci. Technol.* **2002**, *36*, 3504–3511.

50. Scott, T.B.; Allen, G.C.; Heard, P.J.; Randell, M.G. Reduction of U(VI) to U(IV) on the surface of magnetite. *Geochim. Cosmochim. Acta* **2005**, *69*, 5639–5646.

51. Hood, D.; Wen, Y.; Shuller-Nickels, L. Redox behavior of uranium incorporated into hematite. *Mineral. Mag.* **2013**, *77*, doi:10.1180/minmag.2013.077.5.8.

52. Bengtson, A.; Morgan, D.; Becker, U. Spin state of iron in Fe_3O_4 magnetite and h-Fe_3O_4. *Phys. Rev. B* **2013**, *87*, doi:10.1103/PhysRevB.87.155141.

53. Camaioni, D.M.; Schwerdtfeger, C.A. Comment on Accurate experimental values for the free energies of hydration of H^+, OH^-, and H_3O^+. *J. Phys. Chem. A* **2005**, *109*, 10795–10797.

54. Clark, A.E.; Samuels, A.; Wisuri, K.; Landstrom, S.; Saul, T. Thermodynamic and spectroscopic assignment of the aqueous solvation environments of tri- and hexavalent U, Np, and Pu using large hydrated clusters calculated by density functional theory. *Inorg. Chem.* **2014**, submitted for publication.

55. Ferriss, E.D.A.; Ewing, R.C.; Becker, U. Simulation of thermodynamic mixing properties of actinide-containing zircon solid solutions. *Am. Mineral.* **2010**, *95*, 229–241.

56. Duff, M.C.; Coughlin, J.U.; Hunter, D.B. Uranium co-precipitation with iron oxide minerals. *Geochim. Cosmochim. Acta* **2002**, *66*, 3533–3547.

57. Heyd, J.; Scuseria, G.E. Efficient hybrid density functional calculations in solids: Assessment of the Heyd-Scuseria-Ernzerhof screened coulomb hybrid functional. *J. Chem. Phys.* **2004**, *121*, 1187–1192.

58. Remsungnen, T.; Rode, B.M. QM/MM molecular dynamics simulation of the structure of hydrated Fe(II) and Fe(III) ions. *J. Phys. Chem. A* **2003**, *107*, 2324–2328.

59. Remsungnen, T.; Rode, B.M. Molecular dynamics simulation of the hydration of transition metal ions: The role of non-additive effects in the hydration shells of Fe^{2+} and Fe^{3+} ions. *Chem. Phys. Lett.* **2004**, *385*, 491–497.

60. Remsungnen, T.; Rode, B.M. Dynamical properties of the water molecules in the hydration shells of Fe(II) and Fe(III) ions: *Ab initio* QM/MM molecular dynamics simulations. *Chem. Phys. Lett.* **2003**, *367*, 586–592.

61. Cances, E.; Mennucci, B.; Tomasi, J. A new integral equation formalism for the polarizable continuum model: Theoretical background and applications to isotropic and anisotropic dielectrics. *J. Chem. Phys.* **1997**, *107*, 3032–3041.

62. Cossi, M.; Rega, N.; Scalmani, G.; Barone, V. Energies, structures, and electronic properties of molecules in solution with the C-PCM solvation model. *J. Comput. Chem.* **2003**, *24*, 669–681.

63. Cossi, M.; Scalmani, G.; Rega, N.; Barone, V. New developments in the polarizable continuum model for quantum mechanical and classical calculations on molecules in solution. *J. Chem. Phys.* **2002**, *117*, 43–54.

64. Mennucci, B.; Tomasi, J. Continuum solvation models: A new approach to the problem of solute's charge distribution and cavity boundaries. *J. Chem. Phys.* **1997**, *106*, 5151–5158.

65. Marcus, Y. A simple empirical model describing the thermodynamics of hydration of ions of widely varying charges, sizes, and shapes. *Biophys. Chem.* **1994**, *51*, 111–127.

66. Gibson, J.K.; Haire, R.G.; Santos, M.; Marcalo, J.; de Matos, A.P. Oxidation studies of dipositive actinide ions, An^{2+} (An = Th, U, Np, Pu, Am) in the gas phase: Synthesis and characterization of the isolated uranyl, neptunyl, and plutonyl ions $UO_2^{2+}(g)$, $NpO_2^{2+}(g)$, and $PuO_2^{2+}(g)$. *J. Phys. Chem. A* **2005**, *109*, 2768–2781.

67. Marcus, Y. *Ion Solvation*; Wiley: New York, NY, USA, 1985.

5

Testing of Ore Comminution Behavior in the Geometallurgical Context—A Review

Abdul Mwanga *, Jan Rosenkranz and Pertti Lamberg

Minerals and Metallurgical Engineering Laboratory, Luleå University of Technology,
SE-971 87 Luleå, Sweden; E-Mails: jan.rosenkranz@ltu.se (J.R.); pertti.lamberg@ltu.se (P.L.)

* Author to whom correspondence should be addressed; E-Mails: abdul.mwanga@ltu.se

Academic Editor: Stephen E. Haggerty

Abstract: Comminution tests are an important element in the proper design of ore beneficiation plants. In the past, test work has been conducted for particular representative reference samples. Within geometallurgy the entire ore body is explored in order to further identify the variation within the resource and to establish spatial geometallurgical domains that show the differential response to mineral processing. Setting up a geometallurgical program for an ore deposit requires extensive test work. Methods for testing the comminution behavior must therefore be more efficient in terms of time and cost but also with respect to sample requirements. The integration of the test method into the geometallurgical modeling framework is also important. This paper provides an overview of standard comminution test methods used for the investigation of ore comminution behavior and evaluates their applicability and potential in the geometallurgical context.

Keywords: geometallurgy; comminution behavior; metallurgical testing

---

1. Background

Before concentration, metal ores have to be crushed and ground in order for the metal bearing minerals to be liberated. Sufficient size reduction by comminution is not only a prerequisite for any downstream physical separation, but is also the processing step within mineral processing that has the highest energy demand, and which, in practice, is often the limiting factor for plant capacity. Reliable testing of the ore's comminution behavior that provides information about the particle size distribution after fracture identifies the achieved mineral liberation, and determines the comminution energy needed, therefore an integral step in the proper design of ore beneficiation plants.

In the past, the process design has usually been based on particular representative reference samples that were analyzed and tested at different scales but failed to describe all the mineralogical variations occurring in an ore body. The limited picture obtained from this method can lead to insufficient mineral liberation, or, on the flipside, to overgrinding, which leads to low recovery and selectivity in physical separation [1], resulting in poor plant performance and limited production capacity.

During the last few years, geometallurgy has evolved as a new interdisciplinary approach to improving resource efficiency. Geometallurgy combines geological and metallurgical information to create a spatially-based predictive model for mineral processing to be used in production management [2]. Within the geometallurgical approach, the entire ore body is explored in order to identify the variation within the resource and to establish spatial geometallurgical domains that show differential responses to mineral processing. Using these geometallurgical domains in the design of ore beneficiation plants and mine scheduling, allows for higher flexibility in response to changes in the plant feed when mining different parts of the ore [3].

Setting up a geometallurgical program for an ore deposit requires extensive test work, particularly in comminution and physical separation. This can lead to economic benefits, owing to improved plant operation, yet is time consuming and correspondingly costly. The best possible utilization of the available sample material is crucial, since chemical and rock mechanical analyses require samples and thereby limit the available sample amount. Accordingly, the need for efficient geometallurgical testing provides a stimulus for developing new fast comminution test methods or for revising extant test procedures. Improved process models need to be conceived that subsequently make use of the additional information provided for by the geometallurgical approach [4].

Particular requirements resulting from geometallurgical programs make it necessary to analyze the comminution test methods available. Based on published information, this paper provides an overview of standard test methods used for the investigation of ore comminution behavior and evaluates their applicability and potential in the geometallurgical context by defining the corresponding evaluation criteria. Besides the test work itself, the provision of suitable parameters for comminution process modeling is also considered.

2. Criteria of Evaluation

The term "comminution behavior" subsumes the complex interaction of material properties and process parameters. In practice, different comminution test methods are used for different applications. Figure 1 illustrates the dimensions for categorizing the existing comminution test methods. One of the most important dimensions is the particle size. As the comminution properties change with particle size, different tests ranging from crushing to grinding and very fine grinding are employed accordingly.

Furthermore the type of mechanical stress applied in a certain crusher or mill type, i.e., compressive or impact stress, as well as the stress intensity and rate, have to be taken into account in the context of selecting a suitable test method.

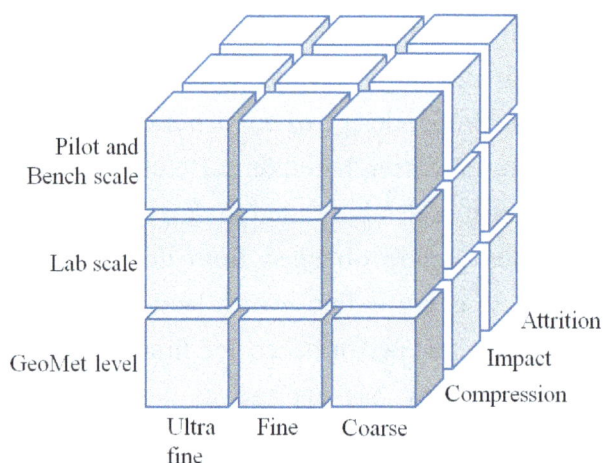

Figure 1. Dimensions in test categorization (GeoMet level = geometallurgical level).

Another dimension is given by the scale of the test, ranging from pilot scale testing down to testing on the geometallurgical level that requires very small sample amounts. The applicability of a comminution test will therefore depend on the project stage and the available sample amount, ranging from drill core sections or hand-picked single particles used in the early stages of the resource evaluation, and at laboratory scale, to pilot scale tests where the metallurgical performance is verified using several tens or hundreds of kilograms of sample. With larger scales the effort in terms of time and cost per test increases. Particularly this third dimension of available sample material will limit the test work to certain methods, *i.e.*, during the different stages of a project tests at different scales need to be considered.

For geometallurgical purposes, the outcome of the experimental work not only is used for resource characterization but has to serve as an input to process modeling, *i.e.*, comminution test results need to be linked to the parameters used in the comminution process models. Within process modeling, different levels of modeling depth and complexity are used. Simple approaches use defined size distribution functions based on single parameters such as energy for grinding or machine-specific size reduction ratios. More sophisticated, rigorous models apply population-balance methods. Here the entire breakage distribution function needs to be constructed based on experimental test work or sampling and back-calculation from continuous comminution tests at an even larger scale [5]. In this context, it must be noted that a methodology that both comprises size reduction and energy for size reduction, and also incorporates mineral liberation within experimental work and process modeling, is still missing.

As discussed above, a geometallurgical program imposes particular requirements on the efficiency and manageability of test methods. Typically, the number of samples needed for realistic testing of the variability within a deposit is relatively large (number of samples > 1000) before an ore body can be divided into domains [6]. A comminution test for geometallurgical purposes should therefore fulfill several technical and economical criteria:

- Simplicity: The test should be relatively simple and easy to implement. It should use instruments that are readily available in conventional analytical and mineral processing laboratories.
- Repeatability: The test should be repeatable and not depend on the individual person conducting the test work.

- Sample preparation: Sample preparation should be possible with low effort and possible to undertake using basic skills or after a short training period.
- Time exposure and costs: The test should be fast and inexpensive, *i.e.*, for execution times of one hour it should be possible to conduct 1000 tests within half a year.
- Sample amount: The amount of sample per test should be < 0.5 kg. Preferably rejects from chemical assays should be sufficient.
- Link to modeling: Tests should provide parameters that can directly be used in process modeling and simulation.
- Mineral liberation: Tests should be easy to extend in order to complement mineral liberation information.

Another criterion is the precision and the statistical quality of the test results. With respect to accuracy, a proper quantification is not an easy task, since the entire chain of sampling and sample preparation, in conjunction with the test and analysis method, needs to be considered. Statistical quality is a parameter that, from the perspective of geometallurgy, not only has to be judged with respect to the repetition of single tests but also in relation to generating a comprehensive data set for the entire geometallurgical program, *i.e.*, a compromise between the quality of single measurements and the overall quantity of measured points must be reached.

3. Review of Existing Test Methods

In the following sections, commonly-used comminution test methods which are of potential use in the geometallurgical context are reviewed against the criteria listed above. The tests are categorized into three groups: (1) rock mechanical tests, (2) particle breakage tests, and (3) bench-scale grindability tests.

3.1. Rock Mechanical Testing

Rock mechanical tests for determining rock strength are usually conducted by means of compressive loading (universal test machines) or simplified instruments. Loading takes place at low velocities relative to other tests. The instrumentation of the test machine allows recording of the load applied to the sample and the displacement over time. Several standard test methods are used that differ in the loading conditions applied [7]. These include uniaxial compressive test, triaxial test, point load test, indirect tensile strength test, and hardness tests.

Samples consist of drill core sections, parts of drill core sections and single particles (grab samples) that are cut to regular shape, and also irregularly-shaped particles. Measured strength parameters depend not only on the modal mineralogy but also on texture and anisotropy [8].

3.1.1. Compression Tests

In compression tests, the drill core sample is pressed between the two parallel platens of the test machine, and is loaded until failure of the test specimen occurs. Data pairs of the applied load and the resulting displacement are recorded over time, yielding the maximum load for calculating the compressive strength. Sample preparation comprises careful cutting of the specimen's top and bottom planes in order to prevent bending effects during the test.

In case the specimen is not further supported, the test is referred to as the *unconfined compressive strength* (UCS) *test* (Figure 2). For triaxial compression tests, the drill core specimens are cut to the required length and then enveloped on the lateral surface by a membrane that seals the specimen from the surrounding pressure medium, usually oil, that provides the radial support. When increasing the axial compressive load, the oil pressure is likewise increased in parallel until failure of the specimen.

A similar experimental setup is used in the *point load test* (PLT), see Figure 3. Instead of using parallel planes, here the compressive load is applied between the tips of two cones, putting a point load on the specimen. These test instruments are quite compact and can even be used in the field.

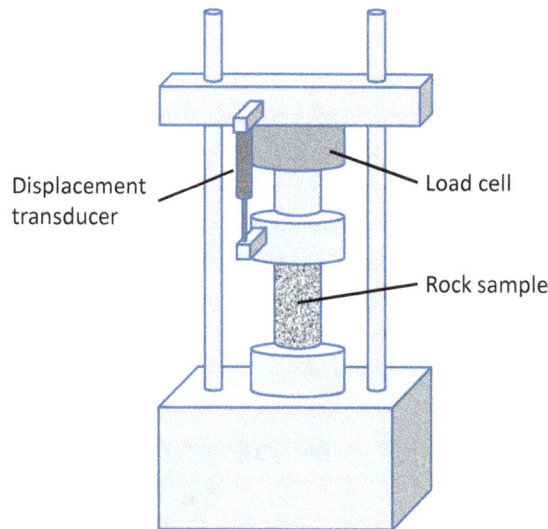

Figure 2. Uniaxial unconfined compression test.

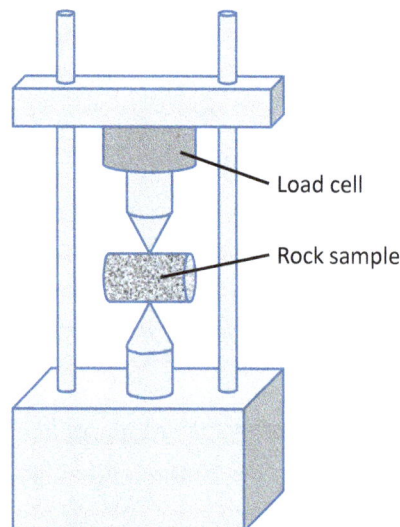

Figure 3. Point load test.

The point load test gives the point load strength index Is_{50} which needs to be corrected to the standard equivalent core diameter D_e of 50 mm if other specimen sizes are used. The Is_{50} is calculated by:

$$Is_{50} = P/D_e^2$$

(1)

where P is the failure load measured in Newton (N) and D_e is in millimeter (mm). The point load strength index can be transferred into uniaxial compressive strength using a linear conversion factor that needs to be determined for a particular ore [9]:

$$UCS = k \times I_{S_{50}}$$ (2)

where k is the ore -specific calibration constant.

3.1.2. Indirect Tensile Strength Test

Using cylindrical specimens, such as slices from drill cores, and loading them radially with compressive forces between two sockets (plates, cushioned plates, and curved clamps) gives an indirect tensile stress and a corresponding deformation in orthogonal direction. This experimental setup is also known as the *Brazilian test* (Figure 4).

The sample is stressed under a linear compressive load that induces tensile stress. Assuming that the material is homogeneous, isotropic, and linearly elastic before failure [10], the failure is expected at the maximum tensile stress. The corresponding tensile strength σ_t (in N/mm^2) is then calculated by

$$\sigma_t = \frac{2 \cdot P}{\pi \cdot D \cdot t}$$ (3)

where P is the failure load (N), D the specimen diameter (mm), and t the thickness of the test specimen (mm).

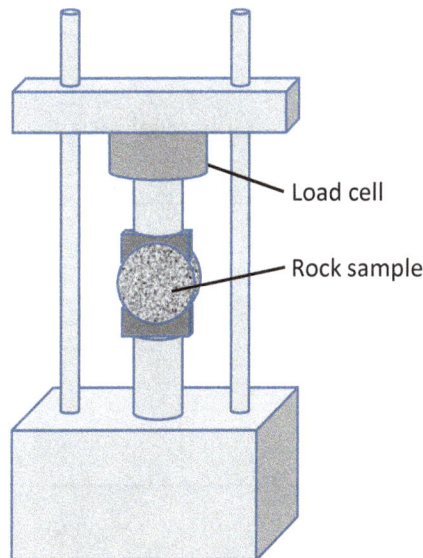

Figure 4. Brazilian test.

3.1.3. Evaluation of Rock Mechanical Tests

Using results from rock mechanical tests to describe comminution behavior is an attractive approach, since no additional test work and sample material are needed. The necessary sample amounts are also small.

All the tests discussed above apply compressive stress in a well-defined way, meaning that the repeatability of the method is ensured, though the interpretation has to take into account the textural effects and the inhomogeneity of samples found in natural mineral resources. Implementation of the test is rather simple and can partly be done in the field. More effort has to be put into proper sample preparation, e.g., when sawing drill cores. Regarding mineral liberation analysis, the fragments generated are usually too coarse to be used in quantitative mineralogical analyses.

It has been shown that the mechanical parameters of rock samples can be used to describe and model comminution processes in the case of crushing [11–13]. For this purpose, mechanical strength, expressed by the maximum load at the point of failure, needs to be transformed into quantities that can be used within the design of crushing stages, e.g., by linking the crusher reduction ratio or crushability index (CI) with compressive strength or impact strength index (ISI). For instance, Toraman et al. [14] suggest the following equation for calculating the CI:

$$CI = -1.21 \times ISI + 165.46 \qquad (4)$$

The empirical correlations that are provided for calculating the crushing index or crusher reduction ratio from UCS or PLT strength values are then ore-specific.

Another application of indirect measurements to determine rock strength properties has been pursued by using non-destructive hardness testers, e.g., the EQUOtip hardness tester [15,16], which provide hardness indices received from drill core samples that can be empirically related to comminution behavior. The EQUOtip measures the impact and rebound velocities which are processed to determine the hardness value. The hardness values show correlation to unconfined compressive strength.

The evaluation of the different rock mechanical tests with respect to the criteria defined in Section 2 is summarized in Table 1. As the criteria are mainly of qualitative nature, the assessment has been based on the three different attributes *adverse*, *acceptable* and *advantageous*.

Table 1. Rock mechanical tests.

Adverse (−), Acceptable (O), Advantageous (+)	UCS	PLT	Brazilian	Hardness Tester
Simplicity	+	+	+	+
Repeatability	O	O	O	O
Sample preparation	−	O	−	+
Time exposure and costs	O	O	O	+
Sample amount	+	+	+	+
Link to modeling	O	O	O	O
Mineral liberation	−	−	−	−

3.2. Particle Breakage Tests

3.2.1. Simple Drop Weight Test

In the drop weight test, a weight is lifted to a certain height and then released to fall on an ore fragment that has been arranged on a rigid anvil underneath (Figure 5).

The breakage product is collected and analyzed in terms of the particle size distribution to determine the breakage distribution. The test sample can comprise single particles, partial or complete drill core section or several coarser particles obtained from pre-crushing. Several test apparati have been conceived [11,17,18], with the JK drop weight test being the most well-known. In order to conduct particle bed fracture tests with finer particles, the anvil is replaced by a die [19].

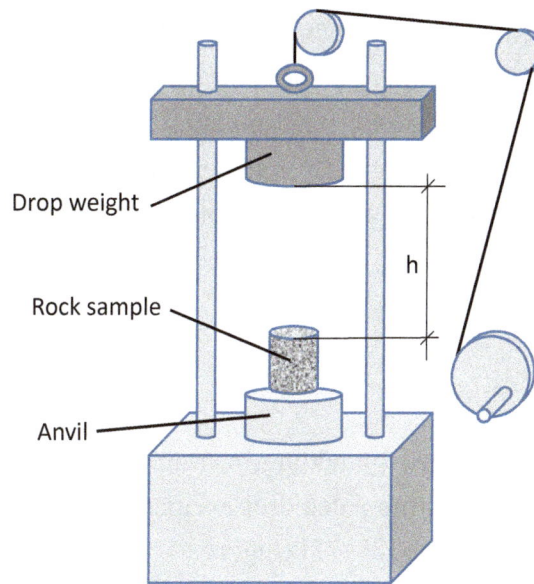

Figure 5. Drop weight tester.

Different impact levels and amounts of specific comminution energy can be obtained by varying the drop weight, the falling height or the fragment mass. The energy provided to the ore fragment can be described by the potential energy (in J/kg) of the drop weight at the initial height:

$$E = m_{dw} \cdot g \cdot h \tag{5}$$

where m_{dw} is the mass of the drop weight (in kg), g is gravitational acceleration (9.81 m/s^2) and h the distance (meters) between the drop weight and the top of the specimen.

In order to evaluate the breakage distribution, a method has been developed by Narayanan *et al.* that links the breakage distribution and comminution energy to the modeling of particle size reduction [19,20]. The percentage passing $1/n$ of the starting particle size can be related to the comminution energy by:

$$t_n = A \cdot \left(1 - e^{-b \cdot E_{spec}}\right) \tag{6}$$

where E_{spec} is the specific energy in kWh/t and A and b are ore-specific dimensionless parameters determined by fitting the model to the experimental data obtained from tests with different specific energies. Typically the t_{10}-value is used as a fineness index to characterize a certain ore sample. In process modeling and simulation an average set of A and b parameters are used assuming that particles of a different size will break in a similar way. For predicting the product size distribution at different grinding times or energy levels the t_n is related to t_{10}. The t_{10} parameter is also known as a function input energy (*i.e.*, impact energy) and material characteristics (*i.e.*, A and b). Therefore, Axb is used to describe the hardness of the ore. A higher value of Axb corresponds to soft material and vice versa.

In order to determine the several t_n curves, five different initial particle sizes in the range of 13.2 to 63 mm and three energy levels need to be investigated, giving a total of 15 tests. The drop weight test as defined by the Julius Kruttschnitt Mineral Research Centre (JKMRC) therefore normally requires 75 to 100 kg of sample material [20].

In order to reduce the sample amount and the number of particle size fractions, an abbreviated drop weight test, called *SAG Mill Comminution* (SMC) *test*, has been developed [21]. Using only a single size fraction, *i.e.*, crushed particles or parts cut from drill cores in the size range 16.2 to 31.5 mm, the necessary sample amount can be reduced to less than 5 kg. The test provides the parameters A and b as well as a Drop Weight Index (DW_i in kWh) but not the t_n family curves.

More recently, a method has been suggested for estimating the ore hardness parameters A and b from the crushing of drill core sections (or naturally fragmented rock) fragments within assay sample preparation [22]. The product size distribution generated from crushing in a laboratory jaw crusher are used to calculate two comminution indices Ci, that can be correlated to Axb and to the Bond ball mill work index (compare Section 3.3), respectively.

3.2.2. Instrumented Drop Weight Test

The drop weight test has been extended by adding instrumentation for measurements during the conduct of the test. A well-established instrumented drop weight test is the *Ultra-fast load cell device* (UFLC) developed at the University of Utah [23–25] (Figure 6).

A sample, consisting of individual particles or a bed of particles, is placed on a vertical steel bar and then impacted by the falling mass. As with the split Hopkinson pressure bar (see Section 3.2.3), the steel bar is instrumented by a pair of strain gauges to detect the impact waves, which allow for the determination of the load actually applied to the sample. The load-time profiles and the calculated deformation are used to estimate the actual transferred energy instead of using the potential energy.

A portable impact load cell, using the same principles as in the Ultra-fast load cell, has been developed by Bourgeois *et al.* [26] for *in situ* quantification of ore breakage properties. The so-called SILC—*Short Impact Load Cell*—is reduced in height and weighs only 30 kg.

Based on the principle design of the simple drop weight test, another instrumented drop weight tester has been designed [27] and has initially been used for the investigation of particle compaction processes. Here, the drop weight itself is instrumented by a load cell and an inductive displacement transducer that yields time-dependent measurement profiles for the entire sequence of primary impact and subsequent rebounds (Figure 7). The comminution energy transferred to the sample is determined using integral calculus for the load-displacement relation.

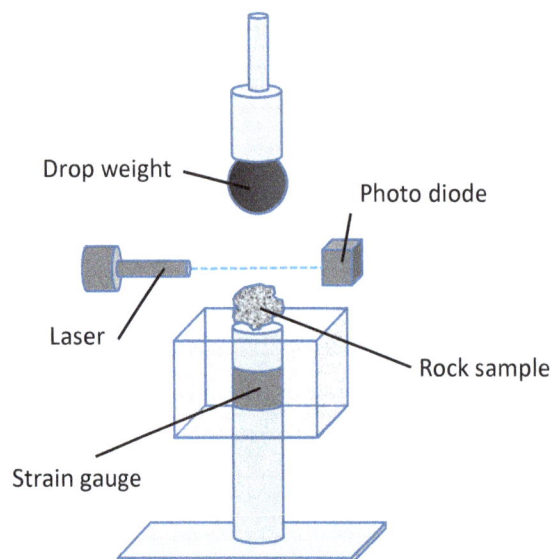

Figure 6. Ultra-fast load cell.

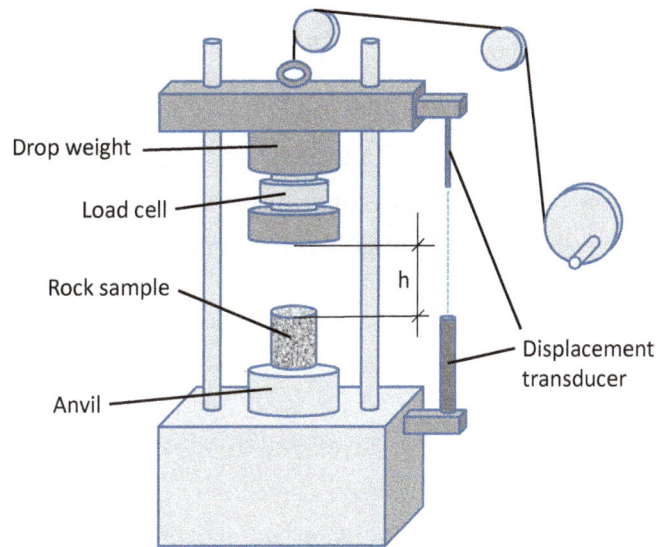

Figure 7. Micro-stamping device.

3.2.3. Twin Pendulum Tests

In twin pendulum tests, a single particle is fractured between two pendulum-mounted hammers that are released from a certain height. Figure 8 shows the experimental setup for the Bond twin pendulum test [28]. A single particle is mounted on a socket and then simultaneously hit by the hammers. The procedure is repeated until the particle breaks, thereby incrementally increasing the deflection angle of the hammers.

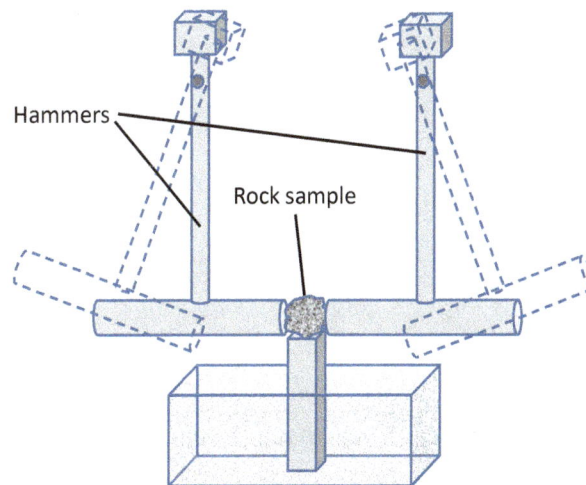

Figure 8. Bond twin pendulum tester.

Bond defined a crushing work index (in kWh/t) by:

$$CW_i = \frac{53.5 \cdot C_B}{\rho_p}$$

(7)

where ρ_p is the particle density in g/cm^3 and C_B the impact energy per particle thickness d_p in J/mm for the last pendulum deflection angle Θ (in degree):

$$C_B = \frac{117 \cdot (1 - \Theta)}{d_p} \qquad (8)$$

The twin pendulum test has been extended by Narayanan by adding instrumentation that allows the recording of the pendulum motion [28–30]. Lifting only one pendulum and mounting the particle on the other allows for determining the energy actually transferred to the sample by evaluating the rebound movements of the two pendulums after collision.

3.2.4. Split Hopkinson Pressure Bar Test

The split Hopkinson pressure bar, originally developed for testing the propagation of stress in materials from the detonation of explosives, has been used for fracturing a sample particle between two horizontally mounted steel bars, called the incident bar and the transmitted bar (Figure 9). The mechanical stress is induced by a loading system, *i.e.*, a gas gun, and the travel of the deformation waves is recorded by means of strain gauges. The signals provide information about load-displacement profiles and allow for energy balancing.

Even though the experiments with the Hopkinson pressure bar are quite laborious, there has been a phase of intensive development of the test method, leading to the Modified Hopkinson Pressure Bar for higher resolution [31] or the CSIRO Hopkinson pressure bar for larger specimen [32] having a vertical assembly like the Ultrafast load cell.

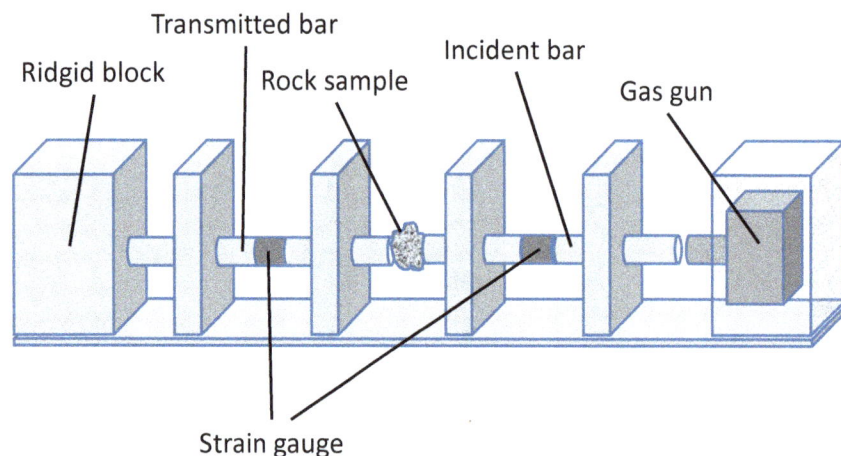

Figure 9. Split Hopkinson pressure bar.

3.2.5. Rotary Single Impact Tester

In single impact tests, the stress applied to a sample results from the collision with a single impact surface. This can be achieved either by accelerating the sample against a plate using gravitational forces or the acceleration forces from a gun, or by advancing the sample with a fast moving tool, commonly realized in a rotor-stator impact system.

Such a rotary impact tester design was first presented by Schönert *et al.* [33] and has meanwhile been commercially adapted also for ore testing by Kojovic *et al.* [34–36]. Figure 10 shows a schematic of the rotor–stator impacting system.

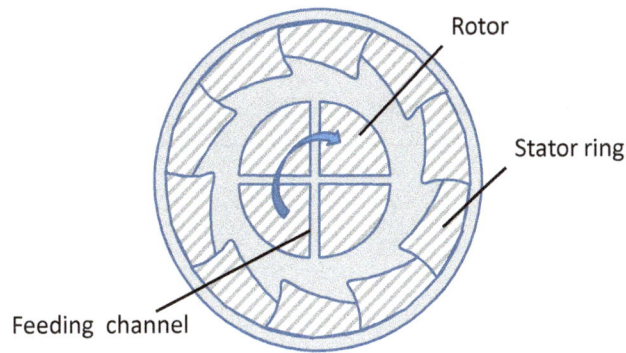

Figure 10. Rotary impact tester.

Particles are fed into the evacuated impact chamber by centrifugal action via several channels in the moving rotor. The surrounding stator has a saw tooth profile that allows for a perpendicular impact of the particles. Compared to other particle breakage tests, significantly more particles and samples can be tested in a given time using this method. The specific kinetic energy (J/kg) used in the comminution test is described by:

$$E_{kin} = \frac{m}{2} \cdot v^2 \bigg/ m = \frac{1}{2} \cdot v^2 \tag{9}$$

where v is impact velocity in m/s and m is the sample mass. Equation (9) shows that the specific breakage energy is independent of the sample mass, which is a unique characteristic of the rotary breakage test.

The results can be used for establishing energy–size reduction relationships within process modeling. One approach is to determine the parameters for population balance models, probability of breakage and breakage distributions. The probability of breakage can, for instance, be described as a function of the impact energy using a Weibull distribution [37]:

$$S = 1 - \exp\left(f_{mat} \cdot k \cdot x \cdot \left(E_{kin} - E_{kin,\min} \right) \right) \tag{10}$$

where f_{mat} is a material constant, x the initial particle size in mm, k the number of impacts, and E_{kin} and $E_{kin,min}$ (in joules) are the kinetic energy impact and the energy threshold without fracture, respectively.

3.2.6. Evaluation of Particle Breakage Tests

The evaluation of the different particle breakage tests with respect to the criteria defined in Section 2 is summarized in Table 2. Particle breakage tests rely on impact stress to break the sample. Except for the rotary breakage test, the mounting of the sample and the conduct of the tests are, in most cases, quite tedious. Repeatability is more affected by the sample characteristics than by the test method.

The necessary sample amounts are small, if using cut drill core samples or fragments. Testing fractionated material from crushing and screening requires significantly more material. This is particularly the case if not only the initial particle size is varied, but also if different energy levels are tested, e.g., when using the concept of t_n-values for process modeling.

Based on the initial particle size, the fragments obtained after breakage are usually too coarse to be used in quantitative mineralogical analyses, i.e., only in the case of ultra-fast load cells, the rotary breakage tester or when using small bars with the split Hopkinson pressure bar test, are sufficiently small particles generated which are meaningful to investigate by mineral liberation analysis.

Table 2. Particle breakage tests.

Adverse (−), Acceptable (O), Advantageous (+)	Drop Weight	SMC	UFLC	Twin Pendulum	Split Hopkinson Bar	Rotary Breakage
Simplicity	O	O	−	−	−	O
Repeatability	O	O	O	−	O	+
Sample preparation	− [1]	O	O	O	−	O
Time exposure and costs	− [1]	O	O	O	−	+
Sample amount	− [1]	O	−	O	−	O
Link to modeling	+	+	+	+	O	+
Mineral liberation	−	−	O [2]	−	O [2]	O [2]

Notes: [1] based on the JK drop weight test; [2] can produce fragments fine enough for liberation analysis.

3.3. Bench-Scale Grindability Tests

3.3.1. Bond Test

The Bond test is used to analyze the grindability of a material. The test applies a standardized ball mill of 305 mm (12 in.), both in diameter and length, with a grinding media charge of certain size distribution and operated at a defined speed [38]. The sample amount is defined by the bulk volume of 0.7 L, consisting of particles smaller than 3.35 mm. The test is conducted as a dry locked-cycle test with sieving of the mill product after each stage. Fines are replaced by an equal amount of fresh feed material, and grinding times are varied in order to reach a simulated circulating load of 250%. Usually samples of up to 10 kg smaller 3.35 mm are required.

From the grinding test the Bond ball mill work index W_i (in kWh/t) is determined [38]:

$$W_i = \frac{1.1 \cdot 44.5}{x_S^{0.23} \cdot G^{0.82} \cdot \left(\dfrac{10}{\sqrt{x_{80,P}}} - \dfrac{10}{\sqrt{x_{80,F}}} \right)} \tag{11}$$

where x_S is the screen aperture in microns, G the grindability (in grams of product per mill revolution), and $x_{80,F}$ and $x_{80,P}$ are the 80% passing particle sizes in μm for the mill feed and product, respectively.

The test results are used to calculate the change in particle size during grinding based on the grinding work input W (in kWh/t) and to size mills to achieve a desired size reduction using the Bond formula, also referred to as Bond's law, as a process model [39]:

$$W = W_i \cdot \left(\frac{10}{\sqrt{x_{80,P}}} - \frac{10}{\sqrt{x_{80,F}}} \right) \tag{12}$$

3.3.2. Variations of the Bond Test

The Bond test was originally developed for determining ball mill grindability but has been adapted for other mill types. Bond also defined a test for rod mill grinding using a 305 mm × 610 mm standard mill requiring up to 20 kg samples. Test conditions differ, e.g., the circulating load is changed to 100%, and also the Bond equation for calculating the rod mill work index is slightly different. In order to describe comminution in autogenous and semi-autogenous mills (AG/SAG) mills and high pressure grinding rolls (HPGR), using the Bond test method requires model extensions by empirical relations. For instance, Barratt provided an empirical formula for a SAG circuit involving three work indices for crushing, rod mill and ball mill grinding [40].

The SPI (SAG Power Index) is another related index used to determine the ore comminution behavior in AG/SAG mills. The index is obtained from a batch test in a mill of 305 mm × 102 mm operated with steel balls of 25 mm in diameter. The feed ore sample is crushed material <12.7 mm. The test was originally developed by Starkey *et al.* in 1994 and reviewed by Amelunxen *et al.* in 2014 [41–43]. Time and power draw are measured for grinding the ore sample to an 80% passing product particle size of 1.7 mm. With respect to modeling, the SPI test has been extended to ore body profiling and the design of semi-autogenous-ball milling circuits (SABC) [44,45] and is used in the CEET software [46].

In the past, several attempts have been made to simplify the Bond procedure. One approach has been to change the test from locked-cycle to a pure single-pass batch test in order to minimize the time, effort and sample amount needed, by developing new test mills, e.g., Outokumpu's Mergan mill [47] or the new size ball mill(NSBM) [48], or by modifying the test procedure, such as by changing the test to wet grinding [49], or by reducing the number of test cycles based on certain assumptions and complemented by simulations for ball mill and rod mill grindability, see [50,51] and JK Bond Ball Mill Lite test [52]. In the *Modbond grindability test*, an open circuit dry batch test run is used for estimating the work index after calibration against the standard Bond ball mill test [53].

3.3.3. Evaluation of Grindability Tests

In grindability tests, a combination of impact and attrition is applied to a bulk of material. The original Bond test is quite tedious, as several grinding cycles are necessary to reach the steady state of the simulated closed circuit. Furthermore, the sample amount required in the test is large, and is more problematic when only drill cores are available. Sample preparation is limited to pre-crushing and screening.

The test is reliable and has a good repeatability if the procedure and test mill comply with the standard. One of the major advantages is surely the huge data base that has been developed during the last six decades. With respect to process modeling, the coupling between the work index and the particle size reduction can be used together with an approximation function for particle size distribution. Attention must be paid to the applicability of the function type for the individual case. The particle size range of the mill product is suitable to perform mineral liberation analyses. Table 3 summarizes the characteristics of the different grindability test methods.

Table 3. Grindability tests.

Adverse (−), Acceptable (O), Advantageous (+)	Original Bond Ball Mill	Original Bond Rod Mill	Simplified Bond e.g., Mergan Mill
Simplicity	O	O	+
Repeatability	+	+	+
Sample preparation	O	O	O
Time exposure and costs	−	−	O
Sample amount	−	−	O
Link to modeling	+	+	+
Mineral liberation	+	+	+

3.4. Pilot and Bench-Pilot Scale Tests

Comminution test work on bench-pilot or pilot scale is done by using different types of crushers and mills depending on the intended process design. Typically these tests include:

- Cone crushers;
- High pressure grinding rolls (HPGR);
- Tumbling mills: ball mills, AG and SAG mills;
- Stirred media mills, e.g., IsaMill or vertical stirred mills.

Stress type and stress rate are based upon the respective machine type.

The tests usually require preparing tens to hundreds of kilograms of sample material and are done in batch or continuous mode. Sample preparation normally comprises pre-crushing and screening to the initial size distribution, as well as sample splitting.

Sampling from the test mill or comminution circuit provides the data necessary for determining breakage probabilities and grinding rates, as well as breakage distributions received from back-calculation by applying population-balance methods. Using the data from liberation analyses allows for describing the particles based on their mineral composition [54].

Pilot and bench-pilot scale tests are used to verify the metallurgical performance of a designed circuit. In the geometallurgical context, these results can be used in calibrating the small scale test results to full scale operation.

3.5. Indirect Methods for Determining Comminution Behavior

Another way of obtaining information about rock mechanical strength is to evaluate the core drilling process with corresponding instrumentation, also referred to as measurement while drilling (MWD). Alternatively, drill cuttings can be evaluated. Variations of the conditions at the drill bit, such as torque, normal or bending forces, result from changes in rock hardness.

Despite the huge data sets that are continuously collected in a large number of operations, the MWD data is seldom used for assessing the comminution properties of ore bodies. One of the obstacles is the lack of reliable on-line information about the condition of the drill bit, which is needed for correcting the recorded down-hole measurements by the dynamic process of drill bit wear.

Petrophysical data from multi-sensor drill core logging have also been used for calibration against measures of ore breakage parameters and grindability obtained from conventional destructive comminution tests [55]. Using density, magnetic susceptibility and seismic wave parameters from Australian copper-gold deposits, the Bond mill work index and the crushability parameters obtained from drop weight testing, could be predicted with acceptable accuracy.

In conjunction with recent advances in quantitative mineralogical analyses, the development in geometallurgical characterization today tends towards the identification of correlations between ore comminution behavior and mineralogical properties. The strategy is to reduce the number of comminution tests necessary for characterizing a deposit and to arrive at a more generic description of mechanical properties based on the occurring minerals [56].

4. Summary of Evaluation Findings

Table 4 summarizes the findings from the evaluation of the different test categories by summing up the attributes from Tables 1–3. None of the tests fulfills all criteria. To best assess the potential of the individual methods for geometallurgical testing, a compromise set of weighting criteria is required to emphasize the most important attributes. The sample amount and testing effort (in terms of time and cost) are considered to be the most significant parameters. In association with known process modeling requirements, the weighting process can be used to identify several methods as relevant and promising for further development of geometallurgical comminution tests, as indicated by mark (v) or v (medium to high potential) in Table 4.

Table 4. Summary table, potential for geometallurgy.

Adverse (−), Acceptable (O), Advantageous (+)	−	O	+	Potential
Unconfined compressive strength test	2	3	2	-
Point load test	1	4	2	(v)
Brazilian test	2	3	2	-
Hardness tester	1	2	4	(v)
Drop weight test [1]	4	2	1	(v)
SMC	1	5	1	v
Ultra-fast load cell test	2	4	1	-
Twin Pendulum test (Bond CWI)	3	3	1	-
Split Hopkinson bar test	4	3	0	-
Rotary breakage test	0	4	3	v
Bond ball mill test (Bond BWI)	2	2	3	(v)
Bond rod mill test (Bond RWI)	2	2	3	(v)
Simplified Bond test, e.g., Mergan mill	0	3	4	v

Note: [1] based on the JK drop weight test.

Table 5 summarizes where the investigated test methods can be applied with respect to the required sample material. In addition, the test methods that have been derived by using the same basic principles as the above-listed tests, have been included. In most cases this refers to tests that are commercially available. The modifications and simplifications made have improved the applicability within the geometallurgical context given some of the defined criteria. In other respects, the shortcomings are more fundamental. Hence, from a more scientific perspective there is still a need and a potential for improvement and/or development of new tests. The authors are aware of recent developments that have not been published yet, as for instance an enhanced version of the rotary breakage test (JKMRC's rotary breakage tester (RBT)"lite" test), which is based on a smaller device and reduced sample amounts.

Figure 11 shows the test methods existing today with their placement in the matrix as introduced in Figure 1, while ignoring the dimension of *stress type*. Considering the other two dimensions *particle size range* and *type of mechanical stress* alone, it can be concluded that it is practically impossible to develop a single universal method for determining ore comminution behavior at all test scales, including the geometallurgical level. Nevertheless, the objective of future research and development should be to reduce the variety of test methods by focusing more on the underlying breakage mechanisms and by linking these to mineral texture and mineral liberation information. The boxes that are not filled indicate directions for further development of geometallurgical comminution testing methods. Replacing the remaining interrogation marks can be done either by developing entirely new comminution test methods or by enhancing the existing methods.

Table 5. Sample requirements

Test method	Drill core/parts	Drill chips	Crushed sample	Sample amount (kg) Required	Consumed
Unconfined compressive strength test	X	-	-	n/a	<1 [1]
Point load test	X	-	-	n/a	<1 [1]
Brazilian test	X	-	-	n/a	<1 [1]
Hardness tester, e.g., EQUOtip [2]	X	-	-	n/a	n/a
Drop weight test [3]	X	X	X	75	25
SMC test	X	X	X	30	5
Ultra-fast load cell test [4]	X	X	X	2	1
Twin Pendulum test (Bond CWI)	X	X	X	25	10
Split Hopkinson bar test [4]	X	X	X	2	1
Rotary breakage test	X	X	X	100 [5]	15
Bond ball mill test (Bond BWI)	-	X	X	10	5
Bond rod mill test (Bond RWI)	-	X	X	15	10
Simplified Bond test (Mergan mill)	-	X	X	10	5
Simplified Bond test (SGS Modbond)	-	X	X	2	1.2
Simplified Bond test (Magdalinovic)	-	X	X	10	5
Simplified Bond test (JK BB lite)	-	X	X	6	<5
SPI grindability test	-	X	X	10	2
GeM comminution index [6]	-	X	X	5	1

Notes: [1] per specimen; [2] Non-destructive method and is limited to sponsors; [3] based on the JK drop weight test; [4] Mainly for research; [5] For the JK RBT lite even smaller sample amounts (<1 kg) have been used; [6] Limited to sponsors.

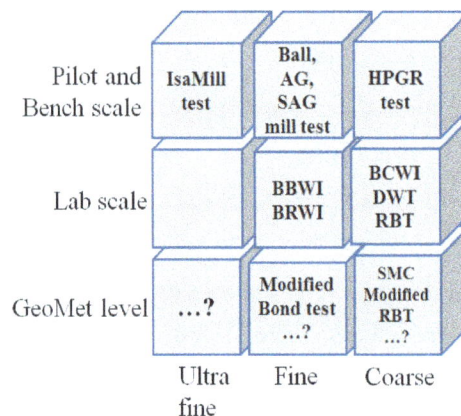

Figure 11. Comminution tests classified according to test design specifications. **HPGR** = High pressure grinding rolls; **BBWI** = Bond ball mill test; **BRWI** = Bond rod mill test; **BCWI** = Bond crusher test; **DWT** = Drop weight test; **RBT** = Rotary breakage tester; **Modified RBT** = Variation of the RBT (e.g., RBT lite) tester; **Modified Bond** = Variations of the Bond test; **SMC** = SAG Mill Comminution test; **GeoMet level** = Geometallurgical level.

5. Conclusions

In the previous sections, different categories of comminution tests have been reviewed against the priorly defined evaluation criteria for geometallurgical testing. For the individual test categories it can be concluded:

- Results from rock mechanical tests should only be used for geometallurgy where already available. As the results cannot be directly used, empirical correlations need to be found for linking mechanical strength or hardness to the description of ore crushing properties. Generally speaking, rock mechanical tests should not be part of a geometallurgical program as they do not provide information about grinding behavior down to the particle liberation level.

- Particle breakage tests have the potential to be used within geometallurgy if they do not require too large sample amounts, or great effort. For instance, the SMC test, based on the standard drop weight test has potential, since only single-size samples are required. Also, the rotary breakage test is promising despite the need for the highly technical and expensive test machines.

- Grindability tests are well established and provide a huge amount of reference data. The Bond equation links comminution energy and resulting particle size reduction, thus already providing a comminution process model. With respect to geometallurgical testing, a clear disadvantage lies in the timely effort needed for conducting the Bond test and the comparatively large sample amount. Steps have been undertaken to simplify the procedure. Future development should target modified Bond tests where the sample amount needed is significantly minimized.

- For ultra-fine grinding new comminution tests need to be developed on geometallurgical level.

- Pilot and bench-pilot scale tests are principally not suitable for mapping of the ore's comminution variability but can be employed in the calibration of small scale tests and at a later stage of a mine development project. Here tumbling mills as well as tests with stirred media mills and HPGR are of relevance.

- Indirect measurements have not been used to a large extent yet but are promising and further development is warranted.

Acknowledgements

The financial support of the Centre for Advanced Mining and Metallurgy (CAMM), a strategic research environment established at Luleå University of Technology funded by the Swedish government, is gratefully acknowledged.

Author Contributions

Abdul Mwanga conducted the literature review and developed the geometallurgy related criteria used in the evaluation. Jan Rosenkranz mainly contributed to the description and systematization of the different comminution test methods. Pertti Lamberg mainly contributed to the test evaluation and the conclusions section.

Conflicts of Interest

The authors declare no conflict of interest.

References

1. Weedon, D.M.; Napier-Munn, T.J.; Evans, C.L. Studies of mineral liberation performance in sulphide comminution circuits. In *Sulphide Deposits—Their Origin and Processing*; Gray, M.J., Bowyer, G.J., Castle, J.F., Vaughan, D.J., Warner, N.A., Eds.; Springer: Berlin, Germany, 1990; pp. 135–154.

2. Lamberg, P. Particles—The Bridge between Geology and Metallurgy. In Proceedings of the Conference in Minerals Engineering, Luleå, Sweden, 8–9 February 2011; pp. 1–16.

3. Alruiz, O.M.; Morrell, S.; Suazo, C.J.; Naranjo, A. A novel approach to the geometallurgical modelling of the Collahuasi grinding circuit. *Miner. Eng.* **2009**, *22*, 1060–1067.

4. Walters, S.; Kojovic, T. Geometallurgical Mapping and Mine Modelling (GeMIII)—The Way of the Future. In Proceedings of the SAG2006 Conference, Vancouver, BC, Canada, 23–27 September 2006; Volume 4, pp. 411–425.

5. Klimpel, R.R. The back-calculation of specific rates of breakage from continuous mill data. *Powder Technol.* **1984**, *38*, 77–91.

6. Williams, S.; Richardson, J. Geometallurgical Mapping: A new approach that reduces technical risk. In Proceedings of the 36th Annual Meeting of the Canadian Mineral Processors, Ottawa, ON, Canada, 20–22 January 2004; pp. 241–268.

7. Russell, A.R.; Muir Wood, A. Point load tests and strength measurements for brittle spheres. *Int. J. Rock Mech. Min. Sci.* **2009**, *46*, 272–280.

8. Shea, W.T.; Kronenberg, A.K. Strength and anisotropy of foliated rocks with varied mica contents. *J. Struct. Geol.* **1993**, *15*, 1097–1121.

9. Rusnak, J.; Mark, C. Using the Point Load Test to Determine the Uniaxial Compressive Strength of Coal Measure Rock. In Proceedings of the 19th International Conference on Ground Control in Mining, Morgantown, WV, USA, 1–3 August 2000; pp. 362–371.

10. Li, D.; Wong, L.N.Y. The Brazilian Disc Test for Rock Mechanics Applications: Review and New Insights. *Rock Mech. Rock Eng.* **2013**, *46*, 269–287.

11. Bearman, R.A.; Briggs, C.A.; Kojovic, T. The Application of Rock Mechanics Parameters to the Prediction of Comminution Behaviour. *Miner. Eng.* **1997**, *10*, 255–264.

12. Koch, P.H.; Mwanga, A.; Lamberg, P.; Pirard, E. Textural Variants of Iron Ore from Malmberget: Characterization, Comminution and Mineral Liberation. In Proceedings of the Exploration Resource and Mining Geology Conference 2013, Cardiff, UK, 21–22 October 2013; pp. 49–50.

13. Olaleye, B.M. Influence of some rock strength properties on jaw crusher performance in granite quarry. *Min. Sci. Technol.* **2010**, *20*, 204–208.

14. Toraman, O.Y.; Kahraman, S.; Cayirli, S. Predicting the crushability of rocks from the impact strength index. *Tech. Note Miner. Eng.* **2010**, *23*, 752–754.

15. Parbhakar-Fox, A.; Lottermoser, B.; Bradshaw, D.J. Cost-Effective Means for Identifying Acid Rock Drainage Risks—Integration of the Geochemistry-Mineralogy-Texture Approach and Geometallurgical Techniques. In Proceedings of the 25th International Mineral Processing Congress, Brisbane, Australia, 6–10 September 2010; pp. 143–154.

16. Verwall, W.; Mulder, A. Estimating rock strength with the EQUOtip hardness tester. *Int. J. Rock Mech. Min. Sci. Geomech.* **1993**, *30*, 659–662.

17. Napier-Munn, T.J.; Morrell, S.; Morrison, R.D.; Kojovic, T. *Mineral Comminution Circuits Their Operation and Optimization*; JKMRC Monograph Series in Mining and Mineral Processing; Julius Kruttschnitt Mineral Research Centre: Brisbane, Australia, 1996; Volume 2.

18. Genc, O.; Ergun, L.; Benzer, H. Single particle impact breakage characterization of materials by drop weight testing. *Physicochem. Probl. Miner. Process.* **2004**, *38*, 241–255.

19. Narayanan, S.S.; Whiten, W.J. Breakage characterization of ores for ball mill modelling. *Proc. AusIMM* **1983**, *286*, 31–39.

20. Narayanan, S.S.; Whiten, W.J. Determination of comminution characteristics from single particle breakage tests and its application to ball mill scale-up. *Trans. Inst. Min. Metall. Sect. C Miner. Process. Extr. Metall.* **1988**, *97*, C115–C124.

21. Morrell, S. A method for predicting the specific energy requirement of comminution circuits and assessing their energy utilization efficiency. *Miner. Eng.* **2008**, *21*, 224–233.

22. Kojovic, T.; Michaux, S.; Walters, S. Development of New Comminution Testing Methodologies for Geometallurgical Mapping of Ore Hardness and Throughput. In Proceedings of the 25th International Mineral Processing Congress, Brisbane, Australia, 6–10 September 2010; pp. 891–899.

23. King, R.P.; Bourgeois, F. Measurement of Fracture Energy during Single-Particle Fracture. *Miner. Eng.* **1993**, *6*, 353–367.

24. Tavares, L.M.; King, R.P. Single-particle fracture under impact loading. *Int. J. Miner. Process.* **1998**, *54*, 1–28.

25. Tavares, L.M. Energy Absorbed in Breakage of Single Particles in Drop Weight Testing. *Miner. Eng.* **1999**, *12*, 43–50.

26. Bourgeois, F.S.; Banini, G.A. A portable load cell for in-situ ore impact breakage testing. *Int. J. Miner. Process.* **2012**, *65*, 31–54.

27. Abel, F.; Rosenkranz, J.; Kuyumcu, H.Z. Stamped coal cakes in coke making technology Part 1—A parameter study on stampability. *Iron Mak. Steelmak.* **2009**, *36*, 321–326.

28. Narayanan, S.S. Development of a Laboratory Single Particle Breakage Technique and Its Application to Ball Mill Scale-Up. Ph.D. Thesis, University of Queensland, Queensland, Australia, 1985.

29. Bond, F.C. Crushing Tests by Pressure and Impact. *Trans. AIME* **1946**, *169*, 58–65.

30. Weedon, D.M.; Wilson, F. Modelling Iron Ore Degradation Using a Twin Pendulum Breakage Device. *Int. J. Miner. Process.* **2000**, *59*, 195–213.

31. Sahoo, R.K.; Weedon, D.M.; Roach, D. Single-Particle Breakage Tests of Gladstone Authority's Coal by a Twin Pendulum Apparatus. *Adv. Powder Technol.* **2004**, *15*, 263–280.

32. Fandrich, R.G.; Clout, J.M.F.; Bourgeois, F.S. The CSIRO Hopkinson Bar Facility for Large Diameter Particle Breakage. *Miner. Eng.* **1998**, *11*, 861–869.

33. Schönert, K.; Marktscheffel, M. Liberation of composite particles by single particle compression, shear and impact loading. In Proceedings of the 6th European Symposium Comminution, Nürnberg, Germany, 16–18 April 1986; pp. 29–45.

34. Shi, F.; Kojovic, T.; Larbi-Bram, S.; Manlapig, E. Development of a rapid particle breakage characterization device—The JKRBT. *Miner. Eng.* **2009**, *22*, 602–612.

35. Briggs, C.A.; Bearman, R.A. An Investigation of Rock Breakage and Damage in Comminution Equipment. *Miner. Eng.* **1996**, *9*, 489–497.

36. Kojovic, T.; Walters, P. Managing your geomet ore characterization needs with the JKRBT. In Proceedings of the GEOMET2012 International Seminar on Geometallurgy, Santiago, Chile, 1–3 December 2012; pp. 48–49.

37. Vogel, L.; Peukert, W. Breakage behaviour of different materials—Construction of a master curve for the breakage probability. *Powder Technol.* **2003**, *129*, 101–110.

38. Bond, F.C. Crushing and grinding calculations. *Br. Chem. Eng.* **1961**, *6*, 378–385.

39. Bond, F.C. The Third Theory of Comminution. *Trans. AIME Min. Eng.* **1952**, *1983*, 484–494.

40. Barratt, D.J. An update on testing, scale up and sizing equipment for autogenous and semi-autogenous grinding circuits. In Proceedings of the International Autogenous and Semi-Autogenous Grinding Technology Conference, Vancouver, BC, Canada, 25–27 September 1989; pp. 25–46.

41. Starkey, J.H.; Dobby, G.; Kosick, G. A New Tool for SAG Hardness Testing. In Proceedings of the 26th Canadian Mineral Processors Annual Meeting, Ottawa, ON, Canada, 18–20 January 1994.

42. Starkey, J.H.; Hindstrom, S.; Nadasdy, G. SAGDesign testing—What is it and Why it Works. In Proceedings of the International Conference on Autogenous and Semi-Autogenous Grinding Technology (SAG 2006), Vancouver, BC, Canada, 23–27 September 2006; Volume 4, pp. 240–254.

43. Amelunxen, P.; Berrios, P.; Rodriguez, E. The SAG grindability index test. *Miner. Eng.* **2014**, *55*, 42–51.

44. Bennett, C.; Dobby, G.S.; Kosick, G. Benchmarking and ore body profiling—The keys to effective production forecasting and SAG circuit optimization. In Proceedings of the International Conference on Autogenous and Semi-Autogenous Grinding Technology (SAG 2001), Vancouver, BC, Canada, 30 September–1 October 2001; Volume 1, pp. 289–300.

45. Dobby, G.; Bennett, C.; Kosick, G. Advances in SAG Circuit Design and Simulation Applied to the Mine Block Model. In Proceedings of the International Conference on Autogenous and Semi-Autogenous Grinding Technology (SAG 2001), Vancouver, BC, Canada, 30 September–1 October 2001; Volume 4, pp. 221–234.

46. Kosick, G.; Dobby, G.; Bennett, C. CEET (Comminution Economic Evaluation Tool) for Comminution Circuit Design and Production Planning. In Proceedings of 2001 SME Annual Meeting, Denver, CO, USA, 26–28 February 2001.

47. Niitti, T. Rapid Evaluation of Grindability by a Simple Batch test. In Proceedings of the International Mineral Processing Congress, Prague, Czechoslovakia, 1–6 June 1970; pp. 41–46.

48. Nematollahi, H. New Size Laboratory Ball Mill for Bond Work Index Determination. *Min. Eng.* **1994**, *46*, 352–353.

49. Tüzün, M.A. Wet Bond Mill Test. *Miner. Eng.* **2001**, *14*, 369–373.

50. Magdalinovica, N. Procedure for Rapid Determination of the Bond Work Index. *Int. J. Miner. Process.* **1989**, *27*, 125–132.

51. Tavares, L.M.; de Carvalho, R.M.; Guerrero, J.C. Simulating the Bond rod mill grindability test. *Miner. Eng.* **2012**, *26*, 99–101.

52. Kojovic, T.; Walters, P. Development of the JK Bond Ball Lite Test (JK BBL). In Proceedings of the GEOMET2012 International Seminar on Geometallurgy, Santiago, Chile, 5–7 December 2012; pp. 46–47.

53. Kosick, G.; Bennett, C. The Value of Orebody Power Requirement Profiles for SAG Circuit Design. In Proceedings of the 31st Annual Meeting of the Canadian Mineral Processors, Ottawa, ON, Canada, 19–21 January 1999; pp. 241–254.

54. Lamberg, P.; Vianna, S.M.S. A technique for tracking multiphase mineral particles in flotation circuits. In Proceedings of the 7th Meeting of the Southern Hemisphere on Mineral Technology, Ouro Preto, Brazil, 20–24 November 2007; pp. 195–202.

55. Vatandoost, A. Petrophysical Characterization of Comminution Behavior. Ph.D. Thesis, University of Tasmania, Tasmania, Australia, 2010.

56. Mwanga, A.; Rosenkranz, J.; Lamberg, P.; Koch, P.-H. Simplified Comminution Test Method for Studying Small Amounts of Ore Samples for Geometallurgical Purposes. In Proceedings of the Exploration Resource and Mining Geology Conference 2013, Cardiff, UK, 21–22 October 2013; pp. 45–48.

6

Use of Phosphate Solubilizing Bacteria to Leach Rare Earth Elements from Monazite-Bearing Ore

Doyun Shin [1,2,*]**, Jiwoong Kim** [1]**, Byung-su Kim** [1,2]**, Jinki Jeong** [1,2] **and Jae-chun Lee** [1,2]

[1] Mineral Resources Resource Division, Korea Institute of Geoscience and Mineral Resources (KIGAM), Gwahangno 124, Yuseong-gu, Daejeon 305-350, Korea;
E-Mails: jwk@kigam.re.kr (J.K.); bskim@kigam.re.kr (B.K.); jinkiz@kigam.re.kr (J.J.); jclee@kigam.re.kr (J.L.)

[2] Department of Resource Recycling Engineering, Korea University of Science and Technology, Gajeongno 217, Yuseong-gu, Daejeon 305-350, Korea

* Author to whom correspondence should be addressed; E-Mail: doyun12@kigam.re.kr

Academic Editor: Anna H. Kaksonen

Abstract: In the present study, the feasibility to use phosphate solubilizing bacteria (PSB) to develop a biological leaching process of rare earth elements (REE) from monazite-bearing ore was determined. To predict the REE leaching capacity of bacteria, the phosphate solubilizing abilities of 10 species of PSB were determined by halo zone formation on Reyes minimal agar media supplemented with bromo cresol green together with a phosphate solubilization test in Reyes minimal liquid media as the screening studies. Calcium phosphate was used as a model mineral phosphate. Among the test PSB strains, *Pseudomonas fluorescens*, *P. putida*, *P. rhizosphaerae*, *Mesorhizobium ciceri*, *Bacillus megaterium*, and *Acetobacter aceti* formed halo zones, with the zone of *A. aceti* being the widest. In the phosphate solubilization test in liquid media, *Azospirillum lipoferum*, *P. rhizosphaerae*, *B. megaterium*, and *A. aceti* caused the leaching of 6.4%, 6.9%, 7.5%, and 32.5% of calcium, respectively. When PSB were used to leach REE from monazite-bearing ore, ~5.7 mg/L of cerium (0.13% of leaching efficiency) and ~2.8 mg/L of lanthanum (0.11%) were leached by *A. aceti*, and *Azospirillum brasilense*, *A. lipoferum*, *P. rhizosphaerae* and *M. ciceri* leached 0.5–1 mg/L of both cerium and lanthanum (0.005%–0.01%), as measured by concentrations in the leaching liquor. These results indicate that determination of halo zone formation was found as a useful method to select high-capacity bacteria in REE leaching. However, as the leaching efficiency determined in our experiments was low, even in the presence of *A. aceti*, further studies are now underway to enhance leaching efficiency by selecting other microorganisms based on halo zone formation.

Keywords: bioleaching; monazite; phosphate solubilizing bacteria; rare earth element

---

1. Introduction

Rare earth elements (REE) have been increasingly used in the fields of optics, permanent magnetism, electronics, superconductor technology, hydrogen storage, medicine, nuclear technology, secondary battery technology, and catalysis [1–3]. The minerals monazite (a phosphate mineral) and bastnasite (a fluorocarbonate mineral) are the main sources of REE in nature. Generally, monazite contains ~70% rare earth metal oxide, with the rare earth fraction comprising 20%–30% Ce_2O_3, 10%–40% La_2O_3, and substantial amounts of neodymium, praseodymium, and samarium. The thorium content is in the range of 4%–12% [2,3].

Caustic soda decomposition and concentrated sulfuric acid digestion have been widely used to decompose monazite for many decades [2,4]. Due to its high chemical and thermal stability, monazite is very difficult to decompose; therefore, it is essential to eliminate the phosphate present in the ore by chemically attacking the mineral with sulfuric acid or sodium hydroxide at high temperature, so as to enhance the capacity to dissolve the REE. The sulfuric acid process results in a loss of phosphate as H_3PO_4, corrosion of the processing facilities, toxic gas and wastewater generation, as well as yielding impure products; the process is therefore no longer in commercial use. Caustic soda decomposition has some advantages in terms of the recovery of unreacted alkali and phosphorous, low energy consumption, and simplicity; however, the process also has limitations, such as the need for high-grade ore sources [5,6].

Biohydrometallurgical technology is an attractive alternative emerging green technology for the recovery of metals due to its environmental friendly, simple, and economic processing. However, very few works have been published on the biological recovery of rare earth metals, in particular, from monazite. Recently thorium, uranium, and REE extraction by microorganisms from monazite concentrate was reported [7,8]. The authors used *Aspergillus ficuum*, organic acid producing fungi, and *Pseudomonas aeruginosa*, organic acid/siderophore producing bacteria. They found that those microorganisms produced citric, oxalic, or 2-ketogluconic acid and dissolved 55% and 47% of REE from monazite by *A. ficuum* and *P. aerunoginosa*, respectively. Another study on REE leaching from phosphate minerals apatite and monazite by organic acids such as citric, oxalic, phthalic, and salicylic was also published, even though chemical organic acids were used in this study, which were not biologically produced [9]. Organic acid producing microorganisms secrete organic acids such as malic, gluconic, or oxalic acids [10,11], and the mechanism of metal dissolution by the microorganisms is both of acidolysis (protons dissociated from the organic acids) and complexolysis (metal-complexing anions from the acids) [12].

In a soil improvement process, phosphate solubilizing bacteria (PSB) have been widely used because they affect the release of phosphorous (P) from inorganic and organic P pools through solubilization and mineralization [13–15] by organic acid production. In the present study, we investigated the feasibility to use PSB in REE leaching from monazite-bearing ore as an exploratory study. Screening studies were performed by determining halo zone formation on agar media and phosphate solubilization in liquid media to predict REE leaching capacity of bacteria. The REE leaching abilities of the PSB from monazite-bearing ore were determined and compared with the screening data.

2. Experimental Section

2.1. Bacterial Strains and Culture Conditions

Ten PSB strains, *Pseudomonas rhizosphaerae* DSM 16299[T] [16], *P. putida* DSM 291[T], *P. fluorescens* DSM 50090[T] [17], *Bacillus megaterium* DSM 32[T] [18], *Paenibacillus polymyxa* DSM 36[T] [14], *Ensifer meliloti* DSM 30135 [19], *Azospirillum brasilense* DSM 1690[T], *A. lipoferum* DSM 1842 [17], *Mesorhizobium ciceri*, and *Acetobacter aceti* DSM 2002 were selected, purchased from Deutsche Sammlung von Mikroorganismen und Zellkulturen (DSMZ; Braunschweig, Germany), and maintained on nutrient agar (NA) at 30 °C. For the organic acid production test, the phosphate solubilization test, and the RE leaching study, Reyes minimal medium (in 1 L: 0.4 g NH_4Cl; 0.78 g KNO_3; 0.1 g NaCl; 0.5 g $MgSO_4\cdot7H_2O$; 0.1 g $CaCl_2\cdot2H_2O$; 0.5 mg $FeSO_4\cdot7H_2O$; 1.56 mg $MnSO_4\cdot H_2O$; 1.40 mg $ZnSO_4\cdot7H_2O$) [20] was used. Glucose or sucrose was added at 30 g/L as a carbon source and $Ca_3(PO_4)_2$ (5.39 g/L), alkaline metal phosphate, was used as a sole phosphate source. The viable bacterial cell number was measured by the plate counting method on NA at 30 °C.

2.2. Comparison of Halo Zone Formation, Organic Acid Production, and Phosphate Solubilization by Phosphate Solubilizing Bacteria

Halo zone formation by test PSB strains were tested using Reyes minimal medium supplemented with 1.5% agar, 5.39 g/L $Ca_3(PO_4)_2$ as a mineral phosphate, 30 g/L glucose or sucrose as a carbon source, and bromocresol green (BCG) as a pH indicator. The pH of the medium was adjusted to 6.5 using 1 N KOH. A 0.5% BCG solution was prepared in 70% ethanol as a stock solution and a 0.5 mL aliquot of the solution was added to 100 mL of Reyes minimal agar medium before autoclaving. Twenty microliters of PSB pregrown in nutrient broth was dot inoculated onto the surface of the agar plate prepared as above, incubated at 30 °C for 48 h. The total halo diameter and colony diameter were measured; halo size was calculated by subtracting the colony diameter from the total halo diameter.

A quantitative estimate of organic acid production by the PSB was performed using a 50-mL conical tube containing 30 mL of Reyes minimal medium supplemented with 0.45 g/L KH_2PO_4 and 30 g/L glucose, and inoculated with the bacterial strain (3 mL inoculum with approximately 10^8 CFU (colony forming unit)/mL). To measure the organic acid producing ability of the test PSB without any interference, KH_2PO_4 was used as an easily edible form of phosphate source because the availability of a phosphate source may affect organic acid production. An autoclaved uninoculated medium served as a control. The vials were incubated for 11 days in a shaking incubator at 30 °C and 180 rpm. At 1-day intervals, individual cultures were centrifuged at 5000 g for 30 min, and the supernatant was filtered with a 0.22-μm pore syringe filter. Concentrations of citric acid ($C_6H_8O_7$), malic acid ($C_4H_6O_5$), tartaric acid ($C_4H_6O_6$), and acetic acid ($C_2H_4O_2$) in the culture were determined by high-performance liquid chromatography (HPLC) (2690 Separation Module, Waters Alliance System, Milford, MA, USA) with 20 mM NaH_2PO_4 (pH 2.7, adjusted with 20% phosphoric acid), using an Atlantis T3 column (5 μm; 4.6 mm × 250 mm), and quantified using a Waters 996 Photo-diode Array Detector at 210 nm.

Phosphate solubilization by the test PSB in liquid media was determined in Reyes minimal medium supplemented with $Ca_3(PO_4)_2$ (5.39 g/L) and glucose (30 g/L). The cultivation procedure was as described above, except for the incubation time (*i.e.*, 5 days). Total calcium concentrations were determined by inductively coupled plasma atomic emission spectrometry (ICP-AES) (Optima 5300DV, Jobin Yvon JY 38, ICAP 6500 DVO; Perkin Elmer, Shelton, CT, USA); phosphate concentrations were determined by ion chromatography (IC) (ICS-3000, Dionex, Sunnyvale, CA, USA). Values reported are the average of three replicates.

2.3. Ore Characterization

The elemental composition of the raw monazite-bearing ore used in this study is shown in Table 1, which was measured by alkali fusion with Na_2CO_3 for silica, permanganametric titration for calcium, and ICP after multi-acid digestion for the other elements [21]. The mineralogical and morphological data of the ore were obtained by X-Ray Diffraction (XRD; Model RTP 300 RC, Rigaku Co., Tokyo, Japan) and a scanning electron microscope (SEM, JEOL 6400; JEOL Ltd., Tokyo, Japan) equipped with energy dispersive spectroscopy (EDS). The semi-quantification spot analysis SEM-EDS was performed using Thin Film Standardless Standard Quantitative Analysis (Oxide).

Table 1. Elemental analysis of monazite-bearing ore used in this study.

Element	Ce	La	Nd	Pr	SiO_2	Al_2O_3	Fe_2O_3	CaO	MgO	K_2O	Sr	TiO_2	P_2O_5
Content (wt%)	3.50	2.09	0.75	0.21	12.77	0.73	39.80	7.52	6.26	<0.001	0.47	0.02	5.33

2.4. Monazite-Bearing Ore Leaching by Phosphate Solubilizing Bacteria

The monazite-bearing ore was ground and then sieved below 150 μm size. A solution of 30 mL Reyes minimal medium (prepared as above) containing 30 g/L of glucose was added to 50-mL conical tubes containing 5 g of the autoclaved ore. Leaching was run for 9 days in a shaking incubator at 30 °C and 180 rpm. On each of the 9 days, samples were taken and prepared independently for further analysis. Total cerium and lanthanum concentrations in the residue and the liquor were determined by ICP and the mineralogical change of the residue was analyzed by SEM-EDS and XRD.

3. Results and Discussion

3.1. Comparison of Phosphate Solubilization by Phosphate Solubilizing Bacteria

Ten PSB strains (as stated above) were selected based on the literature reviews of their potential for phosphate solubilization. The halo zone formation was determined as a screening study to investigate the phosphate solubilizing ability of the PSB. Due to the production of organic acids in the surrounding medium, the halo zone on the agar plate was formed by converting insoluble phosphate into soluble phosphate [22]. The converting efficiencies vary for different forms of phosphate (*i.e.*, $Ca_3(PO_4)_2$, $AlPO_4$, and $FePO_4$); among them, calcium phosphate was most aggressively attacked by PSB. Higher phosphate solubilization from calcium phosphate than from aluminum and iron phosphate minerals has been reported, based on the solubility equilibrium and acidity constants of these compounds [23], thus $Ca_3(PO_4)_2$ was used in this study as a mineral phosphate. Bromocresol green (BCG) was also added in Reyes minimal medium supplemented with glucose and sucrose to improve the clarity and visibility of the yellow-colored halo zone [11]. The test PSB formed different sizes of distinct halo zones on the media depending on the bacterial strain and carbon source (Table 2). The widest halo zone was observed in the presence of *A. aceti* (diameter, 41.0 mm) on media supplemented with glucose; *P. rhizosphaerae*, *P. fluorescens*, *B. megaterium*, and *M. ciceri* also formed halo zones, while *P. polymyxa*, *E. meliloti*, *A. brasilense*, and *A. lipoferum* did not form halo zones on the media supplemented with either glucose or sucrose. Wider halo zones were observed in most samples when glucose was present rather than when sucrose was present; therefore, glucose was used as a carbon source throughout this study. Also, the use of modified Reyes agar medium supplemented with BCG and glucose was the most effective condition to determine halo zone formation.

Table 2. Comparison of tricalcium phosphate solubilization by phosphate solubilizing bacteria (PSB) in Reyes minimal agar medium supplemented with bromocresol green (BCG) and 30 g/L of glucose or sucrose, associated with microbial growth for 24 h.

Bacterial Species	Halo Size (mm) *	
	Glucose	Sucrose
P. rhizosphaerae DSM 16299[T]	13.5	13.4
P. putida DSM 291[T]	-**	7.2
P. fluorescens DSM 50090[T]	5.6	7.8
B. megaterium DSM 32[T]	4.7	12.1
P. polymyxa DSM 36[T]	-	-
E. meliloti DSM 30135[T]	-	-
A. brasilense DSM 1690[T]	-	-
A. lipoferum DSM 1842	-	-
M. ciceri	5.1	4.4
A. aceti DSM 2002	41.0	12.58

* The total halo diameter and colony diameter were measured; halo size was calculated by subtracting the colony diameter from the total halo diameter. ** "-" represents no halo zone formation.

Concentrations of solubilized phosphate and calcium ion in the Reyes minimal medium supplemented with glucose and $Ca_3(PO_4)_2$ were analyzed for a period of five days, to compare the phosphate solubilizing ability of PSB. The viable PSB cell numbers were maintained at approximately 10^7 CFU/mL medium during the experimental periods. The leaching efficiency was calculated by subtracting the mass of calcium and phosphate in the uninoculated control from the mass in the leaching liquor and dividing the leached mass by the total mass. The leaching efficiencies were shown to be highest generally within two to three days of incubation; thereafter, concentrations of leached phosphate and calcium decreased, possibly due to uptake by the PSB or absorption onto bacterial surfaces. The highest leaching efficiencies after three days of leaching were presented by *A. lipoferum*, *P. rhizosphaerae*, and *B. megaterium* (6.4%, 6.9%, and 7.5%, respectively, for calcium; and 3.7%, 6.8%, and 5.7%, respectively, for phosphate (Figure 1); leaching efficiencies by *A. brasilense* and *E. meliloti* were less than 2% for calcium; as expected, *A. aceti* showed the highest leaching efficiency of calcium (~32.5%).

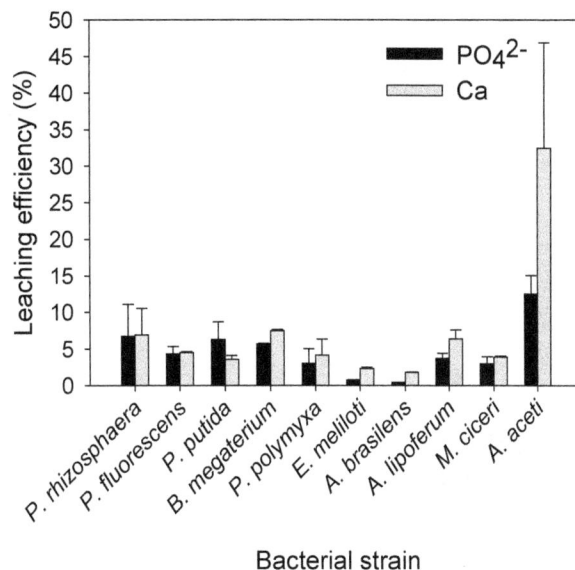

Figure 1. Phosphate and calcium solubilization from $Ca_3(PO_4)_2$ in Reyes basal liquid medium, associated with microbial growth after 3 days.

Inorganic phosphate solubilization by PSB depends on organic acid secretion from PSB [14]. Citric, malic, tartaric, and acetic acids were measured by HPLC in 30-mL of the PSB cultures containing KH_2PO_4 and glucose (Figure 2). The test PSB strains showed organic acid production below ~0.2 mmol, except for *A. aceti*. The maximum malic acid produced by *A. brasilense*, *A. lipoferum*, *M. ciceri*, and *P. rhizosphaerae* occurred at 0.07 mmol at six days, 0.09 mmol at nine days, 0.06 mmol at eight days, and 0.18 mmol at six days, respectively (maximum production). Acetic acid was produced by *A. aceti*, with concentrations reaching a maximum of 15.8 mM (0.48 mmol) at four days of incubation. Hwangbo *et al.* [24] reported that *Enterobacter intermedium* produced gluconic acid from glucose and subsequently converted it to 2-ketogluconic acid, which has a strong ionic strength and therefore easily solubilizes rock phosphate into soluble forms. The same situation was reported when Acetobacter was used to produce gluconic acid [25]. Because of the conversion of gluconic acid as stated above, gluconic acid might not be detected in the microbial cultures in this study. After six days, the amount of malic and acetic acid decreased. The pH of the media during incubation showed similar trends with organic acid production (Figure 3) because the pH decrease is due to the excreted metabolites, which included proton from organic acids, amino acids, and other metabolites. The pH in the medium of *A. aceti* decreased to about pH 3 (corresponding pH to ~15 mM of acetic acid, which was the acid produced) until eight days of incubation, and subsequently increased. The acidities of the media for *A. brasilense*, *A. lipoferum*, and *P. rhizosphaerae* also showed similar trends with acetic acid, with pH decreasing (*i.e.*, 3–4) until six to seven days of incubation and increasing thereafter.

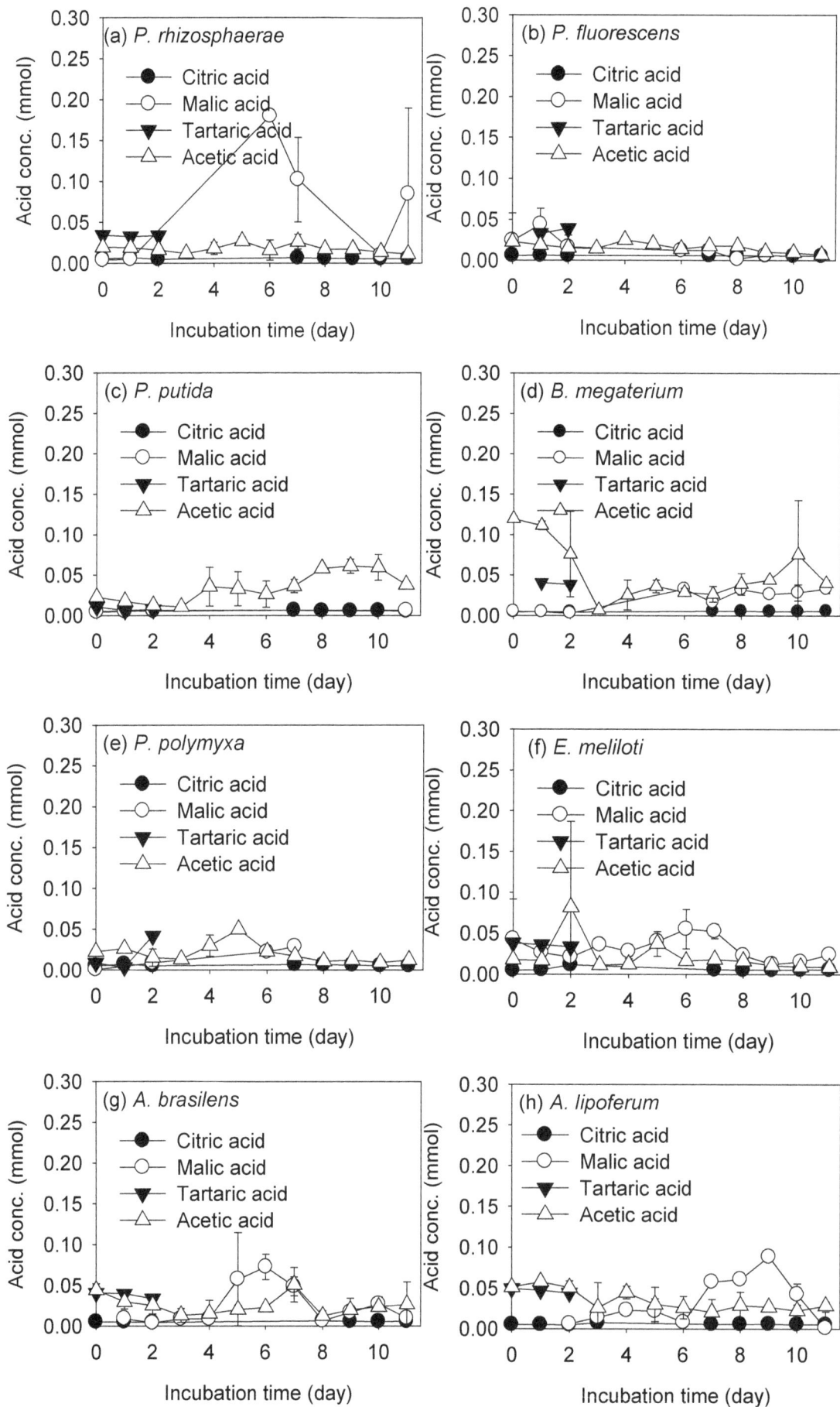

Figure 2. *Cont.*

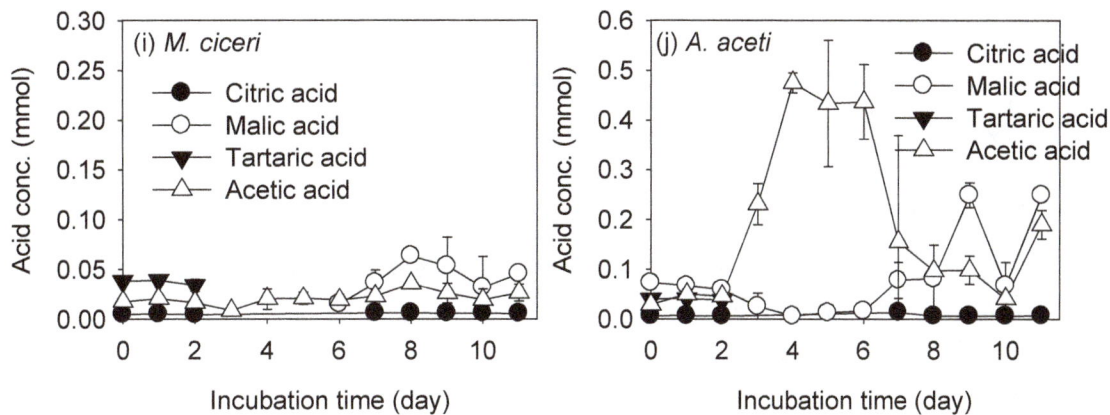

Figure 2. Organic acid production of (**a**) *P. rhizosphaerae* DSM 16299^T, (**b**) *P. fluorescens* DSM 50090^T, (**c**) *P. putida* DSM 291^T, (**d**) *B. megaterium* DSM 32^T, (**e**) *P. polymyxa* DSM 36^T, (**f**) *E. meliloti* DSM 30135, (**g**) *A. brasilense* DSM 1690^T, (**h**) *A. lipoferum* DSM 1842, (**i**) *M. ciceri*, and (**j**) *A. aceti* in Reyes basal liquid medium supplemented with K_2HPO_4, associated with microbial growth for 11 days.

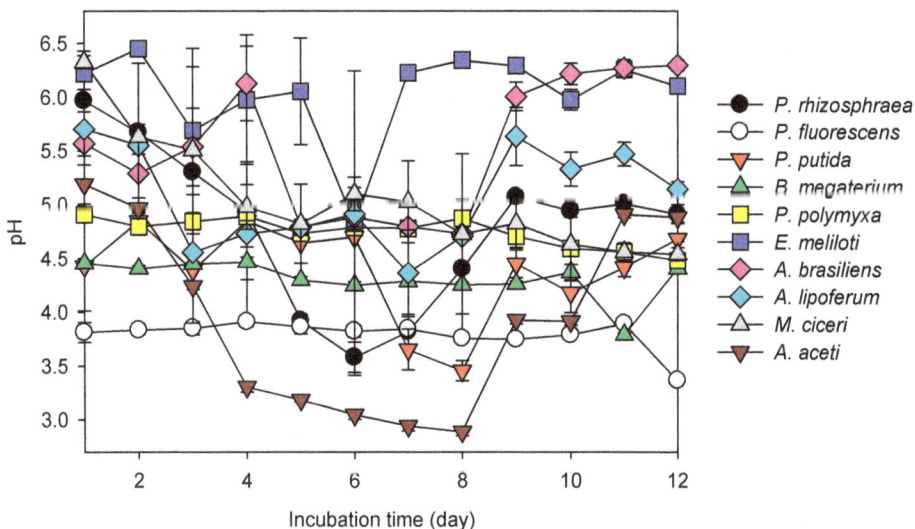

Figure 3. The pH changes induced by phosphate solubilizing bacteria (PSB) during incubation in Reyes minimal medium supplemented with K_2HPO_4.

3.2. REE Leaching by the Test PSB from the Monazite-Bearing Ore and Comparison with the Screening Study Data

Ten PSB were introduced to leach REE from monazite-bearing ore (Figure 4). We detected ~5.7 mg/L of cerium (0.13% of leaching efficiency) and 2.8 mg/L of lanthanum (0.11%) in the leaching liquor of *A. aceti*, and approximately 0.5–1 mg/L of cerium and lanthanum (0.005%–0.01%) in the leaching liquors of *A. brasilense*, *A. lipoferum*, *P. rhizosphaerae*, and *M. ciceri*. Since acetic and malic acid were produced by *A. aceti*, *A. brasilense*, *A. lipoferum*, *P. rhizosphaerae*, and *M. ciceri* as stated above, REE such as cerium and lanthanum may form complexes with these acids. When REE cations (Re_2O_3) are present in a system and acetic or malic acid is fully dissociated in aqueous solution, a complexation reaction may take place (at 25 °C) [26];

$C_2H_4O_2 \leftrightarrow C_2H_3O_2^- + H^+$ (pK_a = 4.756)

$3C_2H_3O_2^- + Re_2O_3 \leftrightarrow Re(C_2H_3O_2)_3$ (REE acetate complex)

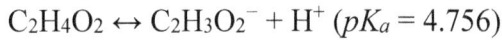

Also, since malic acid is a diprotic acid, the possible complexes of REE reaction with malate anions are as follows;

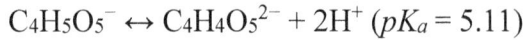

$C_4H_6O_5 \leftrightarrow C_4H_5O_5^- + H^+$ (pK_a = 3.40)

$3C_4H_5O_5^- + Re_2O_3 \leftrightarrow Re(C_4H_5O_5)_3$ (Rare earths malate complex)

$C_4H_5O_5^- \leftrightarrow C_4H_4O_5^{2-} + 2H^+$ (pK_a = 5.11)

$3C_4H_4O_5^{2-} + 2Re_2O_3 \leftrightarrow Re_2(C_4H_4O_5)_3$ (Rare earths malate complex)

The role of low molecular weight organic acids such as acetic, malic, oxalic, and citric acids in metal dissolution has been discussed in many studies [7–9,27,28]. The ligands of the organic acids play a dominant role in dissolving metals, and the differences in the effect of metal dissolution is related to their chemical structure. For instance, citric acid has great chelating ability with metals because citrate (*i.e.*, three carboxylic groups) can form stable chelates with 6-membered ring structures. While malic acid (dicarboxylic acid) is more stable than acetic acid (monocarboxylic acid) [29], REE leaching was observed in the *A. aceti* culture (where most of the leaching agent was acetic acid) because acetic acid production was higher than other organic acids production.

Mineralogical changes of the monazite-bearing ore before and after PSB leaching were also considered. As a variety of minerals that entrap monazite may be present, and these minerals may prevent contact of the organic acids produced by PSB as shown in the XRD patterns and the SEM images (Figures 5 and 6; Table 3). The X-ray pattern of the monazite-bearing ore shows the presence of dolomite ($CaMg(CO_3)_2$), quartz (SiO_2), siderite ($FeCO_3$), pyrite (FeS_2), and monazite ((Ce, La, Pr, Nd, Th, Y)PO_4). The cerium and lanthanum-rich region (spot 1 and 2) had a spotty distribution surrounded by other minerals including Mg, Al, Si, or Fe (spot 3 and 4). High carbon ratio in the spots was because of epoxy impregnation in preparing flat-polished specimens. XRD patterns of the monazite-bearing ore before and after leaching in this study showed a reduction in all peak intensities (Figure 6). High peaks of carbonate minerals (dolomite ($CaMg(CO_3)_2$) and siderite ($FeCO_3$)) were found and these minerals might consume the organic acids produced by PSB and prevent REE dissolution.

Differently from the phosphate solubilization test in liquid media (Figure 1), *P. fluorescens*, *P. putida*, *B. megaterium*, and *P. polymyxa* did not leach REE from the monazite-bearing ore. In the case of *A. brasilense* and *A. lipoferum*, even though they leached REE from the ore in spite of small leaching efficiency (Figure 4) and produced similar amounts of malic acid with *M. ciceri* and *P. rhizosphaerae* (Figure 2), they did not form halo zones (Table 2). This discrepancy sometimes happens because of the difference between cultivation in liquid media and on agar media. A liquid medium is preferred for the production of such organic acids because excreted products are readily available from a liquid culture and the cells are uniformly exposed to conditions of the medium [30]. However, the strongest leaching of REE was observed in the presence of *A. aceti*, as expected by both the results of the halo zone formation and the phosphate solubilization test. In the case of low-capacity bacteria, some errors may occur as stated above in determination of halo zone formation, but this method is still useful to roughly estimate phosphate solubilizing and organic acid producing abilities, as well as REE leaching capacities. Thus we believe that determination of halo zone formation can be used as a simple and rapid screening method to select high-capacity bacteria in REE leaching.

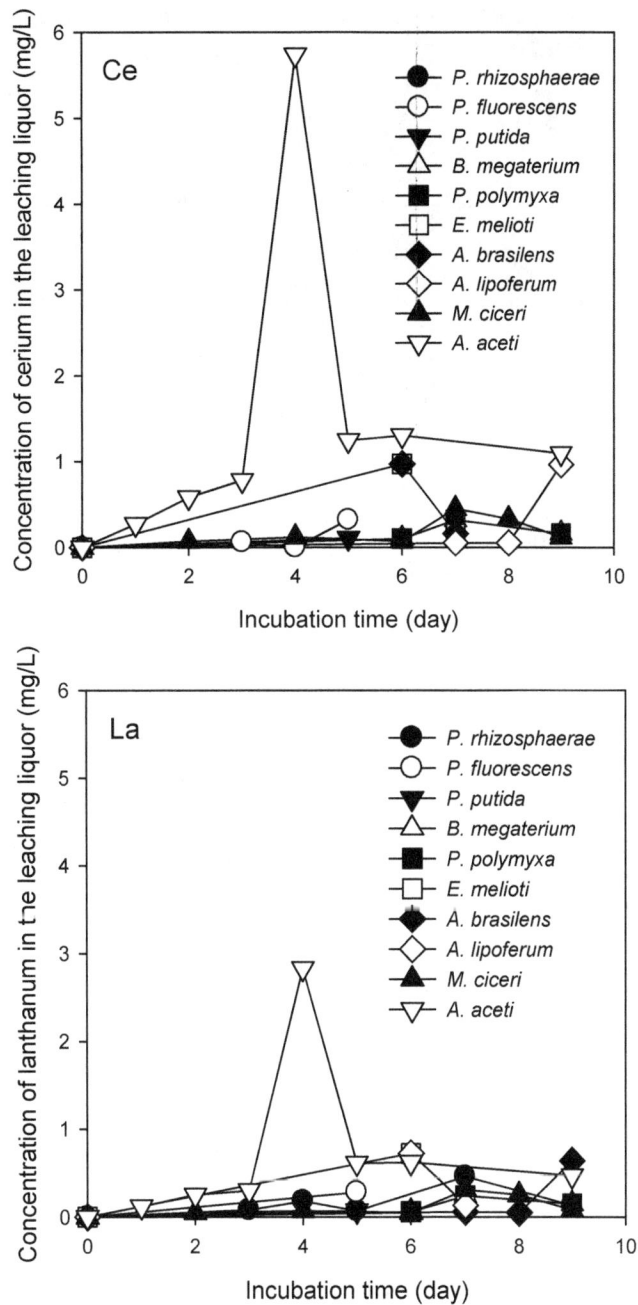

Figure 4. The concentrations of leached cerium and lanthanum from monazite-bearing ore by phosphate solubilizing bacteria (PSB) in Reyes basal liquid medium.

Table 3. Elemental compositions of the spots in Figure 5 by scanning electron microscopy/energy dispersive X-ray spectroscopy (SEM-EDS).

Spot No. in Figure 5	Mass %											
	C	**O**	**Mg**	**Al**	**Si**	**P**	**S**	**Ca**	**Fe**	**La**	**Ce**	**Pr**
1	82.1	4.88	1.67	-	-	-	-	7.06	-	0.18	4.1	-
2	46.0	22.7	1.28	-	5.4	-	8.2	-	-	3.36	12.5	0.62
3	61.3	16.4	1.05	0.24	8.77	2.15	-	-	9.26	-	-	-
4	70.6	15.8	-	4.33	9.3	-	0.56	-	-	-	-	-

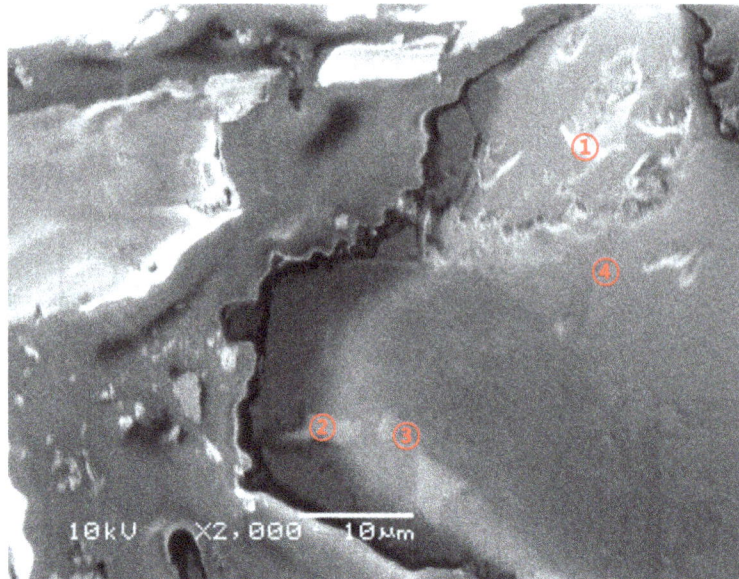

Figure 5. Scanning electron microscopic image of the raw monazite-bearing ore (3000× magnification). The detailed investigations are given in the text and Table 3.

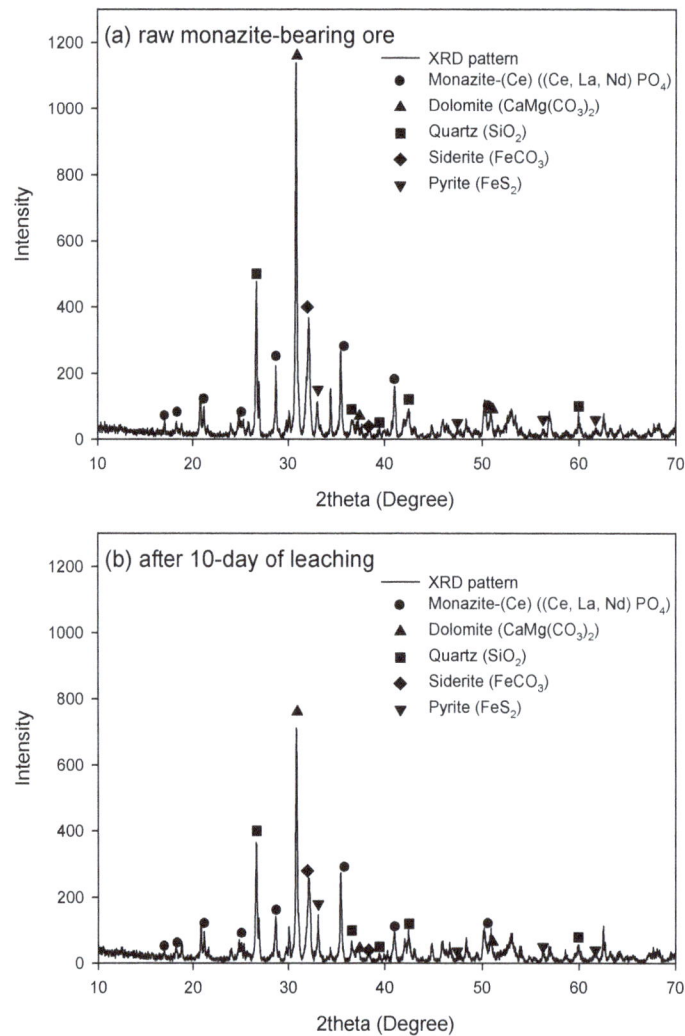

Figure 6. X-Ray diffraction (XRD) patterns of the monazite-bearing ore before and after leaching (10 days).

Biological leaching technology has some advantages over other methods, as it is relatively simple, inexpensive, and environmental friendly. However, even though *A. aceti* showed the highest leaching efficiency of the PSB tested, the efficiency of extraction was only ~0.13% (at least in batch-type experiments), which is still very low. The PSB used in this study have been used in soil improvement applications because they adapt well to soil environments; however, their organic acid producing and phosphate solubilizing ability is not that high. Thus, further studies are underway to explore the use of other microorganisms with good organic acid producing abilities for REE extraction applications, as well as to design continuous leaching systems so as to enhance leaching efficiency because the ore particles might not be exposed or contacted by the organic acid produced by the test PSB strains in a batch culture system.

4. Conclusions

The present study was conducted to use PSB to leach REE from monazite-bearing ore and to predict the REE leaching ability of PSB by determining halo zone formation on agar media and phosphate solubilization in liquid media. Among the ten test PSB strains, the phosphate solubilizing ability of *A. aceti* was the highest, based on results of halo zone formation and the phosphate solubilization test. Based on REE leaching data from the raw monazite-bearing ore by the test PSB, halo zone formation was found to be a useful method to select high-capacity bacteria in REE leaching.

Acknowledgments

This research was supported by the Basic Research Project (GP2015-005 and JP2013-005) of the Korea Institute of Geoscience and Mineral Resources (KIGAM), funded by the Ministry of Science, Information and Communications Technologies, and Future Planning of Korea.

Author Contributions

All of the authors were involved in the laboratory work, sample characterization, writing and revising of all parts of the manuscript.

Conflicts of Interest

The authors declare no conflict of interest.

References

1. British Geological Survey. Rare Earth Elements. Available online: http://www.MineralsUK.com (accessed on 27 March 2015).
2. Gupta, C.K.; Krishnamurthy, N. *Extractive Metallurgy of Rare Earths*; CRC Press: New York, NY, USA, 2005.
3. Maestro, P.; Huguenin, D. Industrial applications of rare earths: Which way for the end of the century. *J. Alloy. Compd.* **1995**, *225*, 520–528.
4. Habashi, F. *Handbook of Extractive Metallurgy*; Wiley-VCH: Heidelberg, Germany, 1997.

5. Zongsen, Y.; Minbo, C. *Rare Earth Elements and Their Applications*; Metallurgical Industry Press: Beijing, China, 1995.

6. Park, H.-K.; Lee, J.-Y.; Cho, S.-W.; Kim, J.-S. Overview on the Technologies for Extraction of Rare Earth Metals. *J. Korean Inst. Resour. Recycl.* **2012**, *21*, 74–83.

7. Desouky, O.A.; El-Mougith, A.A.; Hassanien, W.A.; Awadalla, G.S.; Hussien, S.S. Extraction of some strategic elements from thorium–uranium concentrate using bioproducts of *Aspergillus ficuum* and *Pseudomonas aeruginosa*. *Arabian J. Chem.* **2011**, doi:10.1016/j.arabjc.2011.08.010.

8. Hassanien, W.A.G.; Desouky, O.A.N.; HussienN, S.S.E. Bioleaching of some Rare Earth Elements from Egyptian Monazite using *Aspergillus ficuum* and *Pseudomonas aeruginosa*. *Walailak J. Sci. Technol.* **2014**, *11*, 809–823.

9. Goyne, K.W.; Brantley, S.L.; Chorover, J. Rare earth element release from phosphate minerals in the presence of organic acids. *Chem. Geol.* **2010**, *278*, 1–14.

10. Chung, H.; Park, M.; Madhaiyan, M.; Seshadri, S.; Song, J.; Cho, H.; Sa, T. Isolation and characterization of phosphate solubilizing bacteria from the rhizosphere of crop plants of Korea. *Soil. Biol. Biochem.* **2005**, *37*, 1970–1974.

11. Gadagi, R.S.; Sa, T. New isolation method for microorganisms solulbilizing iron and aluminum phosphates using dyes. *Soil Sci. Plant Nutr.* **2002**, *48*, 615–618.

12. Gadd, G.M. Fungal Production of Citric and Oxalic Acid: Importance in Metal Speciation, Physiology and Biogeochemical Processes. In *Advances in Microbial Physiology*; Poole, R.K., Ed.; Academic Press: Waltham, MA, USA, 1999; Volume 41, pp. 47–92.

13. Chen, Y.P.; Rekha, P.D.; Arun, A.B.; Shen, F.T.; Lai, W.A.; Young, C.C. Phosphate solubilizing bacteria from subtropical soil and their tricalcium phosphate solubilizing abilities. *Appl. Soil Ecol.* **2006**, *34*, 33–41.

14. Rodriguez, H.; Fraga, R. Phosphate solubilizing bacteria and their role in plant growth promotion. *Biotechnol. Adv.* **1999**, *17*, 319–339.

15. Jeong, S.; Moon, H.S.; Nam, K.; Kim, J.Y.; Kim, T.S. Application of phosphate-solubilizing bacteria for enhancing bioavailability and phytoextraction of cadmium (Cd) from polluted soil. *Chemosphere* **2012**, *88*, 204–210.

16. Peix, A.; Rivas, R.; Mateos, P.F.; Martínez-Molina, E.; Rodríguez-Barrueco, C.; Velázquez, E. *Pseudomonas rhizosphaerae* sp. nov., a novel species that actively solubilizes phosphate *in vitro*. *Int. J. Syst. Evol. Microbiol.* **2003**, *53*, 2067–2072.

17. Gholami, A.; Shahsavani, S.; Nezarat, S. The Effect of Plant Growth Promoting Rhizobacteria (PGPR) on Germination, Seedling Growth and Yield of Maize. *Int. J. Biol. Life Sci.* **2009**, *1*, 35–40.

18. Tilak, K.V.B.R.; Ranganayaki, N.; Pal, K.K.; De, R.; Saxena, A.K.; Nautiyal, C.S.; Mittal, S.; Tripathi, A.K.; Johri, B.N. Diversity of plant growth and soil health supporting bacteria. *Curr. Sci.* **2005**, *89*, 136–150.

19. Halder, A.K.; Chakrabartty, P.K. Solubilization of inorganic phosphate by *Rhizobium*. *Folia Microbiol.* **1993**, *38*, 325–330.

20. Reyes, I.; Bernier, L.; Simard, R.R.; Tanguay, P.; Antoun, H. Characteristics of phosphate solubilization by an isolate of a tropical *Penicillium rugulosum* and two UV-induced mutants. *FEMS Microbiol. Ecol.* **1999**, *28*, 291–295.

21. Briggs, P.H.; Meier, A.L. *The Determination of Forty Two Elements in Geological Materials by Inductively Coupled Plasma-Mass Spectrometry*; US Department of the Interior, US Geological Survey: Denver, CO, USA, 1999.

22. Katznelson, H.; Peterson, E.A.; Rouatt, J.W. Phosphate-dissolving microorganisms on seed and in the root zone of plants. *Can. J. Bot.* **1962**, *40*, 1181–1186.

23. Stumm, W.; Morgan, J.J. *Aquatic Chemistry: Chemical Equilibria and Rates in Natural Waters*; John Wiley & Sons: Hoboken, NJ, USA, 2012.

24. Hwangbo, H.; Park, R.D.; Kim, Y.W.; Rim, Y.S.; Park, K.H.; Kim, T.H.; Suh, J.S.; Kim, K.Y. 2-Ketogluconic acid production and phosphate solubilization by *Enterobacter intermedium*. *Curr. Microbiol.* **2003**, *47*, 87–92.

25. Švitel, J.; Šturdik, E. 2-Ketogluconic acid production by *Acetobacter pasteurianus*. *Appl. Biochem. Biotechnol.* **1995**, *53*, 53–63.

26. Dawson, R.M.C. *Data for Biochemical Research*; Clarendon Press: Oxford, UK, 1959.

27. Qin, F.; Shan, X.-Q.; Wei, B. Effects of low-molecular-weight organic acids and residence time on desorption of Cu, Cd, and Pb from soils. *Chemosphere* **2004**, *57*, 253–263.

28. Debela, F.; Arocena, J.M.; Thring, R.W.; Whitcombe, T. Organic acid-induced release of lead from pyromorphite and its relevance to reclamation of Pb-contaminated soils. *Chemosphere* **2010**, *80*, 450–456.

29. Bolan, N.S.; Naidu, R.; Mahimairaja, S.; Baskaran, S. Influence of low-molecular-weight organic acids on the solubilization of phosphates. *Biol. Fertil. Soils* **1994**, *18*, 311–319.

30. Pine, L.; George, J.R.; Reeves, M.W.; Harrell, W.K. Development of a chemically defined liquid medium for growth of *Legionella pneumophila*. *J. Clin. Microbiol.* **1979**, *9*, 615–626.

7

Numerical Simulations of Two-Phase Flow in a Self-Aerated Flotation Machine and Kinetics Modeling

Hassan Fayed [1] and Saad Ragab [2,*]

[1] Numerical Porous Media Center, King Abdullah University of Science and Technology (KAUST), Thuwal 23955-6900, Saudi Arabia; E-Mail: hehady@gmail.com

[2] Department of Engineering Science and Mechanics, Virginia Tech, Blacksburg, VA 24061, USA

* Author to whom correspondence should be addressed; E-Mail: ragab@vt.edu

Academic Editors: Michael G. Nelson and Dariusz Lelinski

Abstract: A new boundary condition treatment has been devised for two-phase flow numerical simulations in a self-aerated minerals flotation machine and applied to a Wemco 0.8 m^3 pilot cell. Airflow rate is not specified *a priori* but is predicted by the simulations as well as power consumption. Time-dependent simulations of two-phase flow in flotation machines are essential to understanding flow behavior and physics in self-aerated machines such as the Wemco machines. In this paper, simulations have been conducted for three different uniform bubble sizes (d_b = 0.5, 0.7 and 1.0 mm) to study the effects of bubble size on air holdup and hydrodynamics in Wemco pilot cells. Moreover, a computational fluid dynamics (CFD)-based flotation model has been developed to predict the pulp recovery rate of minerals from a flotation cell for different bubble sizes, different particle sizes and particle size distribution. The model uses a first-order rate equation, where models for probabilities of collision, adhesion and stabilization and collisions frequency estimated by Zaitchik-2010 model are used for the calculation of rate constant. Spatial distributions of dissipation rate and air volume fraction (also called void fraction) determined by the two-phase simulations are the input for the flotation kinetics model. The average pulp recovery rate has been calculated locally for different uniform bubble and particle diameters. The CFD-based flotation kinetics model is also used to predict pulp recovery rate in the presence of particle size distribution. Particle number density pdf and the data generated for single particle size are used to compute the recovery rate for a specific mean particle diameter. Our computational model gives a figure of merit for the recovery rate of a flotation machine, and as such can be used to assess incremental design improvements as well as design of new machines.

Keywords: minerals flotation machines; two phase flows; flotation kinetics; rate constant; particles size distribution

---

1. Introduction

Mineral flotation machines are classified into two main types; forced air and self-aerated machines. Wemco machines are widely used self-aerated machines where no air pumping mechanism is required, which simplifies flotation plant design and operation. The rate of air flow and power consumption of Wemco machines depend on the flow structure and the hydrodynamics within the pulp volume. For a given machine, rate of airflow depends on the rotor speed (RPM) among other operating conditions such as rotor blades and disperser design. Instantaneous airflow rate in a Wemco machine is not known *a priori* and depends on machine design and operating conditions. Therefore, computational fluid dynamics (CFD) simulation of a Wemco machine is a challenging problem because airflow rate cannot be specified but it has to be an outcome of the simulation. Moreover, rate of air flow may vary significantly with time to the extent that air is temporarily "exhaled" by the standpipe instead of being "inhaled". Computer simulations of such a machine should predict the time-history of the rate of air flow, and the average rate is an output. The unknown rate of air flow and the possibility of "breathing" require careful treatments of the standpipe and pulp–froth interface boundary conditions. Koh and Schwarz [1] conducted CFD simulation of the self-aerated flotation machine Denver-Metso Minerals. In that machine, air flow rate depends on suction pressure created by the impeller, the hydrostatic head of the pulp, and the frictional losses along the delivery shaft from the inlet valve to the impeller. The air motion was not simulated in the standpipe. They predicted the air flow rate iteratively during the simulation by applying pressure loss formula to find the pressure drop in the standpipe. An empirical constant in the formula was adjusted for CFD simulations to match the experimental data.

Computational domains of flotation machines are large, and flow physics are complex involving multi-phase flow turbulence. Even two-phase flow simulations of flotation machines are time consuming and require large computational resources. Some approaches have been used to reduce computational costs for two-phase flow; see, for example, the approach by Tiitinen *et al.* [2], where sector based simulations were used to reduce the number of grid nodes. Bubble size is one of the most important parameters that affect the air holdup of the pulp phase. A spectrum of bubble sizes exists in flotation machines depending on air flow rate and turbulence parameters. To predict such bubble size distribution, another set of equations that describes a population balance can be solved in the course of CFD simulation (Kerdouss *et al.* [3]). This approach increases the computational demands where transport equation for each size group has to be implemented. A more feasible approach is to conduct a parametric study for different uniform bubble sizes to study their effects on air holdup and rate constant.

One of the main characteristics of mechanical flotation machines is to agitate the slurry and disperse air bubbles throughout the pulp volume. In order to assess the performance of flotation machines, it is important to know the spatial distribution of dispersed bubbles within the tank which directly affects air hold up and rate constant. The current CFD simulations are parametric study of two-phase flow in Wemco 0.8 m³ that provide the hydrodynamic data and air volume fraction spatial distribution for uniform bubble size in the pulp phase. Three different bubble sizes—d_b = 0.5, 0.7 and 1.0 mm—are used to investigate the effects of the bubble size on air flow rate, air holdup and rate constant. The paper is organized as follow. Section 2 presents the governing equations for two-phase flow. Machine geometry is presented in Section 3, and simulation results are discussed in Section 4. Flotation model and results are presented and discussed in Section 5, and conclusions are summarized in Section 6.

2. Euler-Euler Two-Fluid Model

A practical approach for two-phase simulations is the Euler-Euler approach in which both phases are modeled by volume-averaged equations [4]. The motion of the two continuous phases is described by the unsteady Reynolds-averaged Navier–Stokes (RANS) equation:

Continuity equation:

$$\frac{\partial(\alpha_i\,\rho_i)}{\partial t} + \nabla.\left(\alpha_i\,\rho_i\,\vec{V}_i\,\right) = 0 \tag{1}$$

Momentum equation:

$$\frac{\partial\left(\alpha_i\,\rho_i\,\vec{V}_i\right)}{\partial t} + \nabla.\left(\alpha_i\,\rho_i\,\vec{V}_i\,\vec{V}_i\,\right) = -\alpha_i\,\nabla P + \nabla.[\mu_{i,\text{eff}}\,(\nabla\,\vec{V}_i\, + \left(\nabla\,\vec{V}_i\,\right)^T)] + S_i + M_i \tag{2}$$

where (i = 1) denotes water phase and (i = 2) denotes gas phase, ∇P is the modified pressure to include the gravity effects, S_i describes any external momentum source and M_i is the interfacial force that acts on phase (i) due to the presence of other phases. In the present simulations, momentum exchanges between the two phases due to drag and buoyancy on bubbles are the only mechanisms that couple the motion of the two phases. Bubbles are deformable fluid particles when moving in high shear rate regions such as in minerals flotation machines. Schiller-Naumann drag model [5] has been used to estimate drag coefficient of air bubbles. Effects of bubbles deformations on the values of drag coefficient are neglected in Schiller-Naumann drag model [5].

Shear stress transport (SST) turbulence model has been used to model turbulence transport where two transport equations are solved. There is no such universal turbulence model for two-phase flow, particularly at high volume fraction [6]. Reynolds stress model is the most adequate model for swirling flows. However, Reynolds stress model (RSM) closes the RANS equations by solving six additional transport equations for averaged Reynolds stress terms, and that require large computational resources. Geometry and flow physics of minerals flotation machines are large and complex. Therefore, using RSM turbulence model is not feasible for such application and SST model has been used in our CFD study.

3. Cell Geometry and Simulations Parameters

The main components of Wemco-0.8 m³ machine (FLSmidth, Salt Lake City, UT, USA) as shown in Figure 1 include a six-blade rotor, a disperser, a draft tube, and a standpipe. Details of the different components are shown in Figure 2. The disperser has 34 holes arranged in two parallel rows. Seventeen semi-circular rods are attached to the inner surface of the disperser. Air is drawn into the machine through a hole in the top of standpipe. The machine is assembled in Figure 3. The simulated Wemco-0.8 m³ model does not have a disperser hood or tank baffles.

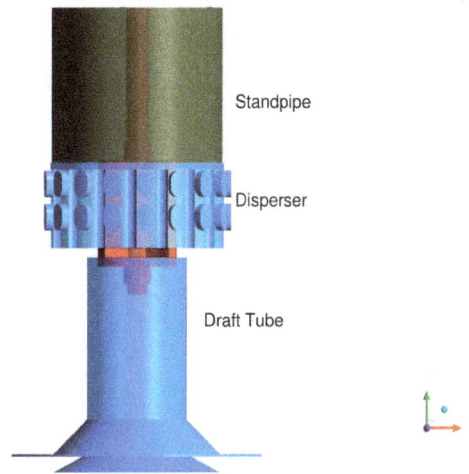

Figure 1. Core of the Wemco-0.8m³ machine: rotor, disperser, draft tube, and standpipe.

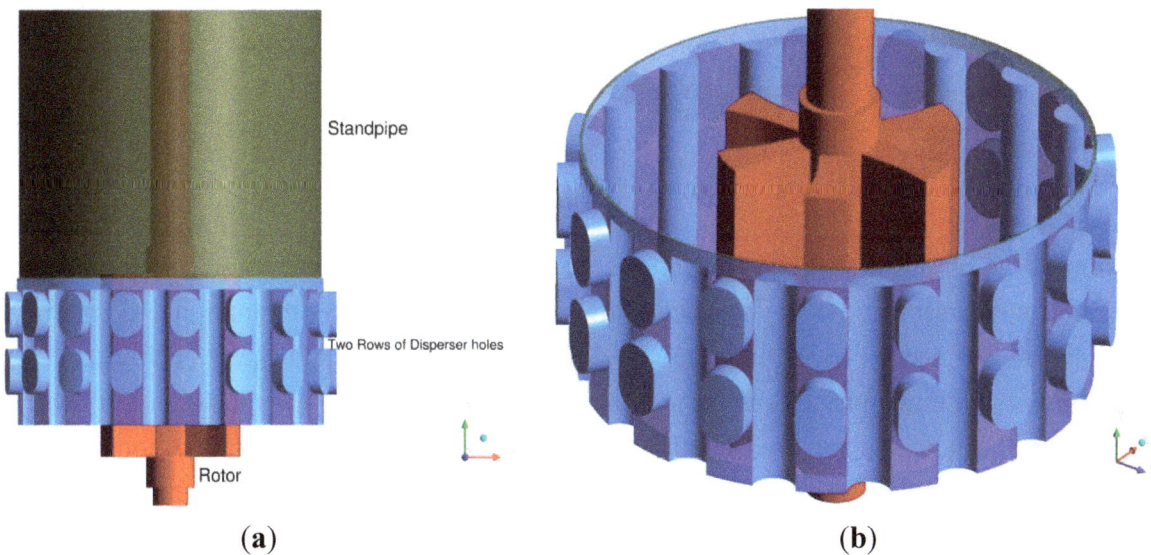

(a)   **(b)**

Figure 2. Details of Wemco-0.8m³ rotor, disperser, and standpipe; (**a**) relative vertical position of rotor and disperse; (**b**) rotor and disperser.

The boundary condition treatment at the air opening in top of the standpipe allows air to enter or exit from the standpipe above the rotor. There is no valve or any other device that obstructs air from flowing back to the ambient through the air inlet hole. If airflow reverses direction it may also carry a small amount of water with it. A container, which is connected to the standpipe only through the air inlet hole, is added. Its function is to retain any water that might be expelled and is to be retrieved when air flows back into the machine. Atmospheric pressure is prescribed at the top of that container. This treatment is unique to the present simulations of a self-aerated machine, and it allows the machine to "breath" for some operating conditions. An important operating assumption used in the present simulations is that the amount of water within the tank remains constant during operation. That is to say no water is allowed to flow over the weir. This condition may not be precisely satisfied in the actual machine operation; but it is a realizable situation. To guarantee that no water exits the tank, the tank wall is extended vertically above the weir edge thereby creating an overflow tank. The overflow tank is a mere vertical extension of the actual tank walls. Atmospheric pressure is prescribed at the top of the overflow tank. The actual tank is initialized with 100% water up to a certain level to be defined later, and the rest of the actual tank and overflow tank are initialized with 100% air.

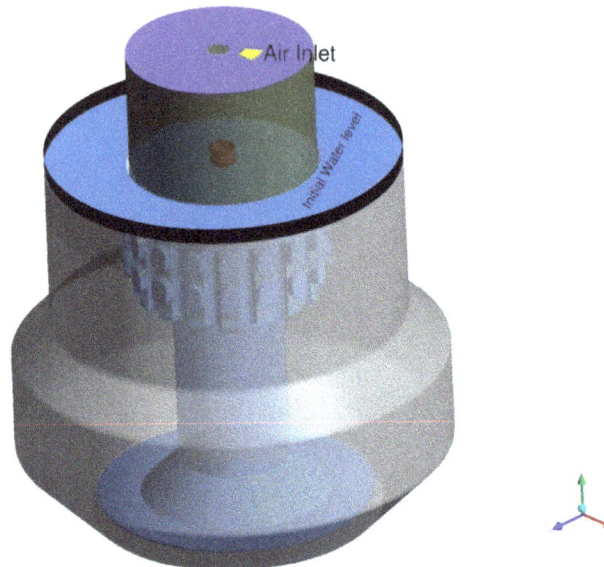

Figure 3. Wemco-0.8 m^3 assembly: tank, air-inlet and initial water level.

The function of the overflow tank is to permit the water level to rise due to the accumulation of air in the pulp, and at the same time it does not allow water to exit the computational domain. No boundary conditions are needed at the interface between the actual tank and the overflow tank. The governing equations of the two-phase flow are solved in both tanks allowing air and water to flow back and forth between the tanks as required by the transport equations. The part of the computational domain below the initial water level is initialized ($t = 0$) with 100% water and the part above that level is initialized by 100% air. A multi block structured grid has been generated for the current simulations using ICEM CFD grid generator. The total number of nodes is around 2.4×10^6. The size of the cells and number of nodes on rotor, disperser walls among other parts have been chosen to resolve boundary layer flow on these walls. Two phase flow simulations have been conducted using ANSYS CFX 12.0 commercial CFD software (ANSYS, Inc., Canonsburg, PA, USA). Governing equations are discretized in space using second order advection scheme and solved in time by using second order backward Euler method. The unsteady calculations have been performed on 10 parallel processors. The time step (Δt) is equal to 0.0007 s, at 620 rpm impeller speed that allows impeller volume to rotate ($\Delta\theta = 1.5$) degrees per time step.

4. Two Phase Flow Results

4.1. Air Flow Rate and Power

The rate of air flow drawn into the machine varies with time as shown in Figure 4a for three assumed bubble diameters. It is evident that the airflow is unsteady and momentarily may reach zero or become slightly negative; air is flowing out of the standpipe. Little effects of the bubble diameter are observed. The power (the computed rotor torque is multiplied by angular velocity) as depicted in Figure 4b also varies with time; it also shows negligible effects of bubble diameter. Air flow rate fluctuates around 0.3 m^3/min and power consumption fluctuate around 1.4 kW. Also, Figure 4a shows that simulations reach steady state after 8 s. The running average of air flow rate and power are defined by:

$$\tilde{Q}_a(t) = \frac{1}{t} \int_0^t Q_a(\tau) \, d\tau \tag{3}$$

$$\tilde{P}(t) = \frac{1}{t} \int_0^t P(\tau) \, d\tau \tag{4}$$

Calculations of the running average include initial transient variations of air flow and power. As shown in Figure 5, the time running averaged air flow rate is converging to 0.35 m^3/min and the power to 1.43 kW. The time-averaged air capacity coefficient $C_a = Q_a/ND^3$ is 0.053 for Wemco 0.8 m^3. Nelson et al. [7] reported measured values of C_a for large Wemco cells (for example, Wemco-160 m^3 SmartCell) in the range of 0.13–0.17. Nelson's et al. [7] work is the only available experimental study for Wemco machines that reports those values. It is notable that the experimental values are much greater than the present CFD prediction for the smaller pilot Wemco 0.8 m^3 model. Values for capacity coefficient are machine dependent, and this discrepancy is due to the differences in machine design as well as operating conditions. In flotation process, air rise velocity in the tank is one of the important design parameters. It is used to determine tank diameter. Minerals flotation engineers usually consider air rise velocity to be around 1.0 m/min for efficient flotation process. This value does not depend on machine type or size. In the current CFD study, the predicted air rise velocity (Q_a/A_c) equals 0.92 m/min which is close to the recommended value. Moreover, predicted air rise velocity lies in the measured range of 0.80–1.24 m/min as reported by Nelson et al. for Wemco-160 m^3.

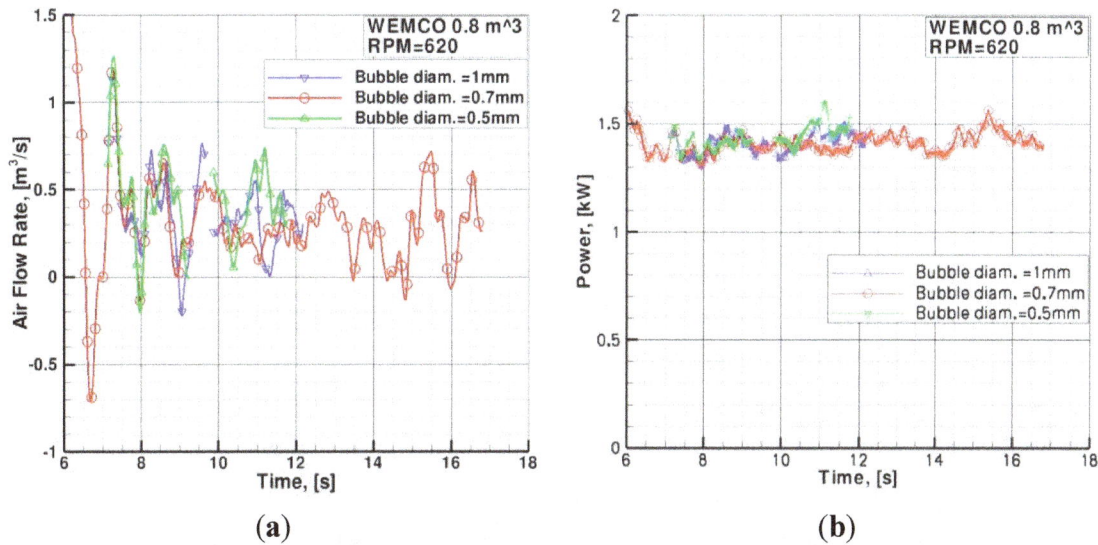

Figure 4. Effects of assumed bubble diameter on airflow rate and power, d_b = 0.5, 0.7 and 1.0 mm; (a) Variation of air volume flow rate with time; (b) Variation of power with time.

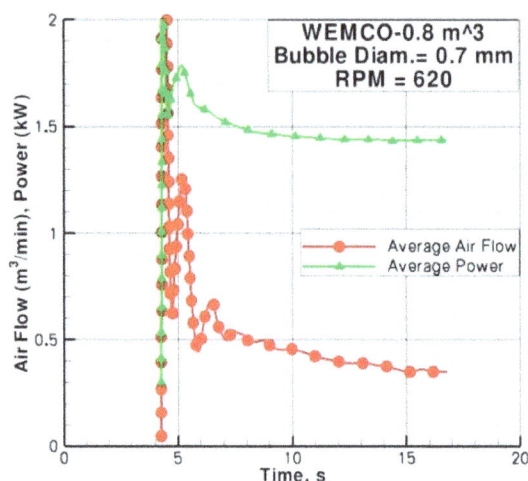

Figure 5. Running average of air flow rate and power.

4.2. Flow Pattern and Velocity Field.

The flow pattern in the Wemco rotor is very complex and unsteady. Analysis of the velocity field and air volume fraction in the rotor region will shed light on the principle of the Wemco machine operation. Water superficial radial velocity contours on horizontal planes that cut the rotor are depicted in Figure 6a–f at time $t = 14.35$ s. In Figure 6c, we see that the two blades at positions twelve-o'clock and six-o'clock have the strongest radial outflow, whereas the two blades in the positions four-o'clock and ten-o'clock have inward radial flow. The radial inward flow is entrained by the rotor from the secondary flow below the disperser. The pumping situation switches at higher elevations as shown in Figure 6e. Here, the radial flow at the blades at twelve-o'clock and six-o'clock is blocked whereas that at the blades at four-o'clock and ten-o'clock is strong and outward. Hence, water exits the rotor non-uniformly both circumferentially and vertically. Lack of flow periodicity in the Wemco rotor has also been predicted by CFD simulations of single-phase flow in Wemco 250 and 300 conducted by the present authors.

The velocity contours in the same planes are shown in Figure 7a–f at $t = 15.4$ s. At this instant, the maximum airflow has been observed and the water content in the rotor is a maximum. The important qualitative difference in the flow at this instant of time and the earlier one is the change in the direction of the vertical velocity at the rotor top. Water flows from the rotor into the standpipe mainly through the gap between rotor and disperser, and then falls down back from the standpipe mixed with air on rotor blades. Then rotor blades pump this mixture through the upper row of disperser holes to the tank. Figure 8a,b show the air volume fraction in a vertical mid plane that passes through the axis of the machine at two different times where the airflow rate is minimum and maximum, respectively. A vortex is formed in the standpipe, and water adheres to the standpipe walls under the effects of centrifugal forces (*i.e.*, water flows up into the standpipe with high circumferential velocity component). Consequently, water head in the standpipe increases continuously until it reaches its maximum level. At this instant, water breaks away from the standpipe walls and falls under gravity to the rotor tips capturing air with it as a mixture as depicted in Figure 8a,b. Water velocity vectors in a mid-plane that passes through the axis of the machine at two different time levels are shown in Figure 9a,b. Figure 9a shows the water velocity vectors at the maximum airflow instant. Water is discharged radially through the upper holes of the disperser where some water falls down on the rotor tips. Water velocity vectors at minimum airflow rate instant are shown in Figure 9b. Secondary flow below the disperser has been noticed for both time instants.

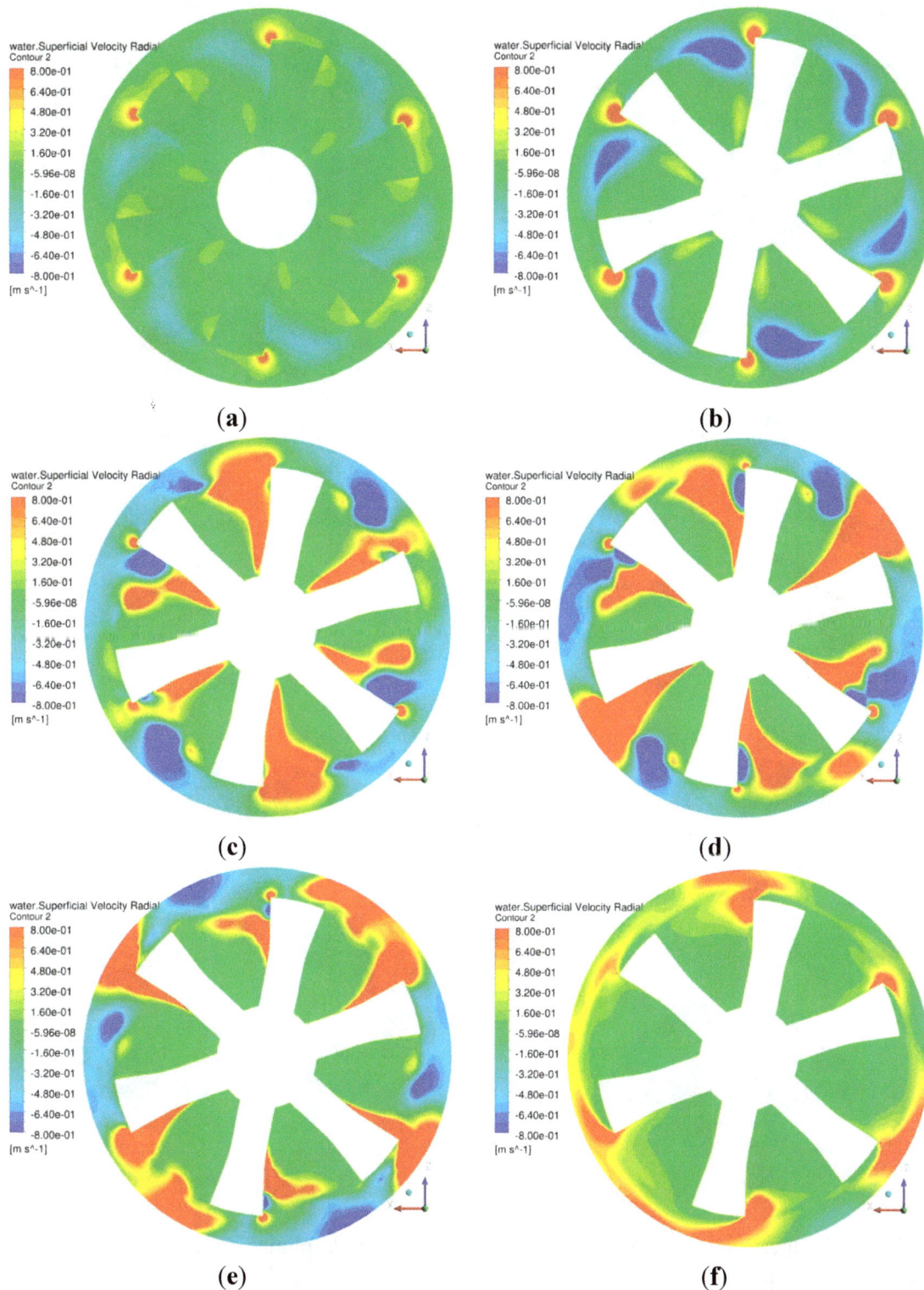

Figure 6. Water superficial radial velocity contours in horizontal planes cutting rotor blades, $t = 14.35$ s, $d_b = 0.7$ mm; (**a**) $y = 0.86$ m; (**b**) $y = 0.87$ m; (**c**) $y = 0.9$ m; (**d**) $y = 0.92$ m; (**e**) $y = 0.96$ m; (**f**) $y = 1.10$ m.

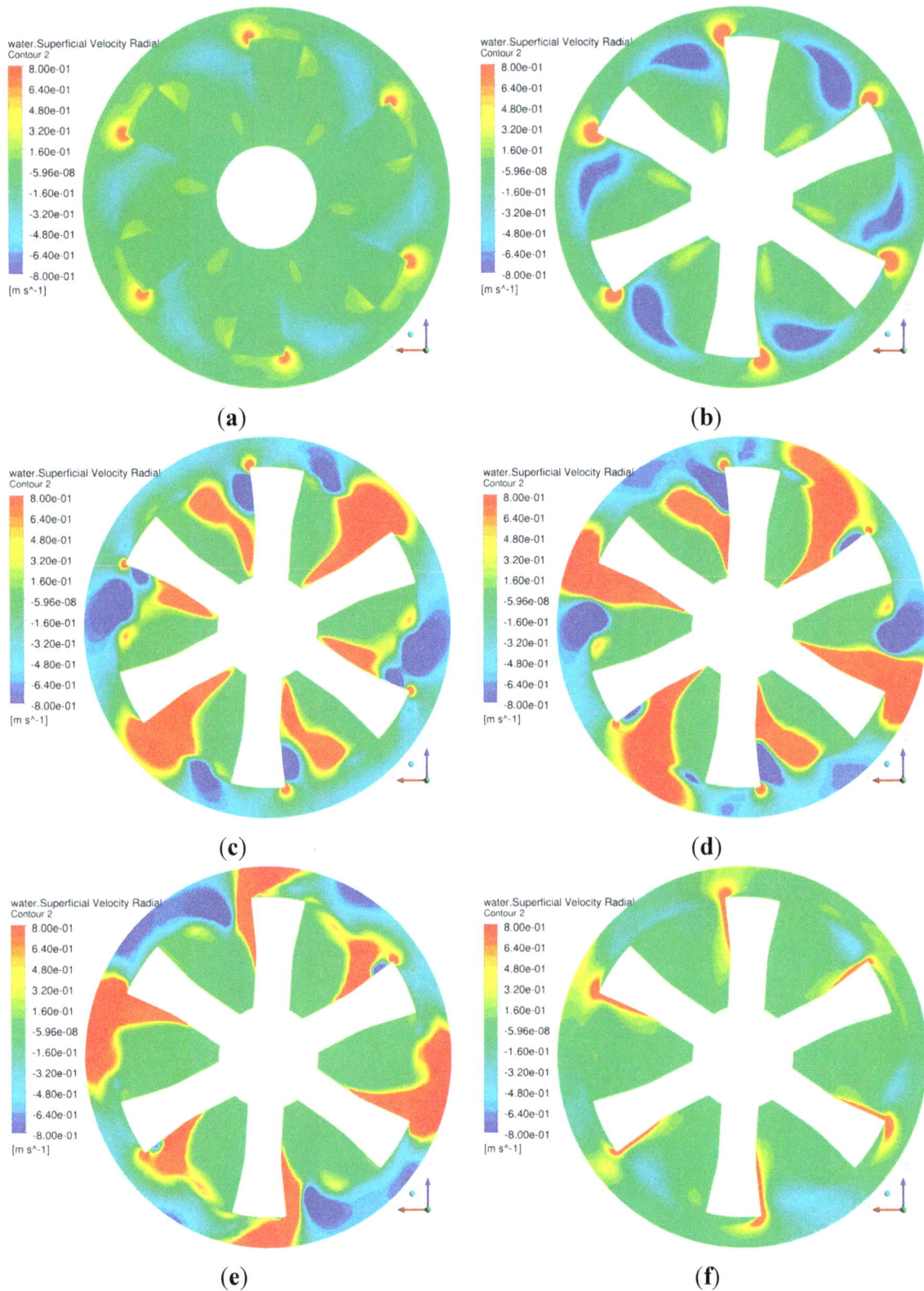

Figure 7. Water superficial radial velocity contours in horizontal planes cutting rotor blades, $t = 15.4$ s, $d_b = 0.7$ mm; (**a**) $y = 0.86$ m; (**b**) $y = 0.87$ m; (**c**) $y = 0.9$ m; (**d**) $y = 0.92$ m; (**e**) $y = 0.96$ m; (**f**) $y = 1.10$ m.

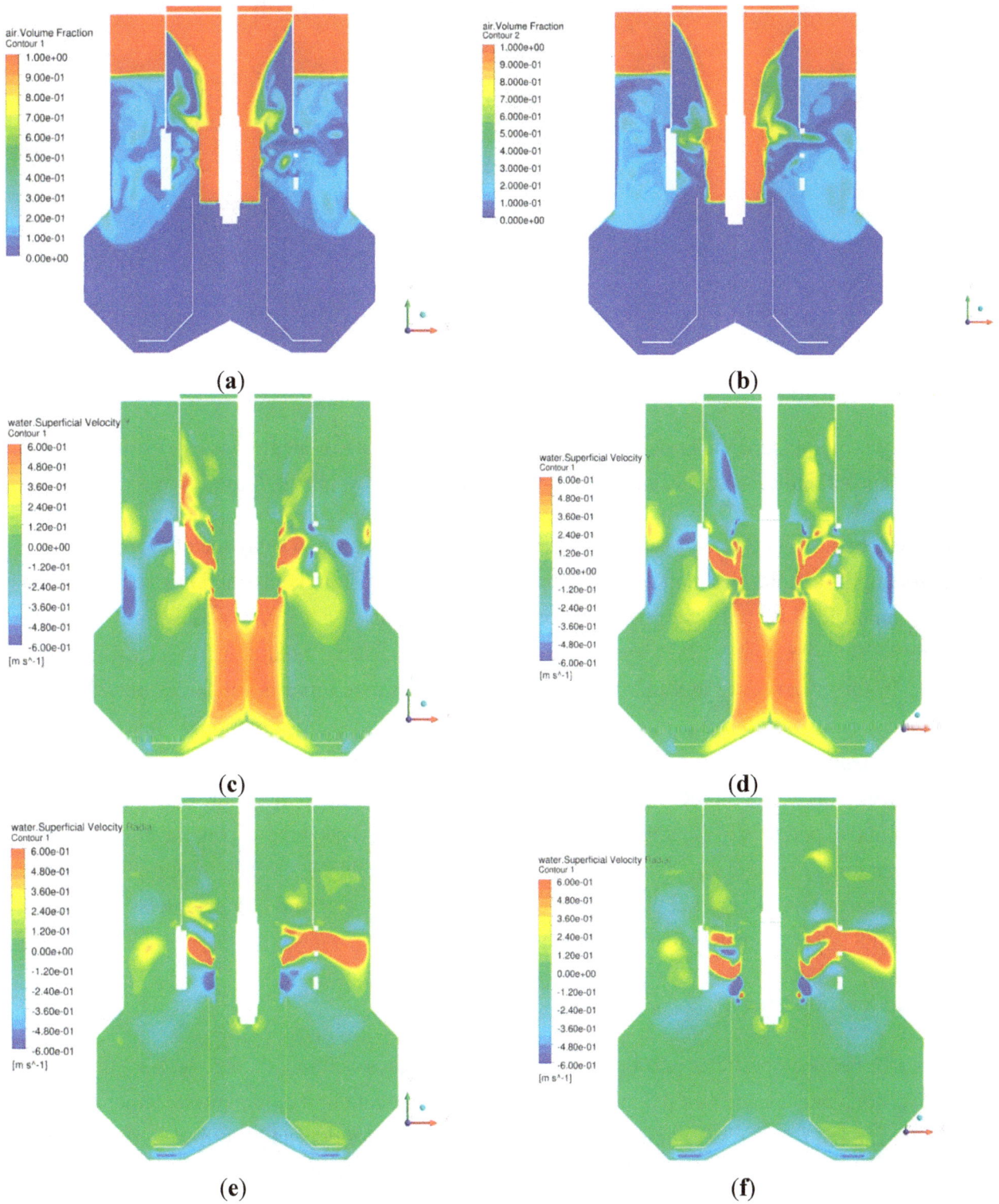

Figure 8. Air volume fraction and velocity contours for, d_b = 0.7 mm (**a**) air volume fraction, t = 14.35 s; (**b**) air volume fraction, t = 15.4 s; (**c**) water superficial vertical velocity, t = 14.35 s; (**d**) water superficial vertical velocity, t = 15.4 s; (**e**) water superficial radial velocity, t = 14.35 s; (**f**) water superficial radial velocity, t = 15.4 s.

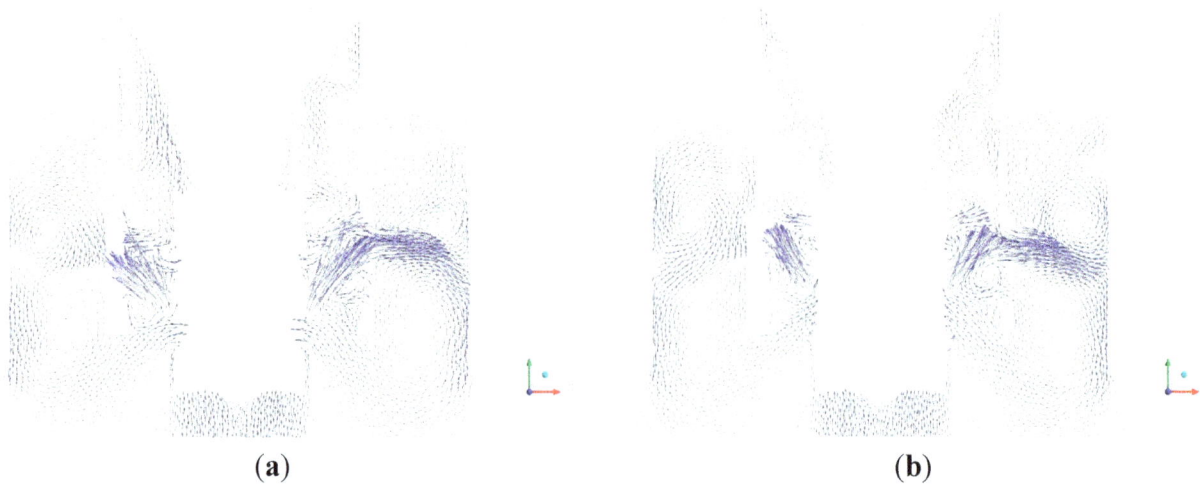

Figure 9. (**a**) Water superficial velocity vectors at maximum rate of air flow, d_b = 0.7 mm; (**b**) water superficial velocity vectors at minimum rate of air flow, d_b = 0.7 mm.

Three uniform bubble diameters of 0.5, 0.7, and 1.0 mm have been used to study the effects of bubble size on air holdup. Air holdup is computed by integration of (local) air void fraction over the pulp volume and then divided by pulp volume. The computed air holdup for 0.5, 0.7, and 1.0 mm bubble diameters are 9.3%, 7.4%, and 5.7%, respectively. It is observed that air holdup increases as bubbles size decrease. In this study, drag and buoyancy are the only considered interfacial forces between air bubbles and the liquid phase. Drag on air bubbles depends on bubble cross sectional area (*i.e.*, d_b^2) while buoyancy depends on the bubble volume (*i.e.*, d_b^3). Therefore, drag is the dominant force for small bubbles while buoyancy is the dominant force for large bubbles. Small bubbles have negligible effective inertia and experience lower slip velocity to the liquid phase. They tend to penetrate deeper into the tank and take longer time (*i.e.*, retention time) to escape out at the free surface of the pulp, which results in higher air holdup. Large bubbles experience higher slip velocity and escape quickly from the tank because of the larger buoyancy that results in lower air holdup.

5. CFD-Based Flotation Model

Collisions rate between particles and bubbles is a critical factor for minerals recovery by flotation machines. Collisions rate depends on the local turbulent dissipation rate in the pulp phase. Other factors such as attachment and detachment rate of particles-bubbles aggregates depend on turbulent dissipation rate as well as surface chemistry of particles and bubbles. Two-phase flow simulations of flotation machines provide spatial distribution of air void fraction and turbulent dissipation rate. A CFD-based flotation model has been developed in this work to predict pulp rate constant of Wemco 0.8 m³ pilot cell. The model provides local values of collisions rate between particles and bubbles as well as probabilities of attachment, collisions and stabilization. Abrahamson collision [8] model is commonly used in estimation for collisions rate in flotation modeling. This paper provides an alternative more accurate model for collisions rate developed by Zaichik *et al.* [9]. In this paper, we present flotation model as a first rate equation. The rate constant in this equation depends on the local collisions frequency, probability of collision, probability of attachment, and probability of stabilization.

Yoon and Lutrell [10], Pyke *et al.* [11], Bloom and Heindel [12] among others developed theoretical flotation models to predict pulp recovery rate. In these models, flotation process is represented mathematically by a first order kinetic ODE Ordinary Differential Equation). These theoretical models depend on the average air void fraction and dissipation rate for a given flotation cell. However, distributions of turbulent dissipation rates and air volume fraction are not uniform and depend on the design of the machine. Distribution of turbulent dissipation rate is the key factor in computing recovery rate. Koh and Schwarz [13] developed a 3D CFD flotation model for a forced air Rushton's impeller flotation tank. Collisions frequency and flotation probabilities have been computed locally. CFD simulations of flotation machines provide the spatial distribution of the rate of turbulent kinetic energy dissipation and air void fraction throughout the cell. Flotation model is based on estimating local recovery rate. Local recovery rate is the multiplication of collision kernel and probabilities of collision, attachment and stabilization. These parameters depend mainly on dissipation rate throughout the machine in addition to particles and bubbles sizes, surface tension and contact angle. The model has been applied to a self-aerated flotation machine (Wemco 0.8) to predict pulp recovery rate. In this paper, a range of particle sizes ($10 \ \mu m \leq d_p \leq 500 \ \mu m$) have been used with three different bubble sizes $d_b = 0.5$ mm, $d_b = 0.7$ mm, $d_b = 1.0$ mm to investigate the effects of particles and bubbles sizes on pulp rate constant. Effects of feed particles size distribution has been studied also in this work.

5.1. Flotation Model: First-Order Rate Equation

The objective of a flotation model is to predict the recovery rate (rate of mass flow of useful minerals collected from a flotation cell). Following Koh and Schwarz [13], we model the flotation kinetics as a first-order rate process, which is given by the fundamental equation:

$$\frac{dN_{p1}}{dt} = -K_1 N_{p1} N_b + K_2 N_a \tag{5}$$

where N_{p1} is the particle number concentration (number of particles per unit volume) of free particles (not attached to bubbles), N_b is the number concentration of bubbles available for attachment (not fully loaded bubbles), N_a is the number of particle-bubble aggregates (bubbles that cannot accept more particles but can lose particles), K_1 is the average particle-bubble attachment rate constant, and K_2 is the average particle-bubble detachment rate constant. We note the difference in the dimensions (units) of K_1 (m^3/s) and K_2 (1/s). The total particle number concentration N_{PT} is:

$$N_{PT} = N_{p1} + N_{p2} \tag{6}$$

where N_{p2} is the number concentration of particles attached to bubbles. All number concentrations are functions of time and location in a flotation cell. A bubble is either fully loaded (cannot accept more particles) or clean (no particles attached). The number of particle-bubble aggregates is proportional to the total number of bubbles,

$$N_a = b \ N_{bT} \tag{7}$$

where b is an average loading parameter, which varies with time and position. The number of clean bubbles (those available for attachment) is

$$N_b = (1 - b)N_{bT} \tag{8}$$

The rate equation can now be written as:

$$\frac{dN_{p1}}{dt} = -k_1(1-b)N_{p1}N_{bT} + k_2 b N_{bT} \tag{9}$$

The number of particles (S) that can be attached to a bubble is given by the ratio of surface area of the bubble to the projected area of the particle.

$$S = 4\left(\frac{d_b}{d_p}\right)^2 \tag{10}$$

Such an estimation is not realistic. In Koh and Schwarz [13] work, only half of that number is assumed as a first approximation.

$$S_{max} = 0.5\,S = 2\left(\frac{d_b}{d_p}\right)^2 = \frac{N_{p2}}{\beta\,N_{bT}} \tag{11}$$

Rearranging, we obtain:

$$b = \frac{N_{p2}}{S_{max}\,N_{bT}} = \frac{N_{p2}}{2\,N_{bT}}\left(\frac{d_b}{d_p}\right)^2 \tag{12}$$

5.2. Attachment Model

Collisions frequency (Z_{pb}) of two groups of dispersed species (e.g., particles and bubbles) is the number of collisions per unit volume per unit time ($m^{-3}\cdot s^{-1}$). It is proportional to the product of the number densities N_b and N_p of bubbles and particles ($Z_{pb} = \beta\,N_p\,N_b$), where β is called the collisions kernel (m^3/s). The attachment rate constant ($k_1, m^3/s$) is defined as the product of particle-bubble collisions kernel times probabilities (or efficiencies). This is because not every collision event leads to successful attachment of a particle to a bubble. Following Koh and Schwarz [13], the attachment rate constant is written as:

$$k_1 = \beta\,P_c\,P_a\,P_s \tag{13}$$

where P_c, P_a, and P_s are probabilities of particle-bubble collision, adhesion, and stabilization. Multiplication by P_c is questionable since it should be considered as a part of the collisions kernel. Apparently including P_c is a remnant of the original model that was developed under quiescent conditions. This collisions probability P_c should be eliminated because collisions kernel is modeled under turbulent conditions, and the assumption of laminar flow is irrelevant. When turbulence is the mechanism of collisions of particles and bubbles, "quiescent" conditions should be defined as the turbulent fluctuations u' and turbulent dissipation rate that maximizes the attachment rate and minimizes detachment.

5.3. Collision Kernel

The collision kernel β is perhaps the most important ingredient of the attachment rate constant k_1. For turbulent flows, two classical models are in use. The first is Saffman and Turner model [14], which is applicable in the limit of zero Stokes particle number. The second is Abrahamson model [8] which is applicable in the limit of infinite particle Stokes number. The form of the later model is usually written (Schubert and Bischofberger 1979) [15] as:

$$\beta = 5 \left(\frac{d_p + d_b}{2} \right)^2 \sqrt{U_p^2 + U_b^2} \tag{14}$$

where U_p and U_b are the turbulent (*rms*) fluctuating velocities of the particles and bubbles relative to the carrier liquid, respectively. Leipe and Mockel's [16] formula (not the original formula by Abrahamson) is used for these velocities,

$$U_i = \frac{0.4 \, \epsilon^{4/9} d^{7/9}}{\nu^{1/3}} \left(\frac{|\rho_i - \rho_f|}{\rho_f} \right)^{2/3} \tag{15}$$

where ϵ is dissipation rate of the turbulent kinetic energy per unit mass (w/kg), ν and ρ_f are the kinematic viscosity and density of the liquid, respectively. ρ_i and d_i are the density and diameter of the colliding particles, $i = p$ for particles and $i = b$ for bubbles. The mass density for a bubble is assumed to be equal $0.5\rho_f$, as if the bubble mass is the virtual mass of a spherical bubble. This equation implies that high dissipation results in higher collisions rates between bubbles and particles, and in this regard high dissipation has favorable effects on flotation. Schubert [15] model have been developed for very high inertia particles. In minerals, flotation particles have a wide spectrum of inertia. Therefore, application of Schubert collisions model results in over prediction of collision kernel as discussed by Fayed [17].

Zaichik *et al.* [9] developed another statistical model for the collisions kernel for particles and bubbles. The model applies to arbitrary values of density ratio and particle sizes. This model has been validated by Fayed and Ragab [18] to study its limitation and it provides more accurate prediction for collisions kernel than Schubert model [15]. The model is not presented here because of the space and interested reader is referred to Zaichik *et al.* [9] and Fayed [17] where the model has been presented and discussed extensively. In this paper, comparison between Schubert model [15] and Zaichik *et al.* model [9] is presented to show the over-estimation of collision kernel by the former model. Hence, Zaichik *et al.* model [9] has been used herein for more accurate prediction of collision kernel.

5.4. Flotation Probabilities

Process of minerals flotation involves two main events which are collisions of particles with bubbles and attachment of the colliding particles with bubbles. The first event happens under turbulent flow conditions where particles and bubbles sizes and turbulent velocity *rms* fluctuations are the main variables that control this event. Upon collision, useful minerals particles have to attach to the bubble and then transported to the froth phase by buoyancy of bubbles. The attachment process is due to the adhesive forces between particles and bubbles and then stabilization of the particle-bubble aggregates. An accurate model for collisions frequency of particles and bubbles has been used in this work [9] to estimate kernel collisions. The second step is to model for the probability of particles' attachment to a bubble. In the literature, considerable efforts have been made to model for the probabilities of collisions, adhesion and stabilization. The concept of probability of collisions has been developed under laminar flow conditions which do not exist in minerals flotation. In this section, we summarize and present models for probabilities of collisions, adhesion and stabilization developed by other authors. These models have been used in our CFD-flotation model to predict pulp recovery rate of a flotation cell. Hopefully, some validation work will be completed on a fundamental level in future to study the limitations of these models and their accuracy.

The probability (efficiency) of collision P_c is given by a formula due to Yoon and Luttrell [10],

$$P_c = \left(\frac{3}{2} + \frac{4}{15} Re_b^{0.72}\right) \frac{d_p^2}{d_b^2} \tag{16}$$

where the bubble Reynolds number is defined by $Re_b = \frac{d_b U_b}{v}$. In the actual calculations P_c is limited to a maximum of 1.

The expression for probability of adhesion P_a is also derived by Yoon and Luttrell [10],

$$P_a = \sin^2\left(2\tan^{-1}\exp\left[\frac{-\left(45 + 8\,Re_b^{0.72}\right) U_b\, t_{\text{ind}}}{15\, d_b \left(\frac{d_b}{d_p} + 1\right)}\right]\right) \tag{17}$$

where t_{ind} is the induction time, which is determined by an empirical formula due to Dai $et\ al.$ [19]:

$$t_{\text{ind}} = \frac{75}{\theta}\, d_p^{0.6} \tag{18}$$

where t_{ind} is measured in seconds, θ is particle-bubble contact angle in degrees, and d_p is the particle diameter in meters.

The formula proposed by Schulze [20] and modified by Bloom and Heindel [12] for the probability of stabilization P_s is used:

$$P_s = 1 - \exp\left[A_s\left(1 - \frac{1}{\min(1, \text{Bo}^*)}\right)\right] \tag{19}$$

where the modified Bond number is defined by:

$$\text{Bo}^* = \frac{d_p^2\left[\Delta\rho g + 1.9\,\rho_p\,\epsilon^{\frac{2}{3}}\,(\frac{d_p}{2} + \frac{d_b}{2})^{-\frac{1}{3}}\right] + 1.5\,d_p\left(\frac{4\,\sigma}{d_b} - d_b\rho g\right)\sin^2(\pi - \frac{\theta}{2})}{\left|6\,\sigma\sin\left(\pi - \frac{\theta}{2}\right)\sin\left(\pi + \frac{\theta}{2}\right)\right|} \tag{20}$$

where $A_s = 0.5$ is an empirical constant suggested by Bloom and Heindel [12], where σ is the surface tension (N/m), $\Delta\rho = \rho_p - \rho_f$, and g is the gravitational acceleration.

5.5. Detachment Model

The particle-bubble detachment rate constant (k_2) is defined by:

$$k_2 = Z_2\, P_d = Z_2(1 - P_s) \tag{21}$$

The probability of detachment P_d is assumed to be equal to $(1 - P_s)$. Koh and Schwarz [13] justified the inclusion of P_s in both k_1 and k_2 because the processes involve different turbulent eddies acting independently of each other. The detachment frequency Z_2 is given by Bloom and Heindel [12] as:

$$Z_2 = \frac{\sqrt{C_1}\, \epsilon^{\frac{1}{3}}}{\left(d_p + d_b\right)^{\frac{2}{3}}} \tag{22}$$

where $C_1 = 2.0$ is an the empirical constant. We note in this equation the unfavorable effects of dissipation that high dissipation results in destabilization of particle-bubble aggregates.

5.6. Particle Size Distribution

Particles in the feed slurry to a flotation machine have a particle size distribution. Several models in the literature have been developed to account for different particles properties in the feed slurry. These models predict the global rate constant of a flotation cell and classified into three groups—discrete rate constant distribution, continuous rate constant distribution and mean rate constant. Several discrete rate constant models have been developed but they differ in the number of fractions (Morris [21], Kelsall [22], Cutting [23], Jowett [24], Imaizumi [25]). Two fraction discrete models developed by (Kelsall [22]) are named as fast float and slow float fractions. Continuous rate constant models assume that rate constants are distributed as a continuous distribution represented by a Gamma-distribution function (Harris [26], Woodburn [27], Loveday [28], Kappur [29]). The mean rate constant model has been proposed by Chen (Chen Z.M. [30,31]). However, prediction of rate constant in all flotation models is strongly affected by operating conditions as well as particles and bubbles properties on a local basis. Flotation modeling using CFD enables more accurate prediction of rate constant and helps to assess the performance of a flotation cell.

The attachment rate k_1 as given above is mainly a function of bubble size, particle size, air/water surface tension and contact angle. An effective pulp recovery rate constant proposed here is defined by:

$$k_1^* = k_1 \frac{\alpha}{v_b}(1 - \alpha) \tag{23}$$

We recognize the factor $\frac{\alpha}{v_b}$ as the local number concentration of bubbles, and the factor $(1 - \alpha)$ is included so that in a region of high air volume fraction ($\alpha \approx 1$) the recovery rate should drop to zero. The local number concentration of particles N_p depends on slurry loading, and we expect it to be proportional to the local water volume fraction, $(1 - \alpha)$. We believe that including the factor $(1 - \alpha)$ gives a better figure of merit that can be used to evaluate different designs in the absence of N_p distribution. The machine average rate constant is obtained by integration over the machine volume,

$$K_1^* = \frac{R_f}{V} \int_V k_1^* (1 - b) dV \tag{24}$$

where R_f is a froth recovery factor.

Minerals particles in the feed slurry have a size distribution and this distribution has great impact on recovery rate. Let $P(dp)$ denote the pdf of particle size distribution. Particle size distribution depends on the grinding process and metallurgical properties of the minerals particles. The current CFD-based flotation kinetic model can be used to determine pulp recovery rate for both bubble size and particle size distributions. We assume a Gamma-distribution for the particles number density function.

$$P(d_p) = \frac{(d_p - \mu_p)^{\gamma - 1}}{B^\gamma \Gamma(\gamma)} \exp\left(-\frac{d_p - \mu_p}{B}\right) \tag{25}$$

where μ_p is the minimum possible particle diameter and Γ is the Gamma function. In this paper, we used $\mu_p = 0$ and $B = 30$. To obtain different particles number distributions, we assign different values for γ starting from 1.0–15, incremented by one. This also changes the mean diameter. The pseudo rate constant for particle size distribution $\overline{K_1^*}$ is computed:

$$\overline{K_1^*} = \int_{d_{p,min}}^{d_{p,max}} P(\lambda) K_1^*(\lambda) \, d\lambda \tag{26}$$

and $d_{p,min}$, $d_{p,max}$ are the minimum and maximum particle diameters present in the feed slurry.

5.7. Flotation Results

In this section, we demonstrate the viability of CFD-based flotation model as a tool to evaluate the performance of flotation machines and provide detailed hydrodynamic and kinetics data that can help improve the design of such machines. Ragab and Fayed [32] developed a CFD-based flotation model, and used it to determine the effects of particle size on the rate constant. We use particle specific gravity $\rho = 4.1$, contact angle $\theta = 40°$, and surface tension $\sigma = 0.06 \, N/m$ as input parameters to the flotation model. Two-phase hydrodynamic simulations provided spatial distributions of dissipation rate, ϵ, and air volume fraction, α (also called void fraction). The number concentration of bubbles is $N_{bT} = \alpha / v_b$, where v_b is the bubble volume and spherical bubbles are assumed. Particles-bubbles collisions rate has been estimated using Zaichik et al.'s model [9]. Zaichik et al.'s model and probabilities models rely on velocity fluctuations, bubble diameter, particle diameter, particle density, air void fraction, contact angle and air-water surface tension coefficient. Velocity fluctuation, u' has been estimated from the local eddy viscosity and dissipation rate. Local air void fraction is obtained from the CFD results of the two-phase flow. Other parameters are user input. CFD-based flotation kinetics model is a post processing program to the two-phase simulations runs of flotation cells such as Wemco 0.8 m³. Effects of particles size distribution on pulp recovery rate are presented here.

As described above, accurate estimation of particle-bubble collision rate is essential for accurate modeling of rate constant. Fayed and Ragab [18] validated Zaichik et al. model [9] and recommended it for use in flotation modeling. A comparison between collisions kernel by Schubert model [15] and Zaichik et al. model [9] are shown in Figure 10a,b. Schubert model overestimates the collisions kernel by an order of magnitude when compared to that of Zaichik et al. model. Based on these results, we recommend Zaichik et al. model in flotation kinetics. Figure 11a–c shows the spatial distribution of probabilities of collision, adhesion and stabilization in a vertical plane that passes through the machine axis. Figure 11a,b shows the favorable effects of turbulent dissipation rate where higher collisions rate and higher probability of collisions in the region between rotor and disperser and within the jet region. However, high local dissipation rates have adverse effects on the probabilities of adhesion and stabilization where low values are observed in the regions between rotor and disperser and within the jet. As described above, flotation process is mainly due to two events—collisions of particles and bubbles and particles attachment to a bubble. That is to say, we can achieve high dissipation rate regions that cause a high collisions rate but lower rates of attachment are obtained which results in a lower rate constant. Therefore, the presented CFD-based flotation model in this paper is a useful tool to optimize for rate constant in flotation machines.

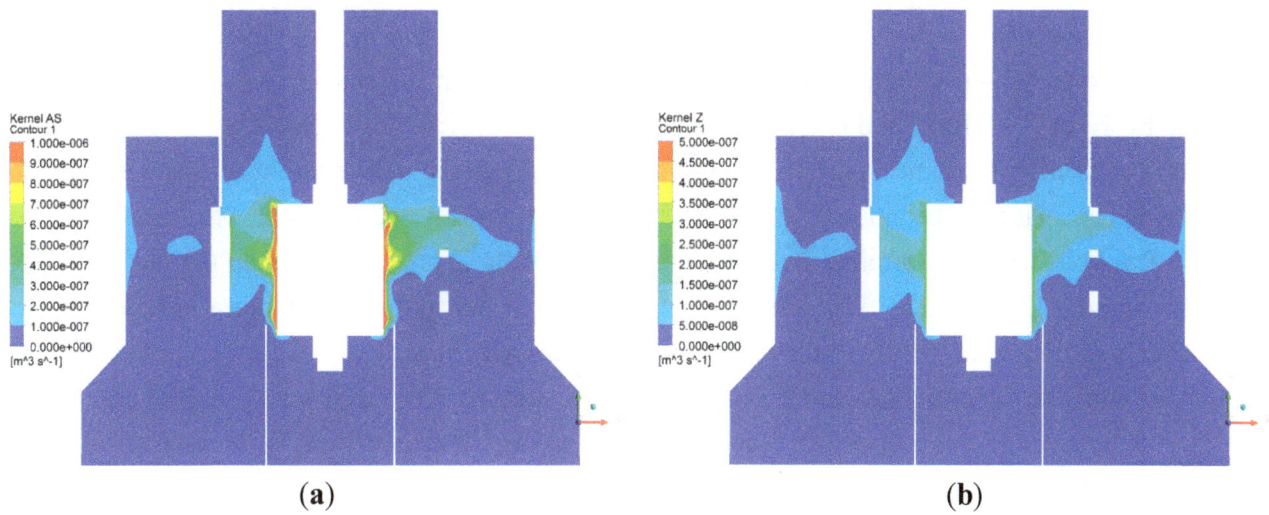

Figure 10. (**a**) Abrahmson-Schubert [6] collisions kernel for d_p = 100 micron; (**b**) Zaichik *et al.* [9] collisions kernel for d_p = 100 micron.

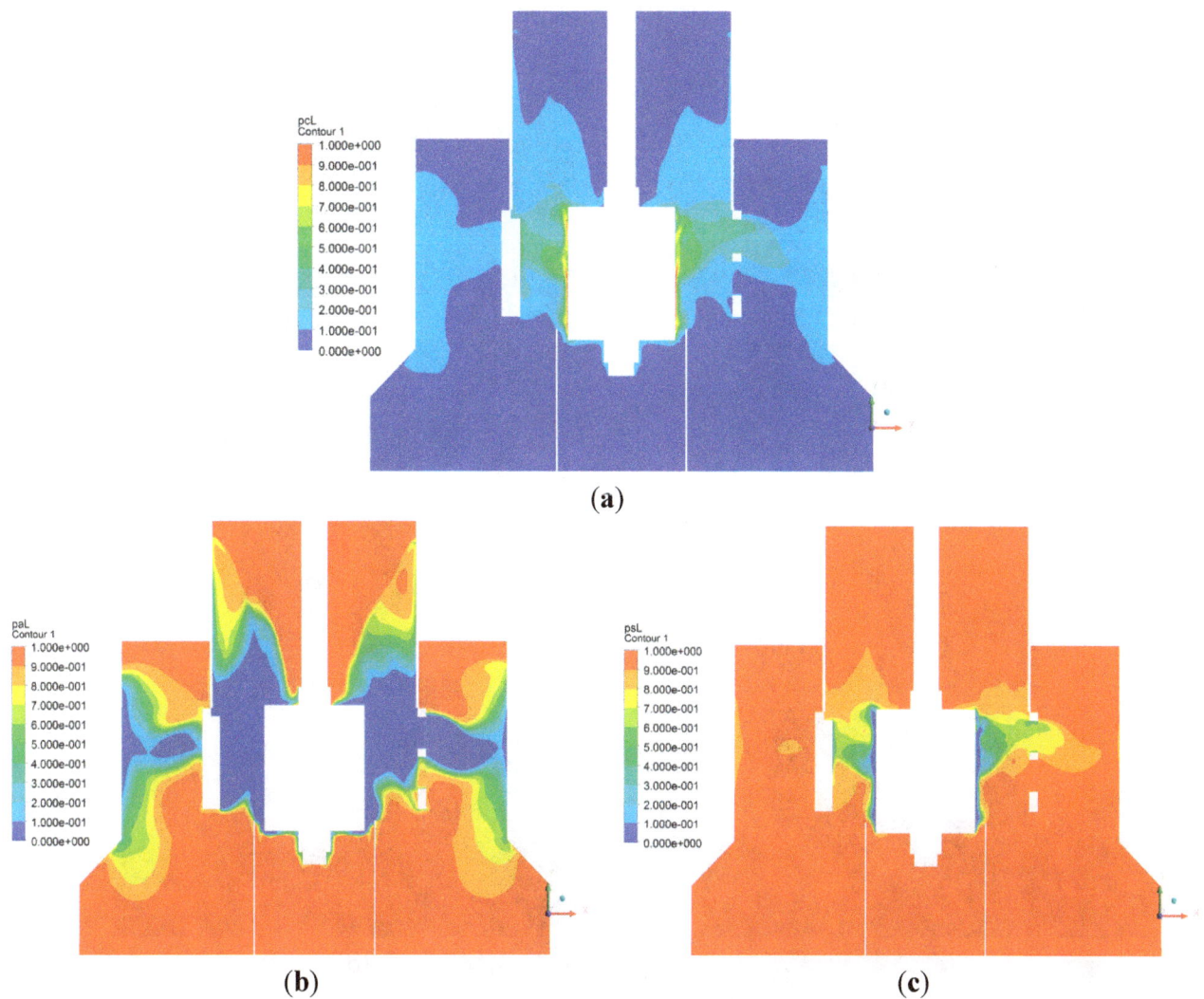

Figure 11. (**a**) Probability of collisions for d_p = 100 micron; (**b**) Probability of adhesion for d_p = 100 micron; (**c**) Probability of stabilization for d_p = 100 micron.

The product of collision kernel and probabilities of collision, adhesion and stabilization gives the local attachment rate k_1. An effective pulp recovery rate constant (in the absence of detachment) is defined here by $k_1 \alpha (1 - \alpha)/v_b$, and corresponding contours are depicted in Figure 12a–d. The factor $(1 - \alpha)$ is included in this definition so that in a region of 100% air ($\alpha = 1.0$) the local recovery rate should be zero. The spatial distribution of $\alpha(1 - \alpha)$, which is zero if $\alpha = 0.0$ or $\alpha = 1.0$ and maximum if $\alpha = 0.5$ (50% void fraction), implies that the local recovery rate is maximized when the air is well dispersed throughout the machine. The actual recovery rate will depend on the balance between the favorable effects of dissipation rate ϵ in increasing the collision frequency against its adverse effects on reducing the attachment rate and increasing detachment rate. Therefore, knowing the spatial distributions of both α and ϵ throughout the machine is essential in understanding the effectiveness of different components (rotor, stator or disperser, jets) on the flotation efficiency. Only through CFD of two-phase simulations (or elaborate experimental measurements) one can determine those spatial distributions. This is a clear advantage of CFD-based flotation models in comparison with models that treat the entire cell as one unit which assumes a single value for ϵ determined by the consumed power and single value for α that is equal to the gas holdup.

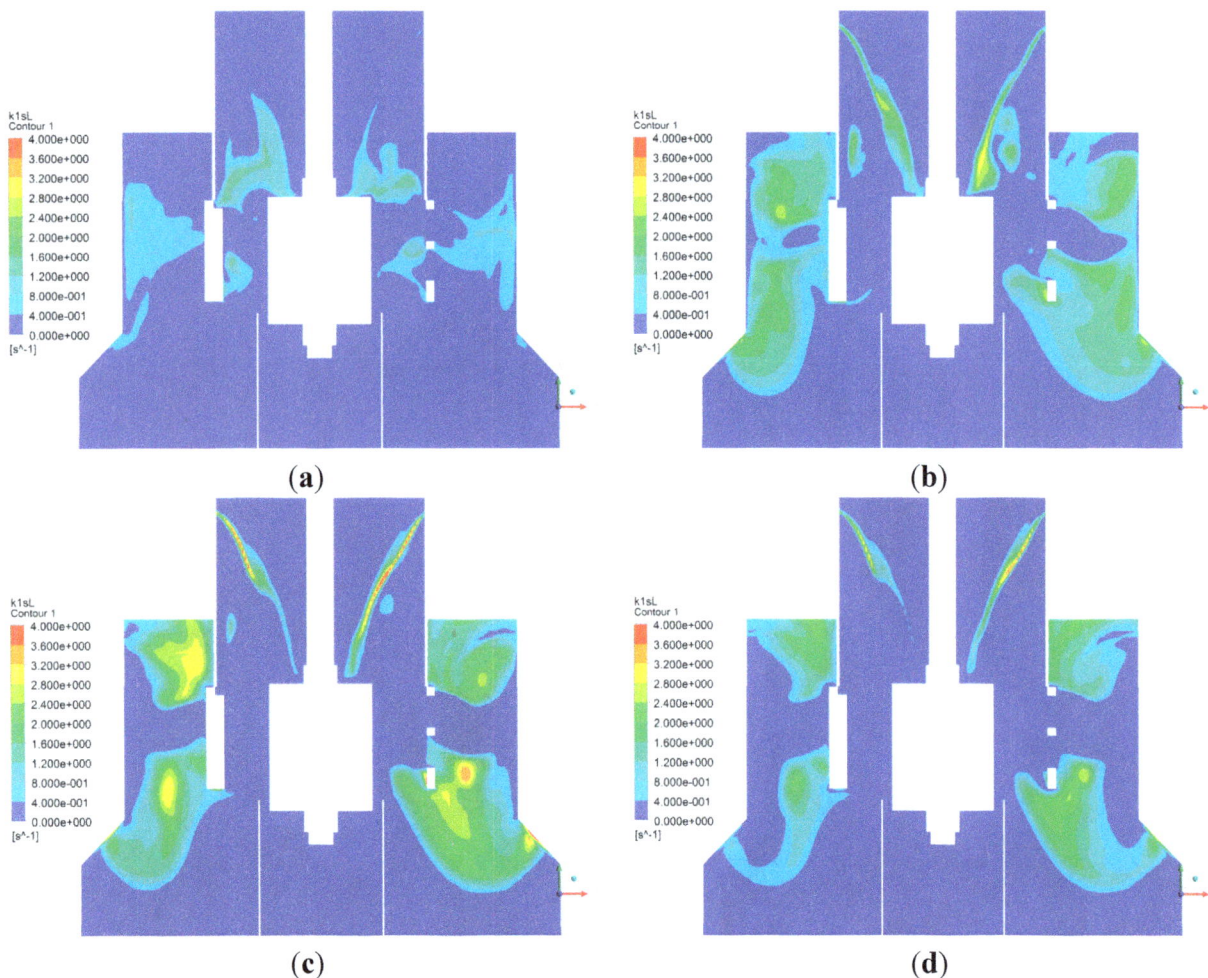

Figure 12. (**a**) Pulp recovery rate k_1^* for 40 µm particle diameter; (**b**) Pulp recovery rate k_1^* for 100 µm particle diameter; (**c**) Pulp recovery rate k_1^* for 200 µm particle diameter; (**d**) Pulp recovery rate k_1^* for 300 µm particle diameter.

Particle diameter has significant effects on distribution of the pulp rate constant k_1^* throughout the Wemco machine. Contours of k_1^* in a vertical mid plane are shown in Figure 12a–d, for particle diameters of 40, 100, 200, and 300 µm. Fine particles $d_p < 50$ µm are efficiently recovered in the high dissipation region between the rotor and disperser. For a medium particle diameter of 100 µm $\leq d_p \leq 200$ µm, recovery happens in the jets out of the disperser. For coarse particles $d_p \geq 300$ µm pulp recovery is more efficient in the moderate dissipation regions outside the disperser below, above and between the disperser jets. In the present work, the pseudo rate constant is computed as a function of particle diameter for two values of contact angle ($\theta = 40°$ and $\theta = 50°$). Values for rate constant are obtained for different uniform bubble size and different uniform particle size. Average flotation rate constant k_1^* can be defined according to the Equation (24) where R_f is a froth recovery factor. We assumed $R_f = 1.0$ and $\beta = 0.0$ for a fully unloaded bubble and studied the effects of bubble diameter, particle diameter, and contact angle on k_1^*. The results are shown in Figure 13a–c. As expected, rate constant increases with the increase of contact angle. Also, higher rate constant is observed for smaller bubble size ($d_b = 0.5$ mm) because air holdup increases for smaller bubble diameter, and hence higher bubble concentration number, as well as a lower rate constant for larger bubble diameter due to the decreases in the air holdup (*i.e.*, local bubbles number concentration). The maximum recovery rate shifts to higher particle diameter with the increase in bubble diameter. The recovery rate shown in these figures is very high relative to experimental data reported in literature because of the assumed values of R_f and β. Maximum rate constant for bubble sizes of 0.5, 0.7 and 1.0 mm occur at particle size of 125, 150 and 175 µm, respectively. This reveals that fact that smaller bubbles sizes ($d_b < 0.5$ mm) are needed to float very tiny particles ($d_p < 100$ µm).

Next, we study the effects of particles size distribution on the mean flotation rate constant. As shown in Figures 13a–c, rate constant is dramatically affected by particle sizes and the feed slurry to a flotation cell contains a wide spectrum of particles sizes. Therefore, modeling for rate constant with particle size distribution is more realistic and useful. Samples of assumed particles' number distribution are shown in Figure 14 for $B = 30$ and three different values for γ ($\gamma = 2, 3$ and 5). Changing values for γ changes the size distribution and mean diameter of the particles. In real samples, this distribution depends on the grinding process as well as metallurgical properties of the ore. The procedure herein is to apply the present CFD flotation model for a single particle size and uniform bubble diameter in the postprocessor of the used CFD package (CFX). Using Equation (23), we obtain the local rate constant for a specific particle size. We used particles sizes ranging from 10–500 µm to be within the practical particle sizes. We used Equation (24) to obtain the average rate constant over the pulp volume and from Equation (26) we calculate the average pulp rate constant for a specific distribution. This process yields a single value for the rate constant that we attribute to the mean value of the proposed particle size. Different particle number distributions pdfs are used (each has different mean diameter) to plot rate constant *versus* particles mean diameter. This deterministic method in the calculation of rate constant for certain particle size distribution is more robust than the other probabilistic models [26–29]. This is because the effects of all variables such as surface tension, contact angle, bubbles size and particles size are explicitly included in the model.

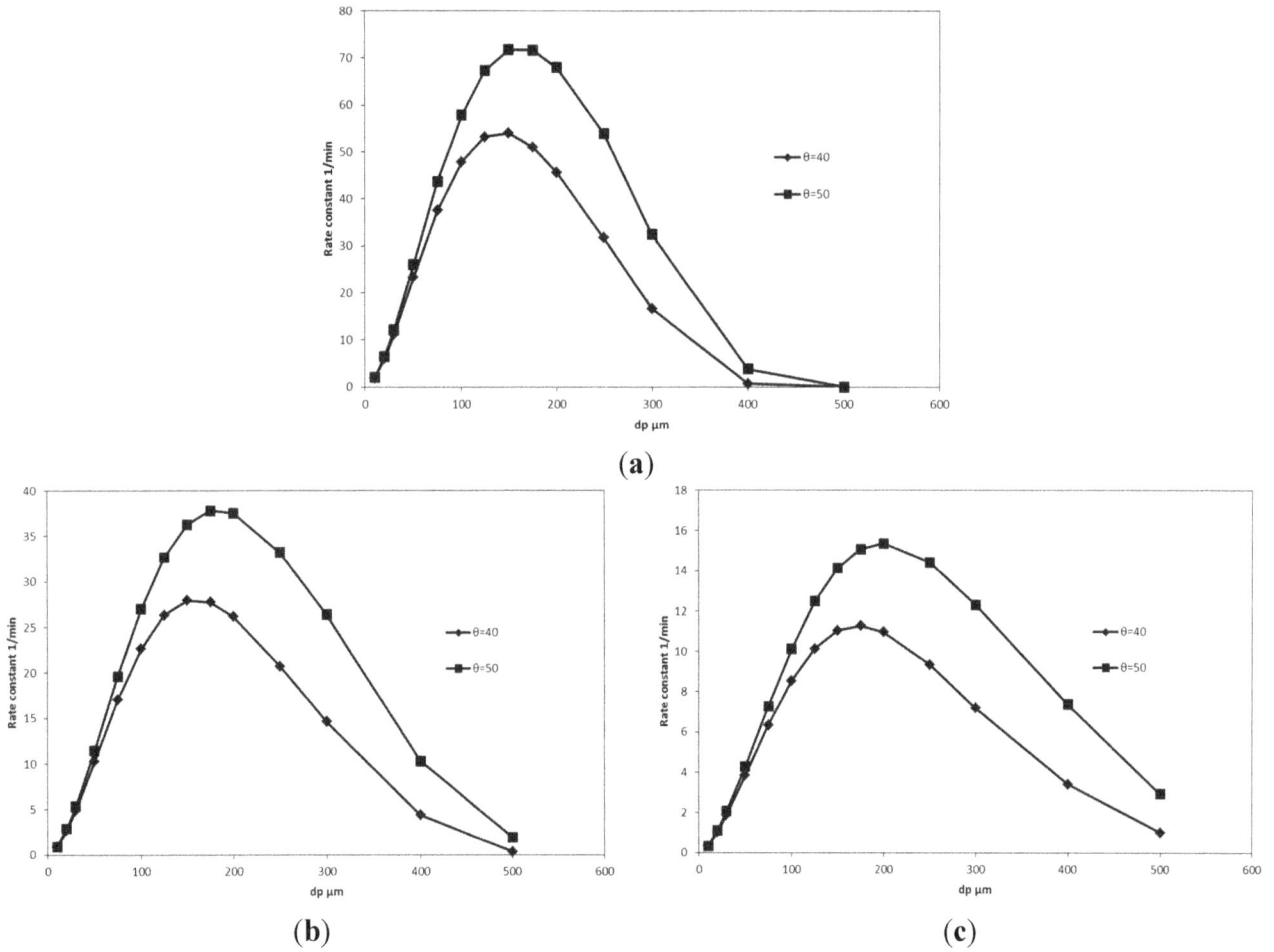

Figure 13. (**a**) Average rate constant for 0.5 mm bubble diameter and for single particle sizes; (**b**) average rate constant for 0.7 mm bubble diameter and for single particle sizes; (**c**) average rate constant for 1.0 mm bubble diameter and for single particle sizes.

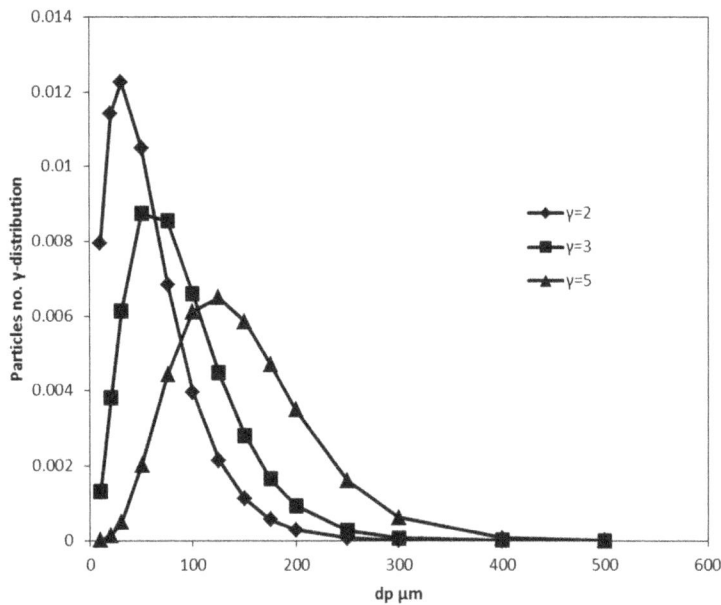

Figure 14. Assumed particles number distribution, $B = 30$.

The effect of particles size distribution on the rate constant has been studied also by an assumed γ-particle number distribution. Effects of different size distribution on mean rate constant are depicted in Figure 15a–c for different bubbles sizes. Figure 15a shows that the maximum rate constant is around the mean particle diameter of 140 μm for a contact angle of 50°, and the particle diameter of 120 μm for a contact angle of 40° where, increasing contact angle increases the possibilities of floating larger particles. A considerable mean rate constant for very small mean diameter ($d_{mean} = 5$ μm) is observed for bubble size of 0.5 mm while very low rate constant exists for the larger bubble size ($d_{mean} = 0.7$ and 1.0 mm) at the same particle mean diameter. Maximum mean rate constant for bubble size ($d_{mean} = 0.7$ mm) is also observed to be around particle mean diameter 140 and 180 μm for bubble size ($d_{mean} = 1.0$ mm) as shown in Figure 15b,c.

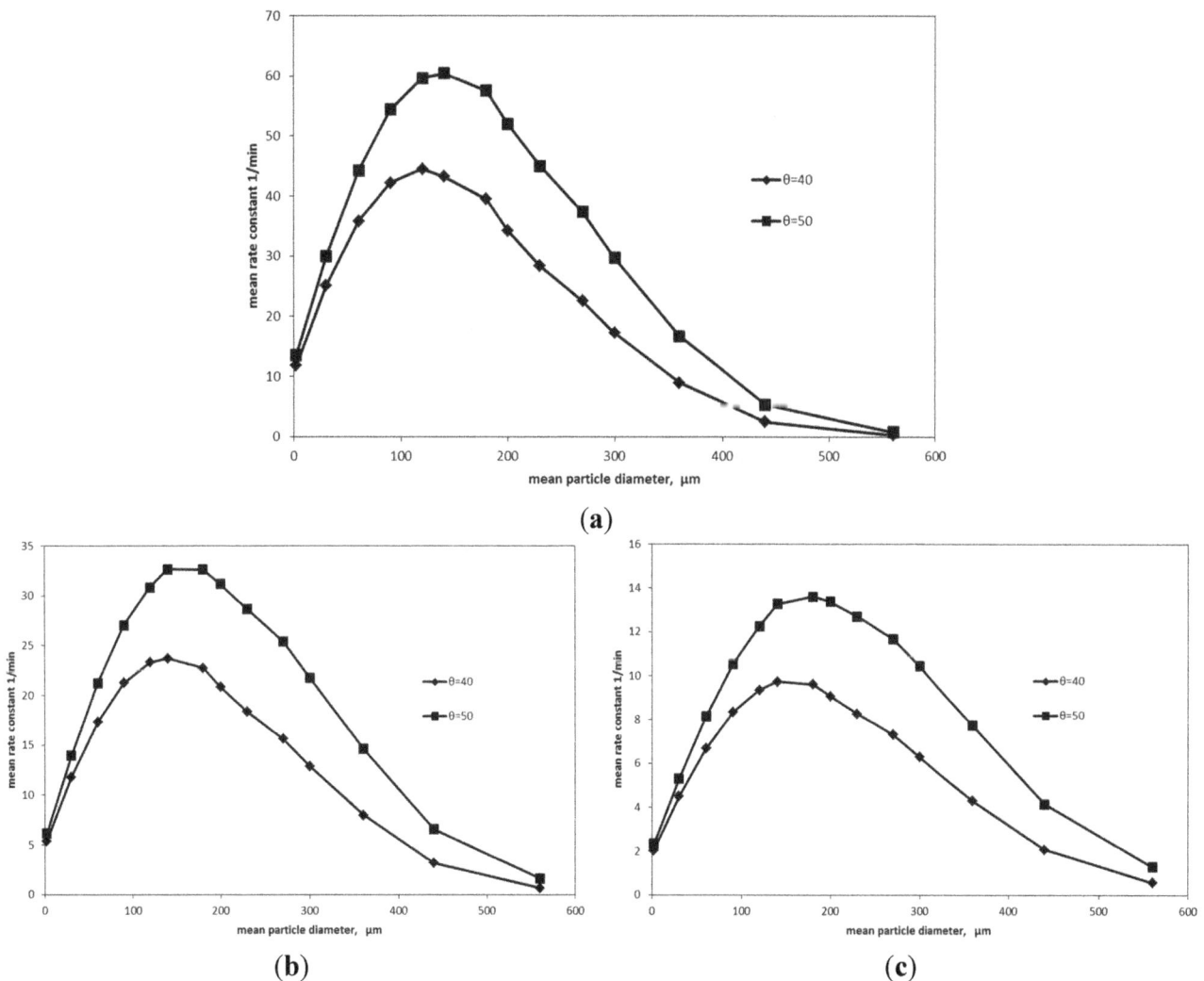

(a)

(b)

(c)

Figure 15. (a) Average rate constant for 0.5 mm bubble diameter and for particle no. γ-distribution; (b) average rate constant for 0.7 mm bubble diameter and for particle no. γ-distribution; (c) average rate constant for 1.0 mm bubble and diameter for particle no. γ-distribution.

6. Conclusions

Two phase simulations of self-aerated flotation Wemco 0.8 m^3 pilot cells have been conducted to study the transient flow structures and predict air flow rate and power consumption as a function of time. The newly devised overflow tank on top of the pulp volume and constant static pressure at the top boundary of this tank and the standpipe opening allowed us to predict the airflow rate and the transient position of the pulp–air interface until it reached a steady position. Also, investigating dynamical behavior of two-phase flow in the standpipe provides us with a detailed picture of the mixing of air and water and pumping this mixture into the tank. The two-phase simulations provide us with essential parameters for flotation modeling such as local turbulent dissipation rate, velocity *rms* fluctuations and air void fraction throughout the whole machine. These parameters have been used to develop a CFD-based flotation model. This model predicts pulp rate constant for an arbitrary particle size and bubble size. It uses a first-order rate equation, where processes of collision, attachment and detachments are described by well-known theoretical and empirical formulae. The model uses local values of the rate of turbulent energy dissipation and air volume fraction. Not only the average pulp recovery rate can be estimated but also the regions of high/low recovery rate can be identified. The CFD-based flotation model presented here is also used to determine the dependence of recovery rate constant at any locality within the pulp on bubble diameter, particle diameter, particle specific gravity, contact angle, and surface tension. Furthermore, we have updated our CFD-based flotation kinetics model to predict pulp recovery rate in the presence of both particles size distributions. The calculations are repeated for many particle diameters selected from a range that covers the anticipated minimum and maximum diameters in the slurry. The particles number density pdf and the data generated for single particle size are used to compute the recovery rate for a specific mean particle diameter. Our computational model gives a figure of merit for the recovery rate of a flotation machine, and as such can be used to assess incremental design improvements and design of new machines. The model is a very useful design tool because it can be used to establish the effects of different components (rotor, stator, or disperser jets) on the pulp recovery.

Acknowledgment

This work has been supported by FLSmidth Minerals Inc., Salt Lake City, Utah.

Author Contributions

Dr. Fayed and Dr. Ragab worked closely on the material presented in this paper during the time Dr. Fayed was working as Graduate Research Assistant in Department of Engineering Science and Mechanics, Virginia Tech, USA. Dr. Fayed constructed geometric models of Wemco machine, generated the mesh, and conducted computer simulations. Dr. Ragab contributed to the theoretical modeling and boundary conditions. Both authors contributed equally to data analysis and drawing conclusions.

Conflicts of Interest

The authors declare no conflict of interest.

References

1. Koh, P.T.L.; Schwarz, P. CFD Model of a Self-Aerating Flotation Cell. *Int. J. Miner. Process.* **2007**, *85*, 16–24.

2. Tiitinen, J.; Koskinen, K.; Ronkainen, S. Numerical modeling of an Outokumpu flotation cell. In Proceedings of the Centenary of Flotation Symposium, Brisbane, Australia, 6–9 June 2005.

3. Kerdouss, F.; Bannari, A.; Proulx, P. CFD modeling of gas dispersion and bubble size in a double turbine stirred tank. *Chem. Eng. Sci.* **2006**, *61*, 3313–3322.

4. Prosperetti, A.; Tryggvason, G. *Computational Methods for Multiphase Flow*; Cambridge University Press: Cambridge, UK, 2007.

5. Clift, R.; Grace, J.R.; Weber, M.E. *Bubbles, Drops and Particles*; Dover Publications: Mineola, NY, USA, 2005.

6. Van den Akker, H.E.A. Toward a truly multiscale computational strategy for simulating turbulent two-phase flow processes. *Ind. Eng. Chem. Res.* **2010**, *49*, 10780–10797.

7. Nelson, M.; Traczyk, F.; Lelinski, D. Design of Mechanical Flotation Machines. In Proceedings of the 2011 SME Annual Meeting, Denver, Colorado, USA, 27 February–2 March 2011.

8. Abrahamson, J. Collision rate of small particles in a vigorously turbulent fluid. *Chem. Eng. Sci.* **1975**, *30*, 1371–1379.

9. Zaichik, L.I.; Simonin, O.; Alipchenkov, V.M. Turbulent collision rates of arbitrary density particles. *Int. J. Heat Mass Transf.* **2010**, *53*, 1613–1620.

10. Yoon, R.H.; Luttrell, G.H. The effect of bubble size on fine particle flotation. *Miner. Process. Extr. Metall. Rev.* **1989**, *5*, 101–122.

11. Pyke, B.; Fornasiero, D.; Ralston, J. Bubble particle heterocoagulation under turbulent conditions. *J. Colloid Interface Sci.* **2003**, *265*, 141–151.

12. Bloom, F.; Heindel, T.J. Modeling flotation separation in semi-batch process. *Chem. Eng. Sci.* **2003**, *58*, 403–422.

13. Koh, P.T.L.; Schwarz, M.P. CFD modeling of bubble-particle attachments in flotation cells. *Miner. Eng.* **2006**, *19*, 619–626.

14. Saffman, P.G.; Turner, T.S. On the collision of drops in turbulent clouds. *J. Fluid Mech.* **1956**, *1*, 16–30.

15. Schubert, H. On the turbulence-controlled microprocesses in otation machines. *Int. J. Miner. Process.* **1999**, *56*, 257–276.

16. Leipe, F.; Mockel, O.H. Untersuchungen zum stoffvereinigen in ussiger phase. *Chem. Technol.* **1976**, *30*, 205–209. (In German)

17. Fayed, H. Particles and Bubbles Collisions in Homogeneous Isotropic Turbulence and Applications to Minerals Flotation Machines. Ph.D. Thesis, Virginia Tech, Blacksburg, VA, USA, 6 December 2013.

18. Fayed, H.E.; Ragab, S.A. Direct Numerical Simulation of Particles-Bubbles Collisions Kernel in Homogeneous Isotropic Turbulence. *J. Comput. Multiph. Flows* **2013**, *5*, 168–188.

19. Dai, Z.; Fornasier, D.; Ralston, J. Particle-bubble attachment in mineral flotation. *J. Colloid Interface Sci.* **1999**, *217*, 70–76.

20. Schulze, H.J. Flotation as a hetrocoagulation process: Possibilities of calculation the probability of flotation. In *Coagulation and Flocculation*; Dobias, B., Ed.; Marcel Dekker: New York, NY, USA, 1993; pp. 321–363.

21. Morris, T.M. Discussion of flotation rates and flotation efficiency. *Min. Eng.* **1952**, *4*, 794–798.

22. Kelsall, D.F.; Stewart, P.S. A critical review of applications of models of grinding and flotation. In Proceedings of the Symposium on Automatic Control Systems in Mineral Processing Plant, Brisbane, Australia, 17–20 May 1971; pp. 213–232.

23. Cutting, G.W.; Devenish, M.A. Steady-state model of froth flotation structures. In Proceedings of the AIME Annual Meeting, New York, NY, USA, 20 February 1975.

24. Jowett, A. Resolution of flotation recovery curves by a difference plot method. *Trans. IMM* **1974**, *70*, 191–204.

25. Imaizumi, T.; Inoue, T. Kinetic considerations of froth flotation. In Proceedings of the 6th International Mineral Processing Congress, Cannes, France, 26 May–2 June 1963; pp. 581–593.

26. Harris, C.C.; Chakravarti, A. Semi-batch froth flotation kinetics; species distribution analysis. *Trans. AIME* **1970**, *247*, 162–172.

27. Woodburn, E.T.; Loveday, B.K. Effect of variable residence time data on the performance of a flotation system. *J. South Afr. Inst. Min. Metall.* **1965**, *65*, 612–628.

28. Loveday, B.K. Analysis of froth flotation kinetics. *Trans. IMM* **1966**, *75*, 219–225.

29. Kapur, P.C.; Mehrotra, S.P. Estimation of the flotation rate distributions by numerical inversion of the Laplace transform. *Chern. Eng. Sci.* **1974**, *29*, 411–415.

30. Chen, Z.M.; Wu, D.C. A study of flotation kinetics. *Nonferr. Met.* **1978**, *10*, 28–33.

31. Chen, Z.M.; Mular, L. A study of flotation kinetics—A kinetic model for continuous flotation. *Nonferr. Met.* **1982**, *3*, 38–43.

32. Ragab, S.; Fayed, H. CFD-Based Flotation Model for Prediction of Pulp Recovery Rate. In Proceedings of the SME Annual Meeting and Exhibit, Seattle, WA, USA, 19–22 February 2012.

Permissions

The contributors of this book come from diverse backgrounds, making this book a truly international effort. This book will bring forth new frontiers with its revolutionizing research information and detailed analysis of the nascent developments around the world.

We would like to thank all the contributing authors for lending their expertise to make the book truly unique. They have played a crucial role in the development of this book. Without their invaluable contributions this book wouldn't have been possible. They have made vital efforts to compile up to date information on the varied aspects of this subject to make this book a valuable addition to the collection of many professionals and students.

This book was conceptualized with the vision of imparting up-to-date information and advanced data in this field. To ensure the same, a matchless editorial board was set up. Every individual on the board went through rigorous rounds of assessment to prove their worth. After which they invested a large part of their time researching and compiling the most relevant data for our readers.

The editorial board has been involved in producing this book since its inception. They have spent rigorous hours researching and exploring the diverse topics which have resulted in the successful publishing of this book. They have passed on their knowledge of decades through this book. To expedite this challenging task, the publisher supported the team at every step. A small team of assistant editors was also appointed to further simplify the editing procedure and attain best results for the readers.

Apart from the editorial board, the designing team has also invested a significant amount of their time in understanding the subject and creating the most relevant covers. They scrutinized every image to scout for the most suitable representation of the subject and create an appropriate cover for the book.

The publishing team has been an ardent support to the editorial, designing and production team. Their endless efforts to recruit the best for this project, has resulted in the accomplishment of this book. They are a veteran in the field of academics and their pool of knowledge is as vast as their experience in printing. Their expertise and guidance has proved useful at every step. Their uncompromising quality standards have made this book an exceptional effort. Their encouragement from time to time has been an inspiration for everyone.

The publisher and the editorial board hope that this book will prove to be a valuable piece of knowledge for researchers, students, practitioners and scholars across the globe.

List of Contributors

Heath D. Watts
Department of Geosciences, The Pennsylvania State University, University Park, PA 16802, USA

Lorena Tribe
Division of Science, The Pennsylvania State University, Berks, Reading, PA 19610, USA

James D. Kubicki
Department of Geosciences, The Pennsylvania State University, University Park, PA 16802, USA
Earth and Environmental Systems Institute, The Pennsylvania State University, University Park, PA 16802, USA

Robert G. Hatfield
College of Earth, Ocean, and Atmospheric Science, Oregon State University, 104 CEOAS Admin Building, Corvallis, OR 97331, USA

Helen R. Watling
CSIRO Mineral Resources Flagship, PO Box 7229, Karawara, WA 6152, Australia

Lindsay C. Shuller-Nickles
Department of Environmental Engineering and Earth Science, Clemson University, 342 Computer Court, Anderson, SC 29625-6510, USA

Will M. Bender
Department of Earth and Environmental Sciences, University of Michigan, 1100 North University Avenue, Ann Arbor, MI 48109-1005, USA

Sarah M. Walker
Department of Earth and Environmental Sciences, University of Michigan, 1100 North University Avenue, Ann Arbor, MI 48109-1005, USA

Udo Becker
Department of Earth and Environmental Sciences, University of Michigan, 1100 North University Avenue, Ann Arbor, MI 48109-1005, USA

Abdul Mwanga
Minerals and Metallurgical Engineering Laboratory, Luleå University of Technology, SE-971 87 Luleå, Sweden

Jan Rosenkranz
Minerals and Metallurgical Engineering Laboratory, Luleå University of Technology, SE-971 87 Luleå, Sweden

Pertti Lamberg
Minerals and Metallurgical Engineering Laboratory, Luleå University of Technology, SE-971 87 Luleå, Sweden

Doyun Shin
Mineral Resources Resource Division, Korea Institute of Geoscience and Mineral Resources (KIGAM), Gwahangno 124, Yuseong-gu, Daejeon 305-350, Korea
Department of Resource Recycling Engineering, Korea University of Science and Technology, Gajeongno 217, Yuseong-gu, Daejeon 305-350, Korea

Jiwoong Kim
Mineral Resources Resource Division, Korea Institute of Geoscience and Mineral Resources (KIGAM), Gwahangno 124, Yuseong-gu, Daejeon 305-350, Korea

Byung-su Kim
Mineral Resources Resource Division, Korea Institute of Geoscience and Mineral Resources (KIGAM), Gwahangno 124, Yuseong-gu, Daejeon 305-350, Korea
Department of Resource Recycling Engineering, Korea University of Science and Technology, Gajeongno 217, Yuseong-gu, Daejeon 305-350, Korea

Jinki Jeong
Mineral Resources Resource Division, Korea Institute of Geoscience and Mineral Resources (KIGAM), Gwahangno 124, Yuseong-gu, Daejeon 305-350, Korea
Department of Resource Recycling Engineering, Korea University of Science and Technology, Gajeongno 217, Yuseong-gu, Daejeon 305-350, Korea

Jae-chun Lee
Mineral Resources Resource Division, Korea Institute of Geoscience and Mineral Resources (KIGAM), Gwahangno 124, Yuseong-gu, Daejeon 305-350, Korea
Department of Resource Recycling Engineering, Korea University of Science and Technology, Gajeongno 217, Yuseong-gu, Daejeon 305-350, Korea

Hassan Fayed
Numerical Porous Media Center, King Abdullah University of Science and Technology (KAUST), Thuwal 23955-6900, Saudi Arabia

Saad Ragab
Department of Engineering Science and Mechanics, Virginia Tech, Blacksburg, VA 24061, USA